The
Greatest
Flying Stories
Ever Told

Also available from The Lyons Press:

The Greatest Adventure Stories Ever Told
The Greatest Baseball Stories Ever Told
The Greatest Boxing Stories Ever Told
The Greatest Cat Stories Ever Told
The Greatest Climbing Stories Ever Told
The Greatest Disaster Stories Ever Told
The Greatest Dog Stories Ever Told
The Greatest Escape Stories Ever Told
The Greatest Exploration Stories Ever Told
The Greatest Fishing Stories Ever Told
The Greatest Football Stories Ever Told
The Greatest Gambling Stories Ever Told
The Greatest Golf Stories Ever Told
The Greatest Hockey Stories Ever Told
The Greatest Horse Stories Ever Told
The Greatest Hunting Stories Ever Told
The Greatest Sailing Stories Ever Told
The Greatest Search and Rescue Stories Ever Told
The Greatest Survival Stories Ever Told
The Greatest Treasure-Hunting Stories Ever Told
The Greatest War Stories Ever Told

The Greatest Flying Stories Ever Told

EDITED WITH INTRODUCTIONS BY
LAMAR UNDERWOOD

Guilford, Connecticut

An imprint of The Rowman & Littlefield Publishing Group, Inc.

4501 Forbes Blvd., Ste. 200

Lanham, MD 20706

www.rowman.com

Distributed by NATIONAL BOOK NETWORK

British Library Cataloguing in Publication Information available

**This book was previously cataloged by the Library of Congress Cataloging-
in-Publication Data**

ISBN 978-1-4930-1969-4 (paperback)

∞™ The paper used in this publication meets the minimum requirements of American
National Standard for Information Sciences—Permanence of Paper for Printed Library Materials, ANSI/NISO Z39.48-1992.

Printed in the United States of America

Contents

High Flight

Its poignant phrases have been engraved in plaques that hang on the walls of countless airport operations offices—"shacks" to the insiders. It has been printed in a number of books and magazine articles. Its words and the story of how they came to be are the star attractions of a number of Internet sites (check the search engine "Google" to find them).

"High Flight" is a piece of verse you may never see studied in the dusty halls of academia where such verses are given distinction as "literature," but in the hearts of aviators John Gillespie Magee's words are immortal.

One of the ironies of "High Flight" is that the story behind the poem requires more words than the verse itself. So be it. For to this editor and all the others who have preceded me in referencing this work, Magee's is a story worth remembering.

He was born in Shanghai, China, in 1922, the son of missionary parents, his father an American and his mother originally British. In September, 1940, he gave up a scholarship to Yale to enlist in the Royal Canadian Air Force. He was posted to England for combat duty in 1941 and went on to fly the Spitfire over France and in the air defense of England against the German Luftwaffe.

In September, 1941, he wrote his parents that he was sending along a poem he had started writing at 30,000 feet and finished when he landed. "High Flight" was scribbled on the back of this letter.

Three days after Pearl Harbor, Magee's Spitfire collided in midair over England with a training plane from a nearby base. He was seen standing in the cockpit, trying to jump from the plane, but was too low for his parachute to open. He was buried at Scopwick Cemetery, Lincolnshire, with full service honors, carried by pilots of his squadron. He was 19 years old.

Oh! I have slipped the surly bonds of earth
And danced the skies on laughter-silvered wings;
Sunward I've climbed, and joined the tumbling mirth
Of sun-split clouds—and done a hundred things
You have not dreamed of—wheeled and soared and swung
High in the sunlit silence. Hov'ring there
I've chased the shouting wind along, and flung
My eager craft through footless halls of air.

Up, up the long, delirious, burning blue
I've topped the wind-swept heights with easy grace
Where never lark, or even eagle flew—
And, while with silent, lifting mind I've trod
The high, untrespassed sanctity of space,
Put out my hand and touched the face of God.

—John Gillespie Magee, Jr.

Introduction

For those who have heard its notes, aviation's siren song is both haunting and irresistible. Whether the seductive pull of airports and airplanes have led you to the field for lessons and licensing, or you're still a dreamer, looking over the fence, you know there's something special going on out there. For those who are young enough, or perhaps settling into a mid-life position where time and finances are manageable, the dream will become reality. They will "slip the surly bonds of earth," as John Gillespie Magee, Jr., so eloquently exalts in the verse "High Flight." Others will have to make peace with themselves—over the dreams that cannot be. Their "flights" will have to come from books and films, from modeling and computer simulations, from whatever rides they can cadge from pilot friends, and from their occasional trips on commercial airlines.

It is almost a hundred years now since Thursday, December 17, 1903, the day Wilbur and Orville Wright stood on a sand dune in North Carolina and successfully launched a motor-driven flying machine into the air. Not a balloon, not a glider. It was a true airplane that soared briefly over the beach at 10:35 that morning. Not much of a plane, not much of a flight. But the Wrights' machine had wings and an engine. Man was finally off the ground. The long, enduring dream of flight had become a reality.

Today, even the most technically awesome and efficient airplanes— those that countless times every day bring people and goods to destinations hundreds and thousands of miles away—stir only a ho-hum attitude from some people. Most people, though, whether grudgingly or with great enthusiasm, will have to admit to harboring a certain amount of curiosity about airplanes and what it takes to fly them. Newspaper editors and television and radio bosses eagerly exploit such sentiments by giving news of aviation accidents, no matter how trivial, coverage far more prominent than other events that are more important but lack the drama of a flight that failed.

Admitting that modern airplanes are technical marvels—which they indeed are!—does not make one an aviation buff. The events in the cockpit, the coping with the skills and devices that enable a pilot to direct his machine through the vastness of the skies and the ever-changing weather, are the trigger that jump-starts the true enthusiast's imagination. Whether real pilots or wannabes, kids or seniors, the aviation faithful are a cult of fascinated on-lookers. No airplane can be ignored—passing overhead, sitting on the ramp, taxiing, taking off. The faithful stare, linked in spirit for the moment not by the machine itself, but by the unseen activity they know to be taking place in the cockpit.

Even though commercial and military aviation are necessarily closed establishments, of little access to the non-professional, the average aviation enthusiast can find plenty of outlets for his passion. They exist in the multifaceted world of general aviation.

You don't have to hang around general aviation airports very long before you realize there's something interesting going on. Peer over the chain-link fence—as I and countless others have done since we were kids—and you will see colorful Pipers, Cessnas, Beechcrafts, and numerous other single- and twin-engine planes you can't quickly name. You will also begin to notice, perhaps with a touch of envy, the people who are allowed through the gate, accessing the ramp and the planes. These are men and women who look . . . well, ordinary. They do not look like movie stars. They are not colorfully dressed, as the early aviators were, with jackets, and boots and scarves that screamed, "I'm a pilot!" These cannot be the steely-eyed heroes you have imagined ruling the skies.

Look there! That guy's a geezer. Seventy, if he's a day. And over there. A teenager is climbing into that Cessna. I can see his pimples from here, for God's sake. And what's the great-looking blonde with a clipboard lecturing that gray-haired guy about? You mean she's an instructor?

Welcome to aviation, my friend. And welcome to freedom. For once you have been in the cockpit of a small plane, known the sheer exhilaration of the takeoff roll, the climb out from the airport, the earth falling away behind you now—experienced these things without fear and trembling but with immense joy and awareness—you will be a slave to your urge to fly. Only at airports will your heart find freedom from a strange sort of longing and emptiness brought on by wanting to get into the cockpit and fly.

Whether you fly a little or a lot or not at all, if your heart is in the cockpit, in this book you have come to the right place to mix and mingle with kindred spirits. In the pages ahead you will be flying with some of the most skilled pilots and talented writers to ever take to the skies and then share their stories with others. From pioneering flights such as Chuck Yeager breaking the sound barrier or Lindbergh crossing the Atlantic; to expressions of the sheer joy of learning to fly such as Diane Ackerman's "On Extended Wings"; to the challenges of building flying skills like Laurence Gonzales's "One, Zero, Charlie"; to the drama of crisis in the cockpit such as Ernest K. Gann's "Dooley"; to chronicles of battles in the skies such as Derek Robinson's "Piece of Cake"—this collection will carry you deep into aviation's diversity and pageantry. The tales chosen to be presented here are not simple reportage of facts. They are intended to put you in the cockpit, to see the weather, the gauges; to hear the roar of engines, the rush of air past the fuselage, the voices of air traffic controllers dedicated to helping you keep your craft from tangling with others who share the skies.

Say what you will about pilots, but the simple fact is that they are not your average walking-around human being. They can fly; not very many people can. Sure, if given the chance and training, they probably could. But they can't.

Being able to do something not very many people can do is probably one of aviation's deeper attractions. And the motivation doesn't stop there. Fact is, some pilots are better than others. They all may have been "created equal," but in the cockpit some individuals rise far above the norm in the skills and understanding of their craft and what it can, and cannot, do. Like it or not, there is a certain amount of drama in the cockpit every time a plane goes up. Some flying tasks may be called "routine," but there is nothing routine about them. Flying is not the same as driving your car to the mall and back. If you buy into that old myth and insist on flying with that kind of attitude, you're going to bust your butt.

The late Ernest K. Gann, legendary pilot and author of such aviation classics as *Fate Is the Hunter*, describes his intimacy with the cockpit:

"For a time I stare at the ceiling of the cockpit which curves only a foot or so above my head. There, at least, everything is familiar . . . If you spend over two thousand hours in the same little room, a room smaller than a prisoner in solitary confinement could tolerate, and if in spite of its cramped peculiarities, then you must know that room and everything within it as a certainty as a miser knows his hoard. For two thousand hours is a long time."

Gann was writing about the airline pilots of the thirties, forties, and fifties in *Fate Is the Hunter*. Today's airline pilots commonly log over 20,000 hours of flying time before reaching mandatory retirement at age sixty. And, of course, the cockpits—Gann's "little rooms"—are intricately more complex.

Reflecting on the difficulties of flying on instruments in the modern cockpit, William Langewiesche in his superb book *Inside the Sky: A Meditation on Flight* (1998), relates: "Yes, the sky at times can seem as familiar as a familiar landscape, but on a dark night and inside the clouds its alien nature reemerges. Again then it becomes a surreal and dangerous place across which we humans may move, but only with care and wonder. The cockpit at such times is like a capsule hurtling through some distant reach of space."

I love that expression, ". . . a surreal and dangerous place across which we humans may move, but only with care and wonder."

Yep, it's just like you thought: It's fascinating up there, but it can be dangerous. And danger, of course, is the clay from which flying stories have been worked, from Orville and Wilbur's time to today's moon shots.

Ernest Gann, the only author with two stories in this collection, reflects on those dangers in our excerpt (called "Dooley") from his classic *Island In the Sky*:

"Just ahead, a scant few feet, the instrument panel offered assorted information of the airspeed, altitude, course, engine temperatures, manifold pressures, and revolutions per minute. They were pleasantly old, familiar hard facts—easy to comprehend and easy to observe. Dooley's twenty years of flying had taught him over and over that he must not be content with them alone.

"It was the things you couldn't see that counted—the hidden, never tangible series of upper world plots and fancies. They invariably joined company with the unwary, always with an air of deceptive innocence. They frequently killed the unwary."

We don't need recent events like young Kennedy's airplane plunging into the Atlantic off Martha's Vineyard, or the Air Alaska jet diving into the Pacific off Los Angeles, to appreciate the poignancy of Gann's words. The history of aviation is replete with tragedy and disaster as it is with vision and discovery. The victims of those devastating losses can best be remembered by the safety measures and new tools of prevention that stemmed from the accidents.

Even before the debris is cleared away from an accident site, people want to know, "What happened?" And today's technology of cockpit voice recorders and aircraft data recorders reveal precisely what did happen. Indeed, such recordings furnish the background of some of the stories in this book that put you in the cockpit during some of the major crashes that led to improvements in flying safety.

Aviators love to tell what they jokingly call "war stories." They all know the clichéd story, beginning, "There I was at 15,000 feet when . . ." You may still hear that expression in hangar talk, but you won't find it here. Our fliers and authors know how to tell a story much better than that. And the stories they have for you are not always about danger and emergencies in the sky, but also include expression of the sheer wonder and joy of flying, or mastering the cockpit to make the airplane go where you want it to go and do what you want it to do.

Real war stories—shooting war stories—are an important and interesting part of aviation literature, and several of the best are included here.

But lest you get the impression that flying is solely about drama and danger, remember too that aviation has its lighter side, as real as service comedy or any other forms of humor. ("A good landing is one you can walk away from. A great landing is one you can taxi away from.") From time to time, you'll see reminders that aviators have a strong sense of humor.

Good thing, too. Aviation professionals, particularly those in general aviation, sure need it at times. Misguided critics and professional naysayers are constantly snapping at the heels of the general aviation community, their twisted minds set on closing down airports and general aviation flying altogether—particularly private flying. The aviation bashers will tell you the skies should be the domain of commercial flights only, and that the folks flying Cessnas and Pipers are fools and a dangerous menace to the public. What rot!

It is my personal opinion that the people who lament the existence of general aviation are strikingly similar to other conspicuous sign-waving "antis." Their targets are also hunters, fishermen, bikers, surfers, mountain climbers, and others pursuing interests in which true freedom and the love of a challenging activity are very visible. Many of these outspoken individuals will throw words like "freedom" and "public interest" around like loose change, while in their hearts wanting a regimented way of life completely devoid of activities that carry the slightest taint of risk. For them, the TV and the mall will do fine.

Both commercial and general aviation today are under a public and government microscope as never before in our history. The events in New York of September 11 have made aviation stories a more-or-less permanent part of daily news. And that's not going to change anytime soon.

To see four of our most remarkable and beloved airplanes used as weapons of war and terror as they were on the morning of September 11 was like witnessing a glimpse of hell itself, of desecration at its ultimate. That morning I had just left Penn Station, headed for my office on Broadway, right across from Madison Square Park where Broadway meets Fifth Avenue. I was following my usual route of walking across 32nd Street, then turning right on Broadway. At about 8:45 I had just made the turn on Broadway, almost directly under the Empire State Building, when the sudden shattering sound of screaming jet engines exploded overhead. Shocked, I looked up to see the dull-colored underbelly of a jet as it roared past, barely missing the Empire State Building, and instantly disappeared over the buildings along Broadway. Jets are familiar sights over Manhattan as they go into and out of LaGuardia. But they are never that low and that fast. I thought I had seen a 737, and as I walked along I could not come to any conclusion about why the plane was flying so low and so fast. I never thought about terrorists. (The plane was actually a 767, American Airlines Flight 11 out of Boston.)

When I reached my office in about five minutes, people were standing by Madison Park, pointing, shouting. As I walked across the Broadway and Fifth Avenue intersection, I heard someone shout, "That jet that just passed over hit the World Trade Center!" I was across the street then, looking at the towers far away downtown. In a little while the second plane hit. You know the rest of the story.

Commercial aviation and general aviation were wounded by the horrible events of September 11 and the subsequent fallout. But the miserable bastards who attacked America cannot destroy what the dreams and sacrifices of an entire century of pilots, engineers, mechanics, and passengers have produced. In the century ahead, our airplanes and pilots will be carrying even more people to their chosen destinations, more efficiently and more safely than ever. And general aviation will be here as well—the hometown airport where young men and women will look over the chain-link fences and dream of the day when an airplane and the sky will be theirs.

In referencing the young and the future of aviation, writer Lane Wallace wrote a piece called "Eyes of a Child," in *Flying* Magazine, February 2000. This portion of that article really hits me hard:

"I'll run my hand over the wing of a small airplane and say to him, 'This plane can teach you more things and give you more gifts than I ever could. It won't get you a better job, a faster car, or a bigger house. But if you treat it with respect and keep your eyes open, it may remind you of some things we used to know—that life is in the moment, joy matters more than money, the world is a beautiful place, and that dreams really, truly are possible.' And then, because airplanes speak in a language beyond words, I'll take him up in the evening summer sky and let the airplane show him what I mean."

With pilots and parents like Lane Wallace around, the future of flying will never be in doubt.

—Lamar Underwood
Spring, 2002

The
Greatest
Flying Stories
Ever Told

PART ONE

Toward the Unknown

Courage is the price that Life exacts for granting peace,
The soul that knows it not, knows no release
From little things,
Knows not the livid loneliness of fear,
Nor mountain heights where bitter joy can hear
The sound of wings.

How can Life grant us boon of living, compensate
For dull grey ugliness and pregnant hate
Unless we dare
The soul's dominion? Each time we make a choice, we pay
With courage to behold the restless day,
And count it fair.

—"Courage" by Amelia Earhart, 1927

(Editor's Note: "Courage" is referenced from the book *The Sound Of Wings: The Life of Amelia Earhart* by Mary S. Lovell. A chapter from that book is excerpted in Chapter Three of this anthology. "Courage" is described by Mary Lovell as appearing in Marion Perkins's "Who Is Amelia Earhart?" *Survey* magazine, July 1, 1928, p. 60.)

The Spirit of St. Louis

BY CHARLES A. LINDBERGH

(Excerpted from *The Spirit of St. Louis*)

Growing up in the second half of her famous father's life, daughter Reeve Lindbergh, the youngest of four brothers and two sisters, had missed the drama and immediate fame resulting from her father's historic flight across the Atlantic. On any occasion when she asked her dad about details of the flight, he would answer, "Read my book."

Fair enough. For Lindbergh's book, a Pulitzer Prize winner, is truly a landmark work of aviation history, written with verve and detail that make reading it an unforgettable experience, worthy of the flight that was one of the most significant events of the twentieth century.

Charles Lindbergh (1902–1974) was 25 when he took off from Roosevelt Field (today a shopping mall) near Garden City, Long Island, just after dawn on Friday, May 20, 1927. His specially prepared single-engine plane, the *Spirit of St. Louis,* was to take him on a voyage fraught with danger—to fly non-stop across the stormy springtime Atlantic to Paris. The same journey in reverse had been attempted earlier that month by the pilots Nungesser and Coli, who took off from Le Bourget field in Paris (Lindbergh's ultimate destination) and flew out over the Atlantic, never to be seen or heard from again.

Lindbergh's successful flight required approximately thirty-three and one-half hours. When he touched down at Le Bourget at 9:52 p.m. Paris time on the evening of May 21, Lindbergh, with no radio aboard, had no idea that tens of thousands of people would be standing there to welcome him. All that Saturday afternoon, however, from the time his plane was spotted upon reaching the Irish coast, radio reports had been coming in marking his progress. Parisians flooded the byways to the airport to witness the culmination of this historic event.

In making the flight, Lindbergh successfully dealt with many dangers known in aviation at that time—wind, ice, clouds, darkness. As a successful air-mail pilot, he had fought many battles with these old enemies before. But on his Atlantic flight, he was forced to take on the most demanding challenge he had ever faced: lack of sleep. He had barely slept a wink the night before he launched the flight, and by Friday evening over the Atlantic he was in extreme danger of falling asleep.

Lindbergh over the Atlantic at night . . . ice in the clouds . . . the need for sleep choking him. What a trip! You know he made it to Paris. But you ought to know what it was really like. And you ought to take the advice he gave his daughter: "Read my book."

I t's nine o'clock. I've reached an altitude of ten thousand feet. Clouds are still rising up to meet me, but the undulating plain they formed in early evening has given way to a foothill country of the sky. Passing over a misty summit, looking down onto a night-filled valley, I wonder what mountains lie ahead. It's cold at this altitude. I zip the flying suit up across my chest. It's cold enough for mittens and my wool-lined helmet, too, but not cold enough to put on flying boots, at least not yet—I'll let them go until later. Too much warmth would make me want, still more, to sleep.

I must straighten out my neck before it cramps permanently in this thrown-back position. I turn from the constellations of the stars to those of the instrument dials. I fix my eyes now on the glowing dots an arm's length before me, now on the points of fire millions of miles away. I travel with their vision back and forth. I feel first the compactness and detailed contents of my cockpit, my dependence on its instruments and levers, the personal proximity of its fabric walls; next, the unlimited expanse and solitude of space. Now, my plane is all-important, and life is vulnerable within it; now, neither it nor life is of any consequence at all, and consciousness seems unbound to either one.

As I fly through the body of night, haze lessens, and I discover that I'm among the cloud mountains themselves—great shadowy forms on every side, dwarfing my plane, dwarfing earthly mountains with their magnitude, awesome in their weird, fantastic shapes. Huge pillars push upward thousands of feet above the common mass. Black valleys and chasms open below me to unfathomed depth.

There's no possibility of flying above those mountains. They look higher than any clouds I ever saw before. How have I come into their midst without knowing they were there? I must have followed a great valley, blinded by the mist. Or did these sky giants draw aside to entice me among them, and close in again now that they have me hopelessly entrapped? Well, if I can't follow the valleys, I'll have to challenge the mountains themselves. Flying through an occasional thunderhead will be less tiring than spending hours on end down in the writhing body of the storm. A few minutes of blind flying followed by relaxation under a star-filled sky is nothing much to dread. It may even be a welcome change, sharpen my dulled senses, break up the monotony of routine flight.

Then I'll hold my course, stay above the stratus layer of the storm, and tunnel through the thunderheads that rise directly on my route.

A pillar of cloud blocks out the stars ahead, spilling over on top like a huge mushroom in the sky. I tighten my belt, push the nose down a bit, and adjust the stabilizer for level flight. In the seconds that intervene while I approach, I make the mental and physical preparation for blind flying.

The body must be informed sternly that the mind will take complete control. The senses must be drafted and lined up in strictest discipline, while logic replaces instinct as commander. If the body feels a wing dropping, and the mind says it is not (because the turn indicator's ball and needle are still centered), the muscles must obey the mind's decision no matter how wrong it seems to them. If the eyes imagine the flicker of a star below where they think the horizon ought to be, if the ears report the engine's tempo too slow for level flight, if the nerves say the seat back's pressure is increasing (as it does in a climb), the hands and the feet must still be loyal to the orders of the mind.

It's a terrific strain on the mind also when it turns from long-proven bodily instincts to the cold, mechanical impartiality of needles moving over dials. For countless centuries, it's been accustomed to relying on the senses. They can keep the body upright on the darkest night. They're trained to catch a stumble in an instant. Deprived of sight, they can still hold a blind man's balance. Why, then, should they be so impotent in an airplane?

The mind must operate as mechanically as the gyroscope which guides it. The muscles must move as unfeelingly as gears. If the senses get excited and out of control, the plane will follow them, and that can be fatal. If the senses break ranks while everything is going right, it may be impossible, with the plane falling dizzily and needles running wild, to bring them back into line, reinstruct them, and force them to gain control while everything is going wrong. It would be like rallying a panicked army under the fire of an advancing enemy. Like an army under fire, blind flying requires absolute discipline. That must be fully understood before it starts.

Wings quiver as I enter the cloud. Air roughens until it jerks the *Spirit of St. Louis* about as though real demons were pulling at fuselage and wings. No stars are overhead now to help, no clouds are below. Everything is uniform blackness, except for the exhaust's flash on passing mist and the glowing dials in my cockpit, so different from all other lights. What lies outside doesn't matter. My world and my life are compressed within these fabric walls.

Flying blind is difficult enough in smooth air. In this swirling cloud, it calls for all the concentration I can muster. The turn and bank indicators, the air speed, the altimeter, and the compass, all those phosphorescent lines and dots in front of me, must be kept in proper place. When a single one strays off, the rest go chasing after it like so many sheep, and have to be caught quickly and carefully herded back into position again.

Remember that flight with the mail last winter—I don't want to go through anything like that up here. I was racing nightfall and a storm to Peoria. Both arrived ahead of me. Caught, skimming treetops, in snow so thick I couldn't see lights half a mile away, I had to decide between cutting the switches and crashing onto whatever lay below, and giving up the earth to climb into the storm. The ceiling wasn't high enough to drop a flare. If I'd been over an area of fields, I might have landed by the vague outlines of late dusk— taken the risk of gullies, scattered trees, and fences. But I was close to the Illinois River, where patches of woodland are thick. And DHs have a reputation for burning when they crash—"Flaming Coffins," they're called. A pilot has about a fifty-fifty chance of living through a landing under such conditions. But to pull up blindly into the storm was almost as dangerous.

I had in my DH a new device for blind flying—a gyroscopic pitch- and-turn indicator. One of the transcontinental pilots had been experimenting with it, and wished to replace it with a more recently constructed mechanism. I'd persuaded our Corporation to buy the instrument from him. Here was the emergency for which I'd wanted it. The trouble was that it had been installed in my plane only a few days before, and I'd had no chance to test it out.

Instrument flying was new. I'd never flown blind, except for a few minutes at a time in high clouds. Could I keep my plane under control? If I turned to the instruments, how could I make contact with the ground again? Night had closed in behind me as well as ahead, and there was a low ceiling over both Springfield and St. Louis. Besides, even if I found a hole in the clouds, even if I saw the lights of a village farther on, how could I tell where I was, how could I locate the Chicago airfield—unless the ceiling there were much higher? "A good pilot doesn't depend on his instruments," I was taught that when I learned to fly.

All these arguments had passed through my mind—harsh impact of earth, splintering wood, ripping fabric, bruised body; versus hurtling, dizzy blindness, lost in the storm. Suddenly, the vague blur of treetops rushing past at eighty miles an hour, a hundred feet below, jumped up in a higher mass—a hillside? I didn't wait to find out. I pulled the stick back, gave up the ground, and turned to my instruments—untrusted needles, rising, falling, leaning right and left.

At a thousand feet, my DH was out of control—skidding, and losing altitude. Altitude was the thing above all else I had to hold. I shoved the throttle wide open, pushed rudder toward the skid, and pulled the stick back farther. The altimeter needle slowed—stopped—started to climb; but air still rushed sideways across my cockpit. It seemed there were more needles than my eyes could watch. There wasn't time for my mind to formulate orders and pass them on to hands and feet. Finally I got the DH headed straight and the wings leveled out. I thought I was getting my plane in hand. The altimeter went up to 1500 feet. But then—she whipped! Loose controls—laboring engine—trembling wings—the final snap as the nose dropped. I shoved the stick forward, but it was too late. While I was concentrating on turn and bank, I'd let my air speed get too low, and the wings stalled.

A whipstall at 1500 feet, with nothing but needles by which to orient myself! Stick neutral—leave the nose down long enough to pick up control—throttle still wide open—let the air speed rise to 80 miles—Stick back firmly—not too fast—Watch the altimeter (before the needle touches zero, you'll be dead)—900 feet—800 feet—700 feet (it steadies)—700—750—it starts to climb.

By then I'd learned that above all else it was essential to keep the turn indicator centered and the air-speed needle high. But in recovering I pulled the stick back too far and held it back too long. The DH whipped again, this time at only 1200 feet. But I made my recovery a little quicker. I'd decided to jump the moment the wings trembled in the stall, if my plane started to whip a third time. But I regained control after the second whip, and climbed slowly, and taught myself to fly by instruments that night.

It's cold up here at—I glance at the altimeter—10,500 feet—*cold*—good Lord, there *are* things to be considered outside the cockpit! How could I forget! I jerk off a leather mitten and thrust my arm out the window. My palm is covered with stinging pinpricks. I pull the flashlight from my pocket and throw its beam onto a strut. The entering edge is irregular and shiny—*ice!* And as far out into darkness as the beam penetrates, the night is filled with countless, horizontal, threadlike streaks. The venturi tubes may clog at any moment!

I've got to turn around, get back into clear air—quickly! But in doing so those instrument needles mustn't move too far or too fast. Mind, not body, must control the turn. My bodily senses want to whip the *Spirit of St. Louis* into a bank and dive it out of the thunderhead, back into open sky:

"Kick rudder hard—no time to lose—the turn indicator's icing up right now."

But the mind retorts, "Steady, steady. It's easy enough to get into a steep bank, but more difficult to get out of one and on your course again. If you turn too fast, you'll lose more time than you save; the plane may get entirely out of control."

"If the turn indicator ices up, it'll get out of control anyway. There's no time—only a few seconds—quick—quick—harder rudder—kick it—"

"Don't do anything of the sort. I've thought all this out carefully and know just what's best to do. You remember, you are to obey my orders!"

"Yes, yes—but just a little faster, then—just a little—"

"No, no faster; turn just the right amount. You're to do exactly what I say; no more, no less!"

"Just a little!"

"No, none!"

I keep pressing rudder cautiously until the turn indicator's needle creeps a quarter-inch to the left. I push the stick over just enough to hold the proper bank—ball high—low—center again—slow and steady movements— mustn't let jerks from the turbulence throw me off—The air speed drops ten miles an hour—The altimeter shows a hundred foot descent—

"Turn faster! You see the air speed's dropping. It's ice doing that! Quick, or it'll be too late!"

"No, it's not ice—at least not very likely. It's probably just the normal slowing down in a bank."

"But the altimeter's dropping too! It's ice, I tell you!"

I open the throttle another 50 revolutions. I don't dare push the stick forward very much to gain speed. The *Spirit of St. Louis* is too close to the top of the main cloud layer. There were less than a thousand feet to spare when I entered the thunderhead. That endless stratus layer is probably full of ice too. If I drop down into it, I may never see the stars again.

The altimeter needle falls 200 feet—300 feet—I push the throttle wide open—I *must* stay above that vast layer of cloud at the thunderhead's base— The bank indicator shows a skid—ball to right of center—a blast of air strikes my cheek—Ease up on the rudder—The air speed rises to 100 miles an hour—The pitch indicator points nose down—Stick back slightly—

I ought to be turned around now—Center the turn indicator—level out the plane—flashlight onto the liquid compass. (It's no time to trust the earth-inductor; it will be working backward anyhow, on a back-track heading.) No, not yet—about 30 degrees more to go—the card's swinging too much to read accurately.

I bank again and glance at the altimeter—10,300 feet. Good—it's gone up a little. I throw my flashlight onto the wing strut. Ice is thicker!

The earth-inductor needle begins moving backward, jumping erratically—Level out wings—About the right heading this time. Now, if the turn indicator doesn't ice up for a few minutes more—I put my hand out the window again—the pinpricks are still there.

Steady the plane. Make the compass card stop swinging—but the air's too rough—Is the turn-indicator getting sluggish—icing?—It seems to move back and forth more slowly—Everything depends on its working till I get outside this cloud—Just two or three more minutes—

My eyes sense a change in the blackness of my cockpit. I look out through the window. Can those be the same stars? Is this the same sky? How bright! How clear! What safety I have reached! Bright, clear, safe? But this is the same hazy air I left, the same fraction of an earthly hour. I've simply been existing in a different frame of space and time. Values are relative, dependent on one's circumstance. They change from frame to frame, and as one travels back and forth between them. Here I've found security where I left danger, flying over a major storm, above a frigid northern ocean. Here's something I never saw before—the brilliant light of a black night.

I was in the thunderhead for ten minutes at most; but it's one of those incidents that can't be measured by minutes. Such periods stand out like islands in a sea of time. It's not the limitless vista of experience, not hours or years that are most important. It's the islands, no matter how small. They impress the senses as they draw the eye at sea. Against them, years roll in and break, as waves upon a coast.

How much ice has accumulated on the plane? I move my flashlight from one spot to another. There's none on the bottom surface of the wing; but a thin layer forward on the strut tells me that it's also clinging to the airfoil's entering edge. I can't see that from my cockpit. There's not enough weight to make much difference, but what about resistance? Will the change in contours have great effect on speed? The air-speed needle shows a five-mile drop. Is it because of ice on the Pitot tube, or the increased drag of the plane? Have those few minutes in the cloud cost me five miles an hour cruising? That seems a heavy penalty to pay.

I turn southward, skirting the edge of the cloud pillar. I'll have to fly around these thunderheads. But can I? There are more masses ahead, and fewer stars. Will they merge into one great citadel of storm? Will I follow up a canyon in the heavens, as I often have on earth, to see it disappear against a mountain ridge? Or can I find real passes in these clouds, as I've found them in the mountains, where a plane can slip between the icy walls?

By day, I could set my course from the edge of one pillar directly to the next ahead, cutting down the angles of my zigzag route. But in the blackness of night, and in the haze that still contaminates the sky, one cloud merges into another miles behind it so I can't distinguish edge from center point. I have to bank as I approach, and follow around the vague wall of mist until I can again take up my compass heading.

Would it be wiser to change my course entirely, and try to fly around the whole storm area, as I'm now flying around its single columns? That's what my fuel reserve is for. Is this the emergency in which to use it? The ship lanes lie three hundred miles to the southward. Clear weather was reported there. But that report is getting old. It's almost fourteen hours since I took off from Roosevelt Field, and most of the weather messages were assembled some time before I left. They told of yesterday's weather, not today's. An unknown storm may have drifted over the ship lanes hours ago. "A high pressure area over the North Atlantic" can't last forever. Stars in the southern sky outline a ridge of cloud fully as high as lies ahead. And to the north? It's the same. I lean to the side of my cockpit and look back at the sky behind. It too is blocked out for thousands of feet above the level where I'm flying.

Great cliffs tower over me, ward me off with icy walls. They belong to mountains of anther world, mountains with forms that change; with summits that overhang; mountains alluring in their softness. There'd be no rending crash if my wing struck one of them. They carry a subtler death. A crash against an earthly mountain is like a sword stroke; one flash and it's over. But to plunge into these mountains of the heavens would be like stepping into quicksand. They enmesh intruders. They're barbaric in their methods. They toss you in their inner turbulence, lash you with their hailstones, poison you with freezing mist. It would be a slow death, a death one would have long minutes to struggle against, trying blindly to regain control of an ice-crippled airplane, climbing, stalling, diving, whipping, always downward toward the sea.

For the first time, the thought of turning back seriously enters my mind. I can climb another five or six thousand feet. The canyons up there may be wider. If they're not, and I can find no passes east or southward, I'll have to turn back—back through the haze-filled valleys behind me, back over Newfoundland, over those ice fields, over Nova Scotia, over New England, over the stretches of ocean between them, back fourteen hundred miles to New York, back for another start from that narrow, muddy runway on Long Island. And when I got back, if I could find my way out through the maze of passageways I've entered, if fog hasn't re-formed along the American coast, if that other storm in Nova Scotia hasn't blocked my route, then, after bucking the wind which has been carrying me along so swiftly, I'd have been about thirty hours in the air—long enough to reach Ireland, if I can keep on heading eastward. Think of flying long enough to reach Ireland, and ending up at Roosevelt Field!

Of course I could try diving quickly to a lower level, where the air may be too warm for ice to form. My mind grasps at the thought of a secret portal to the storm, a passageway deep down in the clouds. Might it lead me through to safety and the light of day again? Hours of blind flying seem less formidable, now—if they're free of ice.

No, it would be a fool's chance; the danger is too great. As far north as Newfoundland and in the cold of night, icing conditions probably extend down to the waves themselves. There's no reason to believe I'd find a ceiling underneath the clouds; and once I got down into their lower levels, ice would clog my instruments long before I could climb back up again.

The pillars of cloud multiply and thicken. I follow narrow canyons between them, weaving in and out around thunderheads, taking always the southward choice for course, edging toward the ship lanes and what I hope is clearer weather. Dark forms blot out the sky on every side, but stars drop down to guide me through the passes.

THE FIFTEENTH HOUR

Over the Atlantic

TIME—9:52 P.M.

Wind velocity	*Unknown*	Visibility	*Night—haze*
Wind direction	*Unknown*	Altitude	10,500 *feet*
True course	66°	Air speed	87 *m.p.h.*
Variation	33° *W.*	Tachometer	1700 *r.p.m.*
Magnetic course	99°	Oil temperature	35° *C.*
Deviation	0°	Oil pressure	60 *lbs.*
Compass course	99°	Fuel pressure	3 *lbs.*
Drift angle	10° *R.*	Mixture	4
Compass heading	89°	Fuel tank	*Fuselage*
Ceiling	*Unlimited above clouds*		

Fourteen hundred miles behind. Twenty-two hundred miles to go. All readings normal. I make the log entries and throw my flashlight onto the wing strut again. The coating of ice is thinner. It's evaporating slowly. When will it all be gone—in an hour, two, or five? I don't know. I've never picked up ice before on a flight long enough to consider its evaporation. When ice collected on our mail planes it stayed with us till we landed. Then we had either to wait until it melted, or peel it off the wings with our hands and beat it off the wires with a stick.

The haze continues to clear. I can see cloud formations farther away, fly closer to their walls, follow a straighter course through their valleys. There's another mushroomed column, miles ahead. Its top silhouettes against a star-brightening sky. I bank toward the southern edge, and settle back in my cockpit.

In keeping his heading by the stars, a pilot must remember that they move. In all the heavens, there is only one he can trust, only one that won't lead him off his course—Polaris, faint star of the northern pole. Those other more brilliant points of light, which at first seem motionless too, sweep through their arcs so rapidly that he can use them only as temporary guides, lining one up ahead, letting it creep to the side, dropping the first to pick a second; then the second for a third; and so on through the night.

As a child, I'd lie on my bed in Minnesota and watch the stars curve upward in their courses—the box-like corners of Orion's Belt—Sirius's piercing brilliance—rising over treetops, climbing slowly toward our roof. I would curl up under my blankets and web the constellations into imaginary scenes of celestial magnitude—a flock of geese in westward flight—God's arrow shooting through the sky. I'd make my wishes on the stars, and drift from wakefulness to sleep as I desired. Dreams of day and dreams of night could merge, while a planet's orbit had no effect on my security. There was no roar of an engine in my ears, no sound above the wind in leaves except the occasional whistle of a train, far away across the river.

The earth-inductor compass needle is halfway to the peg! That constellation I've been following has drawn me south of course. I press left rudder, banking the *Spirit of St. Louis* toward its proper heading. The needle moves up slowly, fluctuating; then overshoots the lubber line and drops down on the other side. I kick right rudder. The needle falters upward. I've never seen it act like that before. Is the earth-inductor failing, or am I half asleep and flying badly?

I throw my flashlight on the liquid compass overhead. It, too, is swinging—probably because of the double change in heading. I center the turn indicator, and hold my nose straight toward a star. But the earth-inductor needle is still top-heavy, and the liquid compass swinging doesn't stop.

Something's seriously wrong. I haven't depended too much on the earth-inductor compass. It's a new and complicated instrument, just past the experimental stage. If it failed completely, I wouldn't be surprised. But the liquid compass!—the whole flight is based on it. It's almost as essential for the liquid compass to work as for the engine to keep running. I never heard of *two* compasses giving out at the same time. The plane must still be turning. There must be something I've overlooked, some simple element I've neglected to consider. I've been sleepy—unobservant—my eyes must be tricking me. But the turn indicator is centered, and the stars confirm that I'm flying straight.

I look at the liquid compass again, as though trying to steady it by concentration alone. The card is rocking through an arc of more than 60 degrees—more than 90 degrees at times. Is it possible that I'm entering a "magnetic storm"? Most pilots scoff at their existence. They say magnetic storms are figments of imagination, like air pockets—just an excuse for getting lost—an attempt to explain away mistakes in navigation. Do magnetic storms really occur, then? Have I found my way through a labyrinth of cloud only to be confronted with this new, unknown danger? How large are magnetic storms? How long do they last? How far off will they throw a compass card? Do they have a permanent effect on the magnets?

The earth-inductor is hopeless. The needle's wobbling back and forth from peg to peg. There's no use paying any attention to it. But the liquid compass hesitates between oscillations, and remains fairly steady for several seconds at a time. I set my heading by these periods of hesitation, and hold it by the stars—except when a thunderhead gets in the way. As long as the stars are there, I can hold a general easterly direction; but there'll be little accuracy to navigation. God only knows where I'll strike the European coast. But if the liquid compass gets any worse, and high clouds shut off the sky, I won't know whether I'm flying north, south, east, or west. I may wander around in circles.

It would be easier to set a course if I could read the liquid compass without a flashlight. But the luminous figures aren't bright enough for that. While the flashlight's on, I can't see the stars; and it's hard to watch the compass overhead and the instruments on the board at the same time.

On what delicate devices flight depends: a magnetized bar of steel, slender as a pencil lead, reaching for the North Pole thousands of miles away, swinging with each bump of air, subject to the slightest disturbance, barely strong enough to point; yet without its directive force the horsepower of the engine, the aerodynamic qualities of the plane, the skills of the pilot, become meaningless. Detouring thunderheads, flying in an unknown wind ten thousand feet above the water, trying to follow a compass that swings thirty degrees and more off course, what hope have I to make my landfall on the southern Irish coast? In fact, can I expect to find a coast at all?

Last night I couldn't go to sleep—Tonight I can barely stay awake—If only I could balance the one against the other—awake—asleep—which is it that I want to be?—But look—there's a great black mass ahead—It *necks out toward the route my compass points for me*—Its ears stick up—its jaws gape wide—It's a cloud—or—maybe it's not a cloud—It could be a dragon, or a tiger—I could imagine it into anything at all—What's that whitish object, moving just beyond the window of my room—no, my cockpit—no, my room—I push goggles up to see more clearly—No, they're bedsheets I'm peeking out between—I'm in the nursery of my Minnesota home, and I'm afraid of the dark! Yes, I know that jungle animals don't jump through windows in the North. I trust my mother and my father when they tell me. But what *is that,* moving slowly, there behind the table? Suppose it *did* leap forth!

Look!—it's turning—it's crawling!—It's about to spring! Nerve and muscle tense against the terror—bed and nursery disappear. I raise my wing and kick left rudder, and watch the compass spin.

Dragons, tigers, jungle animals? How ridiculous! But it's true that I used to be fearful in the dark. It was years before I got completely over it. As a child, I could wander alone, tranquilly, through the most isolated places by the light of day. But at night my mind conjured up drowned bodies on the riverbank, and robbers behind every sumac clump. The reality of life was tame compared to my imagination's fantasies. It was what I couldn't see that frightened me—the python, slithering overhead, the face beyond the curtain. And most of all, the imaginary horrors that took no clear-cut form.

Stars drop lower. Valleys between thunderheads widen. I no longer have to look up through the window on top of the fuselage to find a stable point in space. Clouds tower and slant in bands and layers. Their outlines are sharp—too sharp for clouds at night; and the sky seems lighter. Can it be the first faint warning of the moon's approach? I turn to the south window—the night has a deeper shade. Yes, it must be the morning twilight of the moon—a luminous wash, barely perceptible, on the northeastern wall of night.

But so soon, and so far north! I thought it would rise on the other side of my plane. Have the swinging compasses turned me that many degrees off course? Am I heading for Africa instead of Europe? I hold the *Spirit of St. Louis* straight with the constellation overhead, and watch the liquid compass. No, if the compass is correct during its steady periods, I'm pointed about on course—maybe a little southward, but not over ten degrees, not enough to explain the moon's position unless—unless *both* compasses are *completely* wrong! Maybe the steady periods aren't caused by the magnets pointing toward the pole. Maybe they're the result of some freak vibration. But they can't be very far in error with the North Star in approximate position, high on my left.

I glance at the chart on my lap. Of course I've shortened the night by flying with the earth's rotation. And I've been bending more and more eastward as I follow the great-circle route, cutting each meridian at a greater angle than the last, changing course a degree or two clockwise every hour through the day and night. When I took off from New York, I pointed the *Spirit of St. Louis* northeast; and when the sky cleared, the morning sun beamed down on an angle from my right. But when I reach Paris, I'll be heading south of east, and heavenly bodies will be rising on my left. After all, it's late in May. That's probably where the moon should be for a pilot on the great-circle route.

I'd almost forgotten the moon. Now, like a neglected ally, it's coming to my aid. Every minute will bring improving sight. As the moon climbs higher in the sky, its light will brighten, until finally it ushers in the sun. The stars ahead are already fading. The time is 10:20. There have been only two hours of solid darkness.

Gradually, as light improves, the night's black masses turn into a realm of form and texture. Silhouettes give way to shadings. Clouds open their secret details to the eyes. In the moon's reflected light, they seem more akin to it than to the earth over which they hover. They form a perfect setting for that strange foreign surface one sees through a telescope trained on the satellite of the world. Formations of the moon, they are—volcanoes and flat plateaus; great towers and bottomless pits; crevasses and canyons; ledges no earthly mountains ever knew—reality combined with the fantasy of a dream. There are shapes like growths of coral on the bed of a tropical sea, or the grotesque canyons of sandstone and lava at the edge of Arizona deserts—first black, then gray, now greenish hue in cold, mystical light.

I weave in and out, eastward, toward Europe, hidden away in my plane's tiny cockpit, submerged, alone, in the magnitude of this weird, unhuman space, venturing where man has never been, irretrievably launched on a flight through this sacred garden of the sky, this inner shrine of higher spirits. Am I myself a living, breathing, earth-bound body, or is this a dream of death I'm passing through? Am I alive, or am I really dead, a spirit in a spirit world? Am I actually in a plane boring through the air, over the Atlantic, toward Paris, or have I crashed on some worldly mountain, and is this the afterlife?

Amelia Earhart and the Flight That Failed

BY MARY S. LOVELL

(Excerpted from *The Sound Of Wings: The Life of Amelia Earhart* by Mary S. Lovell)

Amelia Earhart was eleven years old when she saw her first airplane in 1908 during a visit to the Iowa State Fair. She had her first flight in 1919 and began taking lessons in 1921. By 1922, she had set a new altitude record for women (19,000 feet) in the first of what would become a skein of aviation accomplishments. Among these are such milestones as: first woman to fly the Atlantic solo; first person to cross the Atlantic solo twice; first person to fly solo across the Pacific Ocean, Honolulu to Oakland, California. In 1936 she acquired a specially equipped twin-engine plane manufactured by Lockheed and financed by Purdue University and began planning an around-the-world flight to take place in 1937.

On Friday morning, July 2, 1937, Earhart, after completing 22,000 miles of the journey in the previous six weeks, took off from the Lae, New Guinea, bound for Howland Island, a tiny speck of land in the vast Pacific. With Earhart was the flight's very experienced navigator, Fred Noonan. They were estimating a flying time of about 18 hours. The Lockheed Electra carried enough fuel for about twenty to twenty-one hours of flying, depending on fuel management. Only 7,000 miles remained to the California coast now. Once the difficult leg they were now facing was passed, the remainder of the journey would seem to be a certainty.

Radio communications from the plane were heard and answered at various times that day, through the night, and after dawn the following morning when the plane was nearing the vicinity of its destination, where the Coast Guard Cutter *Itasca* was posted and monitoring Earhart's radio calls. The last calls from Earhart seemed to indicate problems in finding the island. Finally, all contact with the plane was lost. The Electra, Amelia Earhart, and Fred Noonan had disappeared. The massive searches that followed did not yield a trace of the plane or its pilots.

Several books and numerous articles surrounding the death of this brave and talented flyer have speculated on what happened that July day. Some tabloid trash went so far as to suggest Earhart had fallen into the hands of the Japanese and was executed. The truth seems likely to be more prosaic, as Mary Lovell reveals in her excellent biography of Amelia Earhart. Noonan and Earhart missed the island, and ran out of fuel while desperately searching for it under the vast Pacific sky that morning. When the Electra went down, they were either killed outright, failed to escape from the body of the plane after a crash landing, or perished in the sea after getting out of the craft.

The brave lady who wrote the poem "Courage," who showed the world more than a thing or two about flying, needed a bit of luck that morning. She needed a break.

She didn't get it.

At precisely 00:00 hours Greenwich Mean time (10:00 A.M. local time at Lae) on Friday, July 2, 1937, Amelia Earhart lined up the Lockheed Electra on the one-thousand-yard runway at Lae and started her takeoff run. It was 12:30 P.M. on July 1 at her destination, Howland Island, which lay two and a half time zones away and across the international date line.

It is believed that the Electra was loaded with one thousand gallons of fuel—less than its tanks held but probably the maximum possible for takeoff on Lae's unpaved runway. This would allow between twenty and twenty-one hours of flying, depending on how efficiently the airplane was flown. Before leaving the United States, Amelia had predicted that the flight from Lae to Howland would take about eighteen and a half hours but, according to *Itasca*'s log, Commander Thompson expected Amelia to arrive at Howland at approximately 19:30 hours (GMT). There is no evidence that Amelia had ever taken off with such a heavy fuel load prior to this. For the Oakland-Honolulu flight, the Electra carried 947 gallons, but on that occasion Paul Mantz had been at her side, working the throttles and retracting the landing gear.

The runway at Lae ended abruptly with a twenty-five-foot drop to the waters of the Huon Gulf and witnesses stated that the takeoff was hair-raising for there was not a breath of wind to help the Electra into the air on that hot, clear morning. Commercial pilot Bert Heath, who was flying into Lae in his trimotored Junkers airplane, watched the Electra's lumbering takeoff run from high above the field. He recalled that close to the seaward end of the runway, a dirt road crossed it. There was a high camber on this road and as the Electra hit the crest, it bounced into the air and "over the drop-off," flying so low over the sea that the propellers were "throwing spray." Mr. Heath noted that the dust kicked up by the Electra over the dirt road hung about in the still air for some time and did not disperse.

A report by James Collopy (district superintendent of Civil Aviation) stated:

> . . . after taking every yard of the 1000 yard runway from the northwest end of the aerodrome towards the sea, the aircraft had not left the ground 50 yards from the end of the runway. When it did leave it sank away but was by this time over the sea. It continued to sink about five or six feet above the water and had not climbed to more than 100 feet before it disappeared from sight.

> In spite of this, however, it was obvious that the aircraft was well-handled and pilots of Guinea Airways who have flown Lockheed aircraft were loud in their praise of the take-off with such an overload.

In the days that Amelia and Noonan had spent in Papua New Guinea, they made contact with Harry Balfour, radio operator at Lae. Between them, they arranged a radio schedule and after the Electra's takeoff, Balfour listened for Amelia's transmissions on her daytime frequency of 6210 kilocycles; within the hour, he received her first communication. From this point on, the flight can be pieced together only by examination of the various radio transmissions that took place over the following twenty or so hours.

The first radio communications between Amelia and Balfour were not logged, but Balfour is on record as stating that he was in contact with her every hour, at which time, position reports and height were received. Amelia's schedule was to transmit at fifteen and forty-five minutes past each hour, and to listen on the hour and half hour to receive messages.

After about five hours, Amelia reported that she was "at 10,000 feet but reducing altitude because of banks of cumulus cloud," and two hours later she told Balfour that she was "at 7,000 feet and making 150 miles per hour." Amelia meant airspeed; the Electra's ground speed would have been significantly less, depending on the strength of the head winds.

At 07:20 hours (GMT), Amelia provided a position: "Latitude: 4 degrees 33.5' South. Longitude: 159 degrees 07' East." This placed the Electra more or less directly on the great circle course for Howland Island, some twenty miles southwest of the Nukumanu Islands—which should have been visible, as it was still daylight—and giving Noonan a perfect opportunity to check his navigation before nightfall. Although this transmission was given only a third of the way into the flight, it was to be Amelia's only position report, and thus it has great significance.

An important factor is that having covered more than 846 miles (735 nautical miles), the Electra had achieved a ground speed of only 118 miles per hour (103 knots). This was considerably less than Amelia would have hoped for with an average airspeed of 150 mph. The shortfall is not unreasonable, however, if one takes into consideration a combination of the excess weight in the early stages of the flight when the Electra was slowly climbing to ten thousand feet, and stronger-than-forecast head winds. The last weather report that Amelia is known to have received before takeoff forecast 12–15 mph east southeast winds, but they were actually 25 mph. Lae had transmitted the increased wind strength of "25 mph to the *Ontario* and 20 mph from there to Howland" but there is uncertainty that Amelia ever received this message due to the twenty-four-hour delay in transmission while cables were routed through American Samoa.

Noonan could have determined the approximate wind strength and its direction, though, by using the drift meter and smoke bombs, and if there had been any doubt at this time about whether they could complete the flight, Amelia would have turned back (as she had at Karachi when conditions were marginal).

From the Electra's reported position near Nukumanu Island, Howland Island is 1,714 miles on the great circle route laid out for Amelia by Clarence Williams. With a decreasing gross weight as the fuel was used up, and head winds forecast to decrease to 20 mph, the Electra was easily capable of averaging 130 mph or better over the remaining distance, but this would give a revised estimated time of arrival (ETA) in the vicinity of Howland of 20:38 GMT, leaving very little margin for error regarding fuel.

At 08:00 hours GMT, Amelia made her final radio contact with Balfour. At this stage, her transmissions were becoming fainter but he was able to make out her message that she was "on course for Howland Island at 12,000 feet." When Amelia informed him that she was changing from her daytime frequency, 6210 kilocycles, to her nighttime one, 3105 kilocycles, Balfour stated, "We asked her to remain on her present frequency but she told me she wished to contact the American Coast Guard cutter *Itasca* so there was nothing to do about it but pass the terminal forecast to her and the upper air report from Ocean Island."

The U.S.S. *Ontario,* designated as Amelia's guard ship on that Lae-Howland flight, was lying 423 miles (368 nautical miles—a nautical mile is equal to 1.151 statute miles), from the Electra's last reported fix near the Nukumanu Islands. The *Ontario's* position—latitude 3 degrees 9' South; longitude 165 degrees 06' East—put the ship less than ten miles north of Amelia's great circle track. Amelia had previously requested by cable that the *Ontario* BROADCAST LETTER N FOR FIVE MINUTES AFTER HOUR GMT FOUR HUNDRED KILOCYCLES WITH OWN CALL LETTERS REPEATED TWICE END EVERY MINUTE. This was done. *Ontario's* captain, Lieutenant H. W. Blakeslee, recalls that it was rainy and overcast that night. The ship had no light switched on but had one ready should they hear a plane.

At 10:30 hours (GMT), three hours and ten minutes after Amelia had reported her position near Nukumanu Island to Balfour, residents on Nauru Island reported hearing Amelia say ". . . ship in sight ahead." Assuming that the ship Amelia reported was the *Ontario,* the Electra was then averaging a ground speed of 125 mph, almost exactly what she should have been making with reported 25-mph head winds and a fuel load approaching recommended operating levels. The reports from Nauru of what Amelia said are open to question, however, for some listeners claim Amelia actually said ". . . lights in sight ahead."

It has been suggested that from the Nukumanu Islands, Noonan may have set a course for Nauru, some 607 miles to the northeast, in order to have one last opportunity to check his astronavigation before setting out over open ocean; the sky was overcast and he may not have been able to take sextant bearings on celestial bodies. Such a detour would have added only thirty-one miles to the flight. It is much more likely, however, that he continued on the direct great circle course for Howland, knowing that some four hundred miles on that track the U.S.S. *Ontario* was waiting, its *specific* duty being to act as a guard ship for the flight.

Captain Blakeslee had no notification that Amelia had left Lae nor any acknowledgment from Amelia that she had heard *Ontario's* Morse signal. A watch was kept, but the crew neither heard nor saw any sign of the Electra. Two hours after the time Amelia was due to pass overhead, the ship, which had been on station for ten days, was ordered to return to port for essential supplies.

Another possibility that has been suggested is that Amelia saw the S.S. *Myrtlebank.* The *Myrtlebank* was headed for Nauru and arrived there at dawn on the following day. At the time of Amelia's report, this ship was lying at latitude 1 degree 40' South; longitude 166 degrees 45' East—some forty-four miles north of Amelia's great circle route.

The suggestion that Amelia flew via Nauru seems unlikely, however, for in order for her to have been within visual distance (thirty-four miles) of the lights on Nauru within three hours and ten minutes after her last reported position, she would have to have been making a ground speed of 170 mph, a significant improvement over the previous calculated performance. On the other hand, from thirty miles off Nauru, she might have been able to see *both* the *Myrtlebank* and the lights on Nauru.

The flight via Nauru would have been possible, of course, if Amelia had transmitted her position to Balfour some time after it was plotted. Her casual attitude toward radio and reporting procedures makes this not unlikely (in the entire trip from Miami to date, she had provided only seven position reports). The position report could have been provided by Noonan at any time following Amelia's previous radio contact, and it is not beyond credibility that, having received it, Amelia waited until her 5:15 scheduled time to report. From the *Itasca* log alone, it is obvious that she was seldom pedantic about keeping to radio schedules.

The wife of the district administrator on Nauru, Mrs. Garsia, noted in her diary for July 2 that Amelia had taken off from Lae and had been notified of the existence of the new Nauru light and local weather. She recorded picking up Amelia's radio transmissions on a shortwave receiver from 6 P.M. [Nauru time] but though the transmissions became increasingly loud, Mrs. Garsia could never make out what was being said. "Amelia was due near or over Nauru at 9:40 P.M. but though we watched she did not come near." It has been said that Nauru could never have picked up Amelia's radio signals if she had not been within fifty miles of the island, but Amelia had broadcast her position to Lae from almost nine hundred miles away; *Itasca,* subsequently, first heard Amelia's faint transmissions at something close to the same distance; and on the following day, a Nauru resident reported receiving Amelia's transmissions when she was in the vicinity of Howland Island.

There is no real evidence as to the precise track of the airplane after Nukumanu. No one saw or heard the plane fly over, and this fact alone has meant that even the most outrageous theories regarding Amelia's course cannot be refuted. Neither can it be refuted that she flew on a direct course for Howland, in which case she must have flown over the Gilbert Islands at about 3 A.M. [local time], but since no assistance had been requested in advance, no one was standing by to receive a transmission.

From 10:30 hours (GMT), there was no radio contact from Amelia for several hours. At 12:42 hours (GMT), *Itasca* signaled coast guard headquarters in San Francisco, where George was waiting for news, to advise that they had not heard from Amelia but saw "no cause for concern as she is still 1000 miles away." It was 14:15 hours (GMT) before the first transmissions from Amelia were received by *Itasca* on 3105 kilocycles.

In the ship's radio room, Chief Radioman Leo G. Bellarts was accompanied by Radiomen Thomas O'Hare and William Galten. A primary radio log, kept by Bellarts, was backed up by a secondary log maintained by one of his operators. In addition to the two radio receivers, there was a loudspeaker. Others waited in the room to hear Amelia: Commander Warner K. Thompson, *Itasca's* captain; three of his officers, Lieutenant Commanders Lee Baker and Frank Kenner, and Ensign William L. Sutter; Richard Black (later Admiral Black) from the Department of the Interior but representing George Putnam regarding administrative details on this leg of the flight; and two press agency reporters, H. N. Hanzlick of the Associated Press and James Carey of United Press International.

At first, no one could make out what Amelia was saying above the static, but Bellarts caught her "low monotone voice" saying: ". . . cloudy and overcast." The two pressmen were familiar with Amelia's voice and agreed that it was she but could not decipher what she was saying. Bellarts sent weather forecasts to KHAQQ [Amelia's call sign] at *Itasca's* scheduled transmission times of 14:30 and 15:00 hours (GMT). Nothing was heard from Amelia during her 14:45 hours slot but this caused no particular concern. She was still a long way from Howland and no one was surprised that reception was poor over such a distance.

At 15:15 hours (GMT), Amelia came in on cue: "*Itasca* from Earhart. Overcast . . . will listen on hour and half hour on 3105." Her voice was clearer this time and understood by others in the radio room. Fifteen minutes later, Bellarts called KHAQQ and gave details of the weather for the Howland Island area before asking, "What is your position? When do you expect to reach Howland? *Itasca* has heard your phone, so go ahead on key. Acknowledge this broadcast next schedule." Each phrase was carefully repeated twice in accordance with the accepted code of practice.

No signal was picked up from Amelia during her next transmission time (15:45 hours GMT), so Bellarts repeated his previous message at 16:00 hours. When Amelia failed to report at 16:15, Bellarts waited eight minutes and transmitted weather by voice and Morse code on frequency 3105 kilocycles.

As Bellarts finished calling, he switched to receive and picked up a faint transmission from KHAQQ that he logged as ". . . partly cloudy . . .," noting after it the code "S1," indicating a very weak reception (signal strengths being measured on a table of 1–5. S1: Very faint; S2: Faint; S3: Fair; S4: Good; S5: Very loud). A minute later, Amelia transmitted again but so faintly that neither Bellarts nor anyone else in the room could make any sense of it.

The *Itasca*'s log reveals that for the next hour and twenty minutes, Bellarts sent weather data and his "What is your position?" message at each scheduled transmission time, but nothing was heard from Amelia. At 17:44 hours (GMT) she broke in at strength 3: "[want] bearing on 3105, on hour, will whistle in mic—" Her transmission broke off, but moments later she spoke again: ". . . about two hundred miles out. Approximately. Whistling now." She whistled briefly but the transmission was too short for a direction finder to home in on the signal.

Dawn was now breaking, recorded in *Itasca*'s log at 17:45 hours GMT (6:15 A.M. ship's time). At 18:00 hours (GMT), Amelia's voice boomed in on schedule at strength S4: "Please take bearing on us and report in half an hour. I will make noise in microphone. About one hundred miles out." Bellarts recorded that, once again, she was on the air too briefly to take direction-finding bearings, but all on board the *Itasca* were heartened as the strength of her calls increased, indicating that she was getting closer.

Amelia's report that she was "about one hundred miles out" fitted in well with the time that *Itasca* expected her to arrive (19:30 hours GMT, according to the ship's log), and they all confidently expected the Electra to appear overhead in an hour or so, but they heard nothing until 19:12 hours (GMT) when Amelia's voice filled the loudspeaker. Signal strength S5 this time—maximum reception. Bellarts logged her call: "KHAQQ calling *Itasca*. We must be on you but cannot see you but gas is running low. Been unable to reach you by radio. We are flying at altitude 1,000 feet." But the secondary log reads: "Earhart says running out of gas. Only half hour left/can't hear us at all/we hear her and are sending on 3105 and 500 same time constantly and listening in for her frequently."

These two accounts have been taken by some researchers as sinister evidence. Why, they have asked, is Amelia sometimes quoted as saying "gas is running low" and at other times, "running out of gas, only half an hour left"?

Reading the pages of these logs, and pages of radio transmissions unrelated to the Earhart mission, the reason is obvious. This mission, while interesting, was really not very special. It was just another job for *Itasca;* she had performed a similar task some months earlier for the Pan Am survey. The men keeping those logs had no reason to believe that their scribbled "radio shorthand" notes would be the subject of countless books. Busy with their duties, they recorded the gist of what they heard, rather than the *precise* wording.

The message received at 19:12 hours GMT (7:42 A.M. ship's time) was actually written in the primary log by Bellarts as follows:

0742 KHAQQ CLING [CALLING] ITASCA WE MUST BE ON YOU BUT CANNOT SEE U BUT GAS IS RUNNING LOW BEEN UNABLE REACH YOU BY RADIO WE ARE FLYING AT ALTITUDE 1,000 FEET.

While the second log, in the same radio room, maintained by another radio operator, reads:

0740: EARHART ON NW [NOW] SEZ RUNNING OUT OF GAS ONLY 1/2 HOUR LEFT CAN'T HR US AT ALL/WE HEAR HER AND ARE SENDING ON 3105 ES 500 SAME TIME CONSTANTLY AND LISTENING IN FOR HER FREQUENTLY

The reported version of what Amelia actually said was made some hours later, after the radio operators and others present in the radio room examined the written log and discussed what *had* been said. The sensational suggestions by several writers that this was done on instructions of the U.S. government to cover up or distort the facts may be discounted if only because of the wealth of information that exists from so many different sources, ranging from the military records to the press (not renowned for keeping secrets) and the inhabitants of various islands. The explanation offered by *Itasca*'s commanding officer is more credible:

The transcript of the radio logs from 0200 until 0930 is necessarily not complete due to the rapidity of events and also due to the Earhart exclusive use of voice, only partially received. At these times tuning was so essential that parts of the actual message may not be given. Officers of appropriate rank and experience were present . . . the radio log stands as it was written at the time and has not been changed or corrected. The transcript is an actual transcript [and] the portions [extracted above] form a true representation of the picture.

Amelia's message that she was low on fuel caused some consternation in the radio room, but there was little that Commander Thompson and his crew could do other than set in motion provisional plans to go to the aid of the fliers if they had to ditch. On Howland Island, a high-frequency direction finder had been installed, manned by Radioman Frank Cipriani. However, when contacted by the *Itasca,* he reported that he was unable to get any fixes on Amelia because her transmissions were too brief. The equipment Cipriani was using was in any case very much experimental and in the "breadboard" stage. It had been taken to Howland by Richard Black and was powered by a dry-cell battery arrangement. In use for the whole night while Cipriani tried in vain to "fix" on Amelia's transmissions, by daybreak—when it was most likely to have been of use—the batteries had run down and were insufficient to power the unit.

Itasca was already making heavy black smoke, which, in the light breeze of under 10 mph, drifted very little. Commander Thompson estimated that the smoke should have been visible for twenty miles or so. Over Howland, the sky was clear and the sun already hot, but to the north and northwest, thirty or forty miles away, *Itasca*'s crew could see formations of large cumulus clouds.

Thompson now decided to break with the radio schedule and transmit and receive on a continuous basis. At 19:13 hours (GMT), within a minute of receiving Amelia's message that she was low on fuel, Bellarts transmitted by voice: "*Itasca* to KHAQQ. Received your message signal strength five. Go ahead please." He followed this with a long stream of letter *A*'s transmitted by Morse-code key on the 3105 frequency. This was repeated at 19:17 (GMT) hours with no response. Two minutes later, he tried again. "*Itasca* to KHAQQ: Your message okay please acknowledge with phone on 3105." Again he sent out a long series of letter *A*'s by key.

Tension was now beginning to build in the radio room. Amelia had not once acknowledged receiving any message from *Itasca* and her messages to them were brief and incomplete. However, their apprehension must have been nothing to that experienced by Amelia and Fred Noonan, who had now been in the air for over nineteen hours. They were physically tired; there was no sight of Howland Island, where Noonan clearly expected it to be; fuel supplies were running low; and Amelia had not been able to make any radio contact. This meant that she was not only unsure that her transmissions were being received but also could not take direction-finder bearings on her DF unit. It is not difficult to imagine the concern that Amelia and Noonan must have been feeling.

At 19:28 hours (GMT), Amelia called *Itasca*. "We are circling but cannot hear you, go ahead on 7500 now or on the schedule time on half hour." Again *Itasca* received her message at signal strength 5. Bellarts, confused by the sudden change of frequency, nevertheless immediately sent out a constant stream of *A*'s on 7500 as requested, and called by voice, "Go ahead on 3105."

To the immense relief of everyone in the radio room, Amelia acknowledged receiving their message almost immediately. "KHAQQ calling *Itasca*. We received your signals but unable to get a minimum. Please take a bearing on us and answer 3105 with voice." She then transmitted a series of long dashes for five seconds or so. The time was 19:30 hours (GMT). It was the only time during the flight that Amelia was in two-way contact with *Itasca* and her reception of the transmission was too poor to enable her to tune her direction finder to the signal.

Five minutes later, having tried and failed to take a bearing on Amelia's transmission, *Itasca* radioed Frank Cipriani on Howland, but he had been no more successful. Bellarts called on 3105 again: "*Itasca* to KHAQQ. Your signals received okay. We are unable to hear you to take a bearing. It is impractical to take a bearing on 3105 on your voice. How do you get that? Go ahead." There was no response, so he tried yet again, sending on 7500 kilocycles, since the only satisfactory contact had been received by Amelia on that frequency. "Itasca to KHAQQ. Go ahead on 3105 or 500 kilocycles."

For more than half an hour, Bellarts sent voice and code signals on all frequencies that he knew Amelia's radio was capable of receiving. The transmissions in the ship's log are interspersed with the laconic "No answer," but the frustration and concern felt by all present were subsequently to be made obvious through the ship's log.

At 20:14 hours (GMT) the listeners in the radio room heard Amelia's voice for the last time. Volume was still strength 5 although it was not quite so loud as the previous transmission. She was sending on 3105 kilocycles and spoke rapidly; some listeners described her transmission as sounding hurried and almost incoherent: "KHAQQ to *Itasca*. We are on the line of position one five seven dash three three seven. Will repeat this message on 6210 kilocycles. We are running north and south."

Bellarts immediately replied: "*Itasca* to KHAQQ. We heard you okay on 3105 kilocycles. Please stay on 3105 do not hear you on 6210. Maintain QSO on 3105." He sent this by voice on 3105 and by key on 7500 kilocycles but nothing further was heard from Amelia on any frequency.

Itasca continued to transmit on all frequencies until 21:30 hours (GMT), when Commander Thompson made the assumption that Amelia had ditched. He then set in motion the search procedure.

Commander Thompson had only the small amount of information Amelia had provided on which to base his search. There seemed no other reason for Amelia's report of flying "at 1,000 feet" than that she was flying beneath a cloud base, for it is easier to see over a greater distance at a higher altitude. Since the only cloud in the vicinity was to the north and northwest of Howland, Thompson concluded that the Electra's most likely position was in that sector. Furthermore, if Amelia had been in the clear skies surrounding and to the south and east of Howland, she ought to have been able to see the pall of black smoke hanging over *Itasca*. Notifying San Francisco of his intention, Thompson gave the command to proceed north at full steam.

Apart from the more sensational claims by writers who believe that Amelia and Noonan were not even in the vicinity of Howland when they disappeared, there has been much conjecture about what might have happened in the cockpit of the Electra during that flight. The main theories are that Amelia and Noonan had a disagreement, causing Amelia to disregard his instructions (as she had done at Dakar); or that Noonan was suffering from a hangover and consequently was not up to the mental effort required, so that the Electra was lost due to his incompetence. The permutations and speculation are seemingly endless, but unless the Electra is found, we will never know the answer, and perhaps not even then.

What is certain is that Amelia and Noonan were exhausted. Even had they been fit and rested when they left Lae, which they were not, they had been awake for more than twenty-four hours, and flying for over twenty, when Amelia was last heard on the radio. The dimensions of the Electra's cockpit were, according to an article written by Amelia before her departure from the United States: "four feet six inches by four feet six inches," and for the entire twenty hours Amelia would have been sitting in the same upright seat, unable to leave it at all. She was surrounded by equipment and would have found it difficult to stretch her limbs, though in smooth conditions she could allow the giro to fly the plane from time to time. The ceaseless engine noise in that confined space, and the vibration in the all-metal fuselage with its empty fuel tanks (also experienced by the crew of the 1967 commemorative flight to Howland in a Lockheed Electra), would have added greatly to the physical stress.

Amelia's only contact with Noonan was on the occasions when he crawled over the fuel tanks and sat in the right-hand copilot's seat in the cockpit, when the two would have had to shout over the engine noise in order to communicate. Otherwise, notes were passed between them on the end of a cleft bamboo pole or by using the pulley system.

Although Fred Noonan was able to stand up and stretch his legs in a limited manner, he would have been working all night in the cramped, poorly lit navigation area and would also have been subject to the constant noise. In these circumstances, it is not unlikely that he could have made a mathematical error at some point, which progressed through his final calculations.

The most likely scenario is that the plane was in the vicinity of Howland and may, at one point, have even been within visual distance but missed it either due to clouds in the line of sight (a relatively small cloud would obscure such a small target) or, if the fliers ever emerged into the clear air north of Howland, possibly missed the tiny island in the glare of the rising sun on the water. This, according to eyewitnesses, reduced visibility to a practical fifteen miles or less, although the *Itasca* at that time was making heavy black smoke, which at one point stretched for ten miles.

What seems most likely is that Amelia ran out of fuel and was forced to land at sea shortly after her last verified message to *Itasca* at 20:14 hours (GMT). Using extensive information gathered during eighteen years of research on the subject, Elgen and Marie Long claim to have pinpointed the most likely position, which is "35 miles west, north west of the island . . . in an area perhaps twenty miles wide and forty miles long."

Coincidentally, and perhaps curiously, using an entirely different line of investigation, British researcher Roy Nesbit (a former Royal Air Force navigator) has reached precisely the same conclusion.

The author's conclusion is that Amelia probably went down within a hundred miles of Howland Island. For lack of proof in any direction, it is not beyond the bounds of *possibility* that Amelia and Noonan were picked up by one of the Japanese fishing boats known to be several hundred miles to the north of Howland. Amelia reported running "north and south on a line of position." This would be a line calculated by Noonan using a sun sight taken earlier in the day, but the Electra might have been *anywhere* on that line, because there was no second coordinate given and one assumes from Amelia's message that they were flying up and down the line within the area where they expected to find the island.

There is substantial testimony to support the fact that a Caucasian couple was seen on the Japanese-mandated Marshall Islands (some six hundred miles to the north of Howland) at some time around 1937, but that this couple was Amelia Earhart and Fred Noonan is open to serious doubt. The testimony of many islanders made more than twenty years later may well have been made in the light of rewards offered for information, coupled with a genuine desire to please the interrogator. There is no really convincing evidence that this was Amelia Earhart's ultimate fate. All the documented facts point to a ditching off Howland Island, which neither flier survived.

It is known that Paul Mantz attempted to verbally prepare Amelia for a ditching in the Pacific in her Vega, prior to her solo transpacific flight; and presumably he may have gone through a similar routine for the Electra. Verbal instructions, however, would have been of minimal help in an actual sea landing.

Amelia would have known that she should keep the undercarriage retracted, but if she was completely out of fuel, as seems likely, she would have had to cope with a dead-stick forced landing on the water. Even over a smooth sea, it is difficult to judge a plane's height above the water surface, and if she stalled too high, the impact alone would have been sufficient to have killed both occupants. In rough seas, it would have been even more difficult to judge and such a stall, or even the effect of the waves in a rough patch of sea, would have caused severe damage to the Electra. When the *Itasca* steamed to the area under heavy cloud to the north of Howland Island, on July 2, she reported a turbulent sea with waves four to six feet high.

Photographs of the Bendix receiver, installed in the Electra immediately before Amelia and Noonan left Miami, were shown to the author by Elgen and Marie Long. They reveal that the unit was sited immediately in front of Amelia's head and was mounted on the top of the instrument panel; its casing was a sharp-cornered metal box. If Amelia had hit her head on this in a ditching process (and it is difficult to see how she might have avoided it, for she had no shoulder harness), it is probable that she would have been knocked unconscious. Noonan may have been in the cockpit, having been warned of a possible emergency landing, or he may have remained aft in the cabin.

Assuming Amelia remained conscious after a forced landing, that everything had gone according to the book, that she had lowered the flaps correctly, glided in perfectly, stalled out at the correct height above the water, and that the plane remained intact, there were other problems to take into account. The floating attitude of the Electra, due to the weight of the engines and empty fuel tanks, was nose down-tail up. With no way of escape through the cockpit hatch (which would have been underwater), Amelia would have had to scramble up the almost vertical slope of the fuel tanks, perhaps with the help of Noonan, or perhaps without his help if he was unconscious.

The life raft and emergency equipment were housed behind the rear cabin bulkhead, at the rear of the plane, and even to have reached this would have required tremendous effort in an upended plane. It would then have been necessary to retrieve this not inconsiderable weight from stowage above head height (due to the plane's nose-down attitude); partially inflate the raft using the CO_2 canister; open the cabin door; throw the life raft into the water and jump after it, continuing the inflation process once safely on it. No trace has ever been found of Amelia's life raft.

There have been at least ten occasions when Lockheed Electras have made forced landings on water. One detailed report reads:

> The pilot experienced engine failure in the right engine and shortly afterwards lost some power in the left engine. He was able to land on a smooth sea, just off the Massachusetts coast into a 10 mph wind with reduced power in one engine. The plane landed with the tail down to allow the easiest deceleration and therefore lessen impact and as a consequence the pilot and all passengers got out safely and were picked up more or less immediately.

The plane floated for eight minutes before sinking.

West With the Night

BY BERYL MARKHAM

(Excerpted from the book *West With the Night*)

". . . she has written so well, and marvelously well, that I was completely ashamed of myself as a writer. I felt that I was simply a carpenter with words, picking up whatever was furnished on the job and nailing them together and sometimes making an okay pig pen. But [she] can write rings around all of us who consider ourselves as writers."

That's pretty heady stuff coming from a writer like Ernest Hemingway, but Beryl Markham was a lady whose life and prose deserve the highest accolades one might imagine. As a settler in Africa, a pioneer aviator, and a writer of vivid prose, Beryl Markham's name stands as an icon among the most outstanding women of the twentieth century.

She was born in England in 1902 and was taken to Africa in 1906 by her father, who was a horse breeder, adventurer, and farmer. From 1931 to 1936, Markham carried mail, passengers, and supplies in her small plane to the most remote outposts on the Sudan, Tanganyika, Kenya, and Rhodesia. She passed away in 1986 at her home in Nairobi, Kenya.

Beryl Markham's book *West With the Night* is filled with engaging stories of her years in Africa, both on the ground and in the air. The title chapter from the book, which we are presenting here, is the story of Markham's attempt to fly the Atlantic solo from east to west. The flight took place in September, 1936.

I have seldom dreamed a dream worth dreaming again, or at least none worth recording. Mine are not enigmatic dreams; they are peopled with characters who are plausible and who do plausible things, and I am the most plausible amongst them. All the characters in my dreams have quiet voices like the voice of the man who telephoned me at Elstree one morning in September of nineteen-thirty-six and told me that there was rain and strong head winds over the west of England and over the Irish Sea, and that there were variable winds and clear skies in mid-Atlantic and fog off the coast of Newfoundland.

'If you are still determined to fly the Atlantic this late in the year,' the voice said, 'the Air Ministry suggests that the weather it is able to forecast for tonight, and for tomorrow morning, will be about the best you can expect.'

The voice had a few other things to say, but not many, and then it was gone, and I lay in bed half-suspecting that the telephone call and the man who made it were only parts of the mediocre dream I had been dreaming. I felt that if I closed my eyes the unreal quality of the message would be re-established, and that, when I opened them again, this would be another ordinary day with its usual beginning and its usual routine.

But of course I could not close my eyes, nor my mind, nor my memory. I could lie there for a few moments—remembering how it had begun, and telling myself, with senseless repetition, that by tomorrow morning I should either have flown the Atlantic to America—or I should not have flown it. In either case this was the day I would try.

I could stare up at the ceiling of my bedroom in Aldenham House, which was a ceiling undistinguished as ceilings go, and feel less resolute than anxious, much less brave than foolhardy. I could say to myself, 'You needn't do it, of course,' knowing at the same time that nothing is so inexorable as a promise to your pride.

I could ask, 'Why risk it?' as I have been asked since, and I could answer, 'Each to his element.' By his nature a sailor must sail, by his nature a flyer must fly. I could compute that I had flown a quarter of a million miles; and I could foresee that, so long as I had a plane and the sky was there, I should go on flying more miles.

There was nothing extraordinary in this. I had learned a craft and had worked hard learning it. My hands had been taught to seek the controls of a plane. Usage had taught them. They were at ease clinging to a stick, as a cobbler's fingers are in repose grasping an awl. No human pursuit achieves dignity until it can be called work, and when you can experience a physical loneliness for the tools of your trade, you see that the other things—the experiments, the irrelevant vocations, the vanities you used to hold—were false to you.

Record flights had actually never interested me very much for myself. There were people who thought that such flights were done for admiration and publicity, and worse. But of all the records—from Louis Blériot's first crossing of the English Channel in nineteen hundred and nine, through and beyond Kingsford Smith's flight from San Francisco to Sydney, Australia— none had been made by amateurs, nor by novices, nor by men or women less than hardened to failure, or less than masters of their trade. None of these was false. They were a company that simple respect and simple ambition made it worth more than an effort to follow.

The Carberrys (of Seramai) were in London and I could remember everything about their dinner party—even the menu. I could remember June Carberry and all her guests, and the man named McCarthy, who lived in Zanzibar, leaning across the table and saying, 'J. C., why don't you finance Beryl for a record flight?'

I could lie there staring lazily at the ceiling and recall J. C.'s dry answer: 'A number of pilots have flown the North Atlantic, west to east. Only Jim Mollison has done it alone the other way—from Ireland. Nobody has done it alone from England—man or woman. I'd be interested in that, but nothing else. If you want to try it, Burl, I'll back you. I think Edgar Percival could build a plane that would do it, provided you can fly it. Want to chance it?'

'Yes.'

I could remember saying that better than I could remember anything—except J. C.'s almost ghoulish grin, and his remark that sealed the agreement: 'It's a deal, Burl. I'll furnish the plane and you fly the Atlantic—but, gee, I wouldn't tackle it for a million. Think of all that black water! Think how cold it is!'

And I had thought of both.

I had thought of both for a while, and then there had been other things to think about. I had moved to Elstree, half-hour's flight from the Percival Aircraft Works at Gravesend, and almost daily for three months now I had flown down to the factory in a hired plane and watched the Vega Gull they were making for me. I had watched her birth and watched her growth. I had watched her wings take shape, and seen wood and fabric moulded to her ribs to form her long, sleek belly, and I had seen her engine cradled into her frame, and made fast.

The Gull had a turquoise-blue body and silver wings. Edgar Percival had made her with care, with skill, and with worry—the care of a veteran flyer, the skill of a master designer, and the worry of a friend. Actually the plane was a standard sport model with a range of only six hundred and sixty miles. But she had a special undercarriage built to carry the weight of her extra oil and petrol tanks. The tanks were fixed into the wings, into the centre section, and into the cabin itself. In the cabin they formed a wall around my seat, and each tank had a petcock of its own. The petcocks were important.

'If you open one,' said Percival, 'without shutting the other first, you may get an airlock. You know the tanks in the cabin have no gauges, so it may be best to let one run completely dry before opening the next. Your motor might go dead in the interval—but she'll start again. She's a De Havilland Gipsy—and Gipsys never stop.'

I had talked to Tom. We had spent hours going over the Atlantic chart, and I had realized that the tinker of Molo, now one of England's great pilots, had traded his dreams and had got in return a better thing. Tom had grown older, too; he had jettisoned a deadweight of irrelevant hopes and wonders, and had left himself a realistic code that had no room for temporizing or easy sentiment.

'I'm glad you're going to do it, Beryl. It won't be simple. If you can get off the ground in the first place, with such an immense load of fuel, you'll be alone in that plane about a night and a day—mostly night. Doing it east to west, the wind's against you. In September, so is the weather. You won't have a radio. If you misjudge your course only a few degrees, you'll end up in Labrador or in the sea—so don't misjudge anything.'

Tom could still grin. He had grinned; he had said: 'Anyway, it ought to amuse you to think that your financial backer lives on a farm called "Place of Death" and your plane is being built at "Gravesend." If you were consistent, you'd christen the Gull "The Flying Tombstone."'

I hadn't been that consistent. I had watched the building of the plane and I had trained for the flight like an athlete. And now, as I lay in bed, fully awake, I could still hear the quiet voice of the man from the Air Ministry intoning, like the voice of a dispassionate court clerk: '. . . the weather for tonight and tomorrow . . . will be about the best you can expect.' I should have liked to discuss the flight once more with Tom before I took off, but he was on a special job up north. I got out of bed and bathed and put on my flying clothes and took some cold chicken packed in a cardboard box and flew over to the military field at Abingdon, where the Vega Gull waited for me under the care of the R.A.F. I remember that the weather was clear and still.

Jim Mollison lent me his watch. He said: 'This is not a gift. I wouldn't part with it for anything. It got me across the North Atlantic and the South Atlantic, too. Don't lose it—and, for God's sake, don't get it wet. Salt water would ruin the works.'

Brian Lewis gave me a live-saving jacket. Brian owned the plane I had been using between Elstree and Gravesend, and he had thought a long time about a farewell gift. What could be more practical than a pneumatic jacket that could be inflated through a rubber tube?

'You could float around in it for days,' said Brian. But I had to decide between the life-saver and warm clothes. I couldn't have both, because of their bulk, and I hate the cold, so I left the jacket.

And Jock Cameron, Brian's mechanic, gave me a sprig of heather. If it had been a whole bush of heather, complete with roots growing in an earthen jar, I think I should have taken it, bulky or not. The blessing of Scotland, bestowed by a Scotsman, is not to be dismissed. Nor is the well-wishing of a ground mechanic to be taken lightly, for these men are the pilot's contact with reality.

It is too much that with all those pedestrian centuries behind us we should, in a few decades, have learned to fly; it is too heady a thought, too proud a boast. Only the dirt on a mechanic's hands, the straining vise, the splintered bolt of steel underfoot on the hangar floor—only these and such anxiety as the face of a Jock Cameron can hold for a pilot and his plane before a flight, serve to remind us that, not unlike the heather, we too are earthbound. We fly, but we have not 'conquered' the air. Nature presides in all her dignity, permitting us the study and the use of such of her forces as we may understand. It is when we presume to intimacy, having been granted only tolerance, that the harsh stick falls across our impudent knuckles and we rub the pain, staring upward, startled by our ignorance.

'Here is a sprig of heather,' said Jock, and I took it and pinned it into a pocket of my flying jacket.

There were press cars parked outside the field at Abingdon, and several press planes and photographers, but the R.A.F. kept everyone away from the grounds except technicians and a few of my friends.

The Carberrys had sailed for New York a month ago to wait for me there. Tom was still out of reach with no knowledge of my decision to leave, but that didn't matter so much, I thought. It didn't matter because Tom was unchanging—neither a fairweather pilot nor a fairweather friend. If for a month, or a year, or two years we sometimes had not seen each other, it still hadn't mattered. Nor did this. Tom would never say, 'You should have let me know.' He assumed that I had learned all that he had tried to teach me, and for my part, I thought of him, even then, as the merest student must think of his mentor. I could sit in a cabin overcrowded with petrol tanks and set my course for North America, but the knowledge of my hands on the controls would be Tom's knowledge. His words of caution and words of guidance, spoken so long ago, so many times, on bright mornings over the veldt or over a forest, or with a far mountain visible at the tip of our wing, would be spoken again, if I asked.

So it didn't matter, I thought. It was silly to think about.

You can live a lifetime and, at the end of it, know more about other people than you know about yourself. You learn to watch other people, but you never watch yourself because you strive against loneliness. If you read a book, or shuffle a deck of cards, or care for a dog, you are avoiding yourself. The abhorrence of loneliness is as natural as wanting to live at all. If it were otherwise, men would never have bothered to make an alphabet, nor to have fashioned words out of what were only animal sounds, nor to have crossed continents—each man to see what the other looked like.

Being alone in an aeroplane for even so short a time as a night and a day, irrevocably alone, with nothing to observe but your instruments and your own hands in semi-darkness, nothing to contemplate but the size of your small courage, nothing to wonder about but the beliefs, the faces, and the hopes rooted in your mind—such an experience can be as startling as the first awareness of a stranger walking by your side at night. You are the stranger.

It is dark already and I am over the south of Ireland. There are the lights of Cork and the lights are wet; they are drenched in Irish rain, and I am above them and dry. I am above them and the plane roars in a sobbing world, but it imparts no sadness to me. I feel the security of solitude, the exhilaration of escape. So long as I can see the lights and imagine the people walking under them, I feel selfishly triumphant, as if I have eluded care and left even the small sorrow of rain in other hands.

It is a little over an hour now since I left Abingdon. England, Wales, and the Irish Sea are behind me like so much time used up. On a long flight distance and time are the same. But there had been a moment when Time stopped—and Distance, too. It was the moment I lifted the blue-and-silver Gull from the aerodrome, the moment the photographers aimed their cameras, the moment I felt the craft refuse its burden and strain toward the earth in sullen rebellion, only to listen at last to the persuasion of stick and elevators, the dogmatic argument of blueprints that said she *had* to fly because the figures proved it.

So she had flown, and once airborne, once she had yielded to the sophistry of a draughtsman's board, she had said, 'There: I have lifted the weight. Now, where are we bound?'—and the question had frightened me.

'We are bound for a place thirty-six hundred miles from here—two thousand miles of it unbroken ocean. Most of the way it will be night. We are flying west with the night.'

So there behind me is Cork; and ahead of me is Berehaven Lighthouse. It is the last light, standing on the last land. I watch it, counting the frequency of its flashes—so many to the minute. Then I pass it and fly out to sea.

The fear is gone now—not overcome nor reasoned away. It is gone because something else has taken its place; the confidence and the trust, the inherent belief in the security of land underfoot—now this faith is transferred to my plane, because the land has vanished and there is no other tangible thing to fix faith upon. Flight is but momentary escape from the eternal custody of earth.

Rain continues to fall, and outside the cabin it is totally dark. My altimeter says that the Atlantic is two thousand feet below me, my Sperry Artificial Horizon says that I am flying level. I judge my drift at three degrees more than my weather chart suggests, and fly accordingly. I am flying blind. A beam to follow would help. So would a radio—but then, so would clear weather. The voice of the man at the Air Ministry had not promised storm.

I feel the wind rising and the rain falls hard. The smell of petrol in the cabin is so strong and the roar of the plane so loud that my senses are almost deadened. Gradually it becomes unthinkable that existence was ever otherwise.

At ten o'clock P.M. I am flying along the Great Circle Course for Harbour Grace, Newfoundland, into a forty-mile headwind at a speed of one hundred and thirty miles an hour. Because of the weather, I cannot be sure of how many more hours I have to fly, but I think it must be between sixteen and eighteen.

At ten-thirty I am still flying on the large cabin tank of petrol, hoping to use it up and put an end to the liquid swirl that has rocked the plane since my take-off. The tank has no gauge, but written on its side is the assurance: 'This tank is good for four hours.'

There is nothing ambiguous about such a guaranty. I believe it, but at twenty-five minutes to eleven, my motor coughs and dies, and the Gull is powerless above the sea.

I realize that the heavy drone of the plane has been, until this moment, complete and comforting silence. It is the actual silence following the last splutter of the engine that stuns me. I can't feel any fear; I can't feel anything. I can only observe with a kind of stupid disinterest that my hands are violently active and know that, while they move, I am being hypnotized by the needle of my altimeter.

I suppose that the denial of natural impulse is what is meant by 'keeping calm,' but impulse has reason in it. If it is night and you are sitting in an aeroplane with a stalled motor, and there are two thousand feet between you and the sea, nothing can be more reasonable than the impulse to pull back your stick in the hope of adding to that two thousand, if only by a little. The thought, the knowledge, the law that tells you that your hope lies not in this, but in a contrary act—the act of directing your impotent craft toward the water—seems a terrifying abandonment, not only of reason, but of sanity. Your mind and your heart reject it. It is your hands—your stranger's hands—that follow with unfeeling precision the letter of the law.

I sit there and watch my hands push forward on the stick and feel the Gull respond and begin its dive to the sea. Of course it is a simple thing; surely the cabin tank has run dry too soon. I need only to turn another petcock . . .

But it is dark in the cabin. It is easy to see the luminous dial of the altimeter and to note that my height is now eleven hundred feet, but it is not easy to see a petcock that is somewhere near the floor of the plane. A hand gropes and reappears with an electric torch, and fingers, moving with agonizing composure, find the petcock and turn it; and I wait.

At three hundred feet the motor is still dead, and I am conscious that the needle of my altimeter seems to whirl like the spoke of a spindle winding up the remaining distance between the plane and the water. There is some lightning, but the quick flash only serves to emphasize the darkness. How high can waves reach—twenty feet, perhaps? Thirty?

It is impossible to avoid the thought that this is the end of my flight, but my reactions are not orthodox; the various incidents of my entire life do not run through my mind like a motion-picture film gone mad. I only feel that all this has happened before—and it has. It has all happened a hundred times in my mind, in my sleep, so that now I am not really caught in terror; I recognize a familiar scene, a familiar story with its climax dulled by too much telling.

I do not know how close to the waves I am when the motor explodes to life again. But the sound is almost meaningless. I see my hand easing back on the stick, and I feel the Gull climb up into the storm, and I see the altimeter whirl like a spindle again, paying out the distance between myself and the sea.

The storm is strong. It is comforting. It is like a friend shaking me and saying, 'Wake up! You were only dreaming.'

But soon I am thinking. By simple calculation I find that my motor had been silent for perhaps an instant more than thirty seconds.

I ought to thank God—and I do, though indirectly. I thank Geoffrey De Havilland who designed the indomitable Gipsy, and who, after all, must have been designed by God in the first place.

A lighted ship—the daybreak—some steep cliffs standing in the sea. The meaning of these will never change for pilots. If one day an ocean can be flown within an hour, if men can build a plane that so masters time, the sight of land will be no less welcome to the steersman of that fantastic craft. He will have cheated laws that the cunning of science has taught him how to cheat, and he will feel his guilt and be eager for the sanctuary of the soil.

I saw the ship and the daybreak, and then I saw the cliffs of New-foundland wound in ribbons of fog. I felt the elation I had so long imagined, and I felt the happy guilt of having circumvented the stern authority of the weather and the sea. But mine was a minor triumph; my swift Gull was not so swift as to have escaped unnoticed. The night and the storm had caught her and we had flown blind for nineteen hours.

I was tired now, and cold. Ice began to film the glass of the cabin win-dows and the fog played a magician's game with the land. But the land was there. I could not see it, but I had seen it. I could not afford to believe that it was any land but the land I wanted. I could not afford to believe that my navi-gation was at fault, because there was no time for doubt.

South to Cape Race, west to Sydney on Cape Breton Island. With my protractor, my map, and my compass, I set my new course, humming the ditty that Tom had taught me: 'Variation West—magnetic best. Variation East—mag-netic least." A silly rhyme, but it served to placate, for the moment, two warring poles—the magnetic and the true. I flew south and found the lighthouse of Cape Race protruding from the fog like a warning finger. I circled twice and went on over the Gulf of Saint Lawrence.

After a while there would be New Brunswick, and then Maine—and then New York. I could anticipate. I could almost say, 'Well, if you stay awake, you'll find it's only a matter of time now'—but there was no question of stay-ing awake. I was tired and I had not moved an inch since that uncertain mo-ment at Abingdon when the Gull had elected to rise with her load and fly, but I could not have closed my eyes. I could sit there in the cabin walled in glass and petrol tanks, and be grateful for the sun and the light, and the fact that I could see the water under me. They were almost the last waves I had to pass. Four hundred miles of water, but then the land again—Cape Breton. I would stop at Sydney to refuel and go on. It was easy now. It would be like stopping at Kisumu and going on.

Success breeds confidence. But who has a right to confidence except the Gods? I had a following wind, my last tank of petrol was more than three-quarters full, and the world was as bright to me as if it were a new world, never touched. If I had been wiser, I might have known that such moments are, like innocence, short-lived. My engine began to shudder before I saw the land. It died, it spluttered, it started again and limped along. It coughed and spat black exhaust toward the sea.

There are words for everything. There was a word for this—airlock, I thought. This had to be an airlock because there was petrol enough. I thought I might clear it by turning on and turning off all the empty tanks, and so I did that. The handles of the petcocks were sharp little pins of metal, and when I had opened and closed them a dozen times, I saw that my hands were bleeding and that the blood was dropping on my maps and on my clothes, but the effort wasn't any good. I coasted along on a sick and halting engine. The oil pressure and the oil temperature gauges were normal, the magnetos working, and yet I lost altitude slowly while the realization of failure seeped into my heart. If I made the land, I should have been the first to fly the North Atlantic from England, but from my point of view, from a pilot's point of view, a forced landing was failure because New York was my goal. If only I could land and then take off, I would make it still . . . if only, if only . . .

The engine cuts again, and then catches, and each time it spurts to life I climb as high as I can get, and then it splutters and stops and I glide once more toward the water, to rise again and descend again, like a hunting sea bird.

I find the land. Visibility is perfect now and I see land forty or fifty miles ahead. If I am on my course, that will be Cape Breton. Minute after minute goes by. The minutes almost materialize; they pass before my eyes like links in a long slow-moving chain, and each time the engine cuts, I see a broken link in the chain and catch my breath until it passes.

The land is under me. I snatch my map and stare at it to confirm my whereabouts. I am, even at my present crippled speed, only twelve minutes from Sydney Airport, where I can land for repairs and then go on.

The engine cuts once more and I begin to glide, but now I am not worried; she will start again, as she has done, and I will gain altitude and fly into Sydney.

But she doesn't start. This time she's dead as death; the Gull settles earthward and it isn't any earth I know. It is black earth stuck with boulders and I hang above it, on hope and on a motionless propeller. Only I cannot hang above it long. The earth hurries to meet me, I bank, turn, and sideslip to dodge the boulders, my wheels touch, and I feel them submerge. The nose of the plane is engulfed in mud, and I go forward striking my head on the glass of the cabin front, hearing it shatter, feeling blood pour over my face.

I stumble out of the plane and sink to my knees in muck and stand there foolishly staring, not at the lifeless land, but at my watch.

Twenty-one hours and twenty-five minutes.

Atlantic flight. Abingdon, England, to a nameless swamp—nonstop.

A Cape Breton Islander found me—a fisherman trudging over the bog saw the Gull with her tail in the air and her nose buried, and then he saw me floundering in the embracing soil of his native land. I had been wandering for an hour and the black mud had got up to my waist and the blood from the cut in my head had met the mud halfway.

From a distance, the fisherman directed me with his arms and with shouts toward the firm places in the bog, and for another hour I walked on them and came toward him like a citizen of Hades blinded by the sun, but it wasn't the sun; I hadn't slept for forty hours.

He took me to his hut on the edge of the coast and found that built upon the rocks there was a little cubicle that housed an ancient telephone— put there in case of shipwrecks.

I telephoned to Sydney Airport to say that I was safe and to prevent a needless search being made. On the following morning I did step out of a plane at Floyd Bennett Field and there was a crowd of people still waiting there to greet me, but the plane I stepped from was not the Gull, and for days while I was in New York I kept thinking about that and wishing over and over again that it had been the Gull, until the wish lost its significance, and time moved on, overcoming many things it met on the way.

The Sound Barrier

BY GENERAL CHUCK YEAGER AND LEO JANOS

(Excerpted from *Yeager* by
General Chuck Yeager and Leo Janos)

"I was always afraid of dying. Always. It was my fear that made me learn everything I could about my airplane and my emergency equipment, and kept my flying respectful of my machine and always alert in the cockpit. Death is the great enemy and robber in my profession, taking away so many friends over the years, all of them young."

That's quite a statement, in my opinion, coming from *any* pilot. But the fact that it comes from the autobiography of Chuck Yeager gives these words the weight of authority and experience. For veteran or wannabe aviators, the message cries out to be heeded and remembered.

Long before Tom Wolfe wrote the now-famous—and deservedly so—*The Right Stuff*, epitomizing the flying skills of test pilots and astronauts, Chuck Yeager was an aviation icon for flying the first plane to break the sound barrier. Wolfe's phenomenally successful book, and the movie which followed, made Yeager's fame as a fighter pilot in World War Two and as a test pilot after the war known in households throughout the world.

The movie version of *The Right Stuff*, in my opinion, particularly in the scenes just prior to depicting Chuck Yeager breaking the sound barrier, strays far, far into the realm of BS by suggesting Yeager took the mission in an overnight gesture. Tom Wolfe got it right in his book. Hollywood didn't—and obviously didn't care, either.

We're going to visit with Tom Wolfe's great book in the next chapter, which tells of John Glenn becoming the first American to orbit the earth. But for a description of what happened in the Bell X-1 on October 14, 1947, we have the man who flew the bullet-like plane at speeds no human had ever attained and lived to tell about it—Chuck Yeager himself.

First Powered Flight: August 29, 1947

Shivering, you bang your gloved hands together and strap on your oxygen mask inside the coldest airplane ever flown. You're being cold-soaked from the hundreds of gallons of liquid oxygen (LOX) fuel stored in the compartment directly behind you at minus 296 degrees. No heater, no defroster; you'll just have to grit your teeth for the next fifteen minutes until you land and feel that wonderful hot desert sun. But that cold saps your strength: it's like trying to work and concentrate inside a frozen food locker.

That cold will take you on the ride of your life. You watched the X-1 get its 7:00 A.M. feeding in a swirling cloud of vapor fog, saw the frost form under its orange belly. That was an eerie sight; you're carrying six hundred gallons of LOX and water alcohol on board that can blow up at the flick of an igniter switch and scatter your pieces over several counties. But if all goes well, the beast will chug-a-lug a ton of fuel a minute.

Anyone with brain cells would have to wonder what in hell he was doing in such a situation—strapped inside a live bomb that's about to be dropped out of a bomb bay. But risks are the spice of life, and this is the kind of moment that a test pilot lives for. The butterflies are fluttering, but you feed off fear as if it's a high-energy candy bar. It keeps you alert and focused.

You accept risk as part of every new challenge; it comes with the territory. So you learn all you can about the ship and its systems, practice flying it on ground runs and glide flights, plan for any possible contingency, until the odds against you seem more friendly. You like the X-1; she's a sound airplane, but she's also an experimental machine, and you're a researcher on an experimental flight. You know you can be hammered by something unexpected, but you count on your experience, concentration, and instincts to pull you through. And luck. Without luck . . .

You can't watch yourself fly. But you know when you're in sync with the machine, so plugged into its instruments and controls that your mind and your hand become the heart of its operating system. You can make that airplane talk, and like a good horse, the machine knows when it's in competent hands. You know what you can get away with. And you can be wrong only once. You smile reading newspaper stories about a pilot in a disabled plane that maneuvered to miss a schoolyard before he hit the ground. That's crap. In an emergency situation, a pilot thinks only about one thing—survival. You battle to survive right down to the ground; you think about nothing else. Your concentration is riveted on what to try next. You don't say anything on the radio, and you aren't even aware that a schoolyard exists. That's exactly how it is.

There are at least a dozen different ways that the X-1 can kill you, so your concentration is total during the preflight check procedures. You load up nitrogen gas pressures in the manifolds—your life's blood because the nitrogen gas runs all the internal systems as well as the flaps and landing gear. Then you bleed off the liquid oxygen manifold and shut it down. All's in order.

Half an hour ago we taxied out to takeoff in the mother ship. Because of the possibility of crashing with so much volatile fuel, they closed down the base until we were safely off the ground. That's the only acknowledgment from the base commander that we even exist. There's no interest in our flights because practically nobody at Muroc gives us any chance for success. Those bastards think they have it all figured. They call our flights "Slick Goodlin's Revenge." The word is that he knew when to get out in one piece by quitting over money.

One minute to drop. Ridley flashes the word from the copilot's seat in the mother ship. We're at 25,000 feet as the B-29 noses over and starts its shallow dive. Major Cardenas, the driver, starts counting backwards from ten.

C-r-r-ack. The bomb shackle release jolts you up from your seat, and as you sail out of the dark bomb bay the sun explodes in brightness. You're looking into the sky. *Wrong!* You should be dropped level. The dive speed was too slow, and they dropped you in a nose-up stall. You blink to get your vision, fighting the stall with your control wheel, dropping toward the basement like an elevator whose cable snapped. You're three thousand pounds heavier than in those glide flights. Down goes that nose and you pick up speed. You level out about a thousand feet below the mother ship and reach for that rocket igniter switch.

The moment of truth: if you are gonna be blown up, this is likely to be when. You light the first chamber.

Whoosh. Slammed back in your seat, a tremendous kick in the butt. Nose up and hold on. Barely a sound; you can hear your breathing in the oxygen mask—you're outracing the noise behind you—and for the first time in a powered airplane you can hear the air beating against the windshield as the distant dot that is Hoover's high chase P-80 grows ever bigger. You pass him like he's standing still, and he reports seeing diamond-shaped shock waves leaping out of your fiery exhaust. Climbing faster than you can even think, but using only one of four rocket chambers, you turn it off and light another. We're streaking up at .7 Mach; this beast's power is awesome. You've never known such a feeling of speed while pointing up in the sky. At 45,000 feet, where morning resembles the beginning of dusk, you turn on the last of the four chambers. God, what a ride! And you still have nearly half your fuel left.

Until this moment, you obeyed the flight plan to the letter: firing only one chamber at a time, to closely monitor the chamber pressures; if you use two or more, there's too much to watch. If you fire all four, you may accelerate too rapidly, be forced to raise your nose to slow down, and get yourself into a high-speed stall.

Now the flight plan calls for you to jettison remaining fuel and glide down to land. But you're bug-eyed, thrilled to your toes, and the fighter jock takes over from the cautious test pilot. Screw it! You're up there in the dark part of the sky in the most fabulous flying machine ever built, and you're just not ready to go home. The moment calls for a nice slow roll, and you lower your wing, pulling a couple of Gs until you're hanging upside down in zero Gs and the engine quits. As soon as the X-1 rights itself it starts again, but you've been stupid. At zero Gs the fuel couldn't feed the engine, and you might have been blown up. But the X-1 is forgiving—this time.

You know what you're supposed to do, but you know what you're gonna do. You turn off the engine, but instead of jettisoning the remaining fuel, you roll over and dive for Muroc Air Base. We blister down, shit-heavy, .8 Mach in front of the needle, a dive-glide faster than most jets at full power. You're thinking, "Let's show those bastards the real X-1."

Below 10,000 feet is the danger zone, the limit for jettisoning fuel with enough maneuver time to glide down to a safe landing. But we're below 5,000, lined up with Muroc's main runway. And we're still in a dive.

We whistle down that main runway, only 300 feet off the ground, until we are parallel with the control tower. You hit the main rocket switch. The four chambers blow a thirty-foot lick of flame. Christ, the impact nearly knocks you back into last week. That nose is pointed so straight up that you can't see the blue sky out the windshield. We are no longer an airplane: we're a sky-rocket. You're not flying. You're holding on to the tiger's tail. Straight up, you're going .75 Mach! In one minute the fuel is gone. By then you're at 35,000 feet, traveling at .85 Mach. You're so excited, scared, and thrilled that you can't say a word until the next day.

But others said plenty. The NACA team thought I was a wild man. Dick Frost chewed me out for doing that slow roll. Even Jack Ridley shook his head. He said, "Any spectators down there knew damned well that wasn't Slick rattling those dishes. Okay, son, you got it all out of your system, but now you're gonna hang tough." Colonel Boyd fired a rocket of his own. "Reply by endorsement about why you exceeded .82 Mach in violation of my direct orders." I asked Ridley to write my reply. "Bullshit," he said. "You did it. You explain it."

I wrote back: "The airplane felt so good and flew so well that I felt certain we would have no trouble going slightly above the agreed speed. The violation of your direct orders was due to the excited state of the undersigned and will not be repeated."

A few days later, the old man called me. "Damn it, I expect you to stick to the program and do what you are supposed to. Don't get overeager and cocky. Do you want to jeopardize the first Air Corps research project?"

"No, sir."

"Well, then obey the goddamn rules."

From then on I did. But on that first powered flight I wanted to answer those who said we were doomed in the attempt to go faster than sound. My message was, "Stick it where the sun don't shine."

Going out to .85 Mach put the program out on a limb because it carried us beyond the limits of what was then known about high-speed aerodynamics. Wind tunnels could only measure up to .85 Mach, and as Walt Williams of NACA was quick to point out to me, "From now on, Chuck, you'll be flying in the realm of the unknown." Ridley and I called it, "the Ughknown."

Whatever happened, I figured I was better off than the British test pilots who had attempted supersonic flights in high-powered dives. If they got into trouble, that was it—especially in a tailless airplane like *The Swallow*. All my attempts would be made in climbs—the power of the rocket over the jet—and that way, if I encountered a problem, I could quickly slow down. But the price of rocket power was flying with volatile fuel. Running four chambers, my fuel lasted only two and a half minutes; it lasted five minutes on two chambers and ten minutes on one. Each minute of climbing we got lighter and faster, so that by the time we had climbed up and over at 45,000 feet, we were at max speed.

Who would decide the max speed of a particular flight? This was an Air Corps research project, but the seventeen NACA engineers and technicians used their expertise to try to control these missions. They were there as advisers, with high-speed wind tunnel experience, and were performing the data reduction collected on the X-1 flights, so they tried to dictate the speed in our flight plans. Ridley, Frost, and I always wanted to go faster than they did. They would recommend a Mach number, then the three of us would sit down and decide whether or not we wanted to stick with their recommendation. They were so conservative that it would've taken me six months to get to the barrier.

I wanted to be careful, but I also wanted to get it over with. Colonel Boyd sided with NACA caution, going up only two-hundredths of a Mach on each consecutive flight. Once I flew back with Hoover to see if I could get the old man to agree to speed things up. We met in the evening at his home. But Bob led off by trying to explain why he had been forced to crash-land a P-80 a few days before. I could tell the old man wasn't buying Bob's explanations; those thick eyebrows were bunching up. But ol' Hoover pushed on, becoming emotional to the point where he accidentally spat a capped tooth onto the old man's lap. I decided to have my say at another time.

So I flew in small increments of speed. On October 5, I made my sixth powered flight and experienced shock-wave buffeting for the first time as I reached .86 Mach. It felt like I was driving on bad shock absorbers over uneven paving stones. The right wing suddenly got heavy and began to drop, and when I tried to correct it my controls were sluggish. I increased my speed to .88 Mach to see what would happen. I saw my aileron vibrating with shock waves, and only with effort could I hold my wing level.

The X-1 was built with a high tail to avoid air turbulence off the wings; the tail was also thinner than the wings, so that shock waves would not form simultaneously on both surfaces. Thus far, the shock waves and buffeting had been manageable, and because the ship was stressed for eighteen Gs, I never was concerned about being shaken apart. Also, I was only flying twice a week, to give NACA time to reduce all the flight data and analyze it. Special sensing devices pinpointed the exact location of shock waves on any part of the airframe. The data revealed that the airplane was functioning exactly as its designers planned.

But on my very next flight we got knocked on our fannies. I was flying at .94 Mach at 40,000 feet, experiencing the usual buffeting, when I pulled back on the control wheel, and Christ, nothing happened! The airplane continued flying with the same attitude and in the same direction.

The control wheel felt as if the cables had snapped. I didn't know what in hell was happening. I turned off the engine and slowed down. I jettisoned my fuel and landed feeling certain that I had taken my last ride in the X-1. Flying at .94, I lost my pitch control. My elevator ceased to function. At the speed of sound, the ship's nose was predicted to go either up or down, and without pitch control, I was in a helluva bind.

I told Ridley I thought we had had it. There was no way I was going faster than .94 Mach without an elevator. He looked sick. So did Dick Frost and the NACA team. We called Colonel Boyd at Wright, and he flew out immediately to confer with us. Meanwhile, NACA analyzed the telemetry data from the flight and found that at .94 Mach, a shock wave was slammed right at the hinge point of the elevator on the tail, negating my controls. Colonel Boyd just shook his head. "Well," he said, "it looks to me like we've reached the end of the line." Everyone seemed to agree except for Jack Ridley.

He sat at a corner of the conference table scribbling little notes and equations. He said, "Well, maybe Chuck can fly without using the elevator. Maybe he can get by using only the horizontal stabilizer." The stabilizer was the winglike structure on the tail that stabilized pitch control. Bell's engineers had purposely built into them an extra control authority because they had anticipated elevator ineffectiveness caused by shock waves. This extra authority was a trim switch in the cockpit that would allow a small air motor to pivot the stabilizer up or down, creating a moving tail that could act as an auxiliary elevator by lowering or raising the airplane's nose. We were leery about trying it while flying at high speeds; instead, we set the trim on the ground and left it alone.

Jack thought we should spend a day ground testing the hell out of that system, learn everything there was to know about it, then flight test it. No one disagreed. There was no other alternative except to call the whole thing quits, but Jack got a lot of "what if" questions that spelled out all the risks. What if the motor got stuck in a trim up or trim down position? Answer: Yeager would have a problem. What if the turbulent airflow at high speed Mach overwhelmed the motor and kept the tail from pivoting? Answer: Yeager would be no worse off than he was during the previous mission. Yeah, but what if that turbulent air ripped off that damned tail as it was pivoting? Answer: Yeager better have paid-up insurance. We were dealing with the Ughknown.

Before returning to Wright, Colonel Boyd approved our ground tests. We were to report the results to him, and then he'd decide whether to proceed with a flight test. Then the old man took me aside. "Listen," he said, "I don't want you to be railroaded into this deal by Ridley or anyone else. If you don't feel comfortable with the risks, I want you to tell me so. I'll respect your decision. Please don't play the hero, Chuck. It makes no sense getting you hurt or killed."

I told him, "Colonel Boyd, it's my ass on the line. I want us to succeed but I'm not going to get splattered doing it."

So, Ridley and I ground tested that stabilizer system every which way but loose. It worked fine, and provided just enough control (about a quarter of a degree change in the angle of incidence) so that we both felt I could get by without using the airplane's elevator. "It may not be much," Ridley said, "and it may feel ragged to you up there, but it will keep you flying." I agreed. But would the system work at high Mach speed? Only one way to find out. Colonel Boyd gave us the go ahead.

No X-1 flight was ever routine. But when I was dropped to repeat the same flight profile that had lost my elevator effectiveness, I admit to being unusually grim. I flew as alert and precisely as I knew how. If the damned Ughknown swallowed me up, there wasn't much I could do about it, but I concentrated on that trim switch. At the slightest indication that something wasn't right, I would break the record for backing off.

Pushing the switch forward opened a solenoid that allowed high-pressure nitrogen gas through the top motor to the stabilizer, changing its angle of attack and stabilizing its upward pitch. If I pulled back, that would start the bottom motor, turning it in the opposite direction. I could just beep it and supposedly make pitch changes. I let the airplane accelerate up to .85 Mach before testing the trim switch. I pulled back on the switch, moving the leading edge of the stabilizer down one degree, and her nose rose. I retrimmed it back to where it was, and we leveled out. I climbed and accelerated up to .9 Mach and made the same change, achieving the same result. I retrimmed it and let it go out to .94 Mach, where I had lost my elevator effectiveness, made the same trim change, again raising the nose, just as I had done at the lower Mach numbers. Ridley was right: the stabilizer gave me just enough pitch control to keep me safe. I felt we could probably make it through without the elevator.

I had her out to .96 at 43,000 feet and was about to turn off the engine and begin jettisoning the remaining fuel, when the windshield began to frost. Because of the intense cabin cold, fogging was a continual problem, but I was usually able to wipe it away. This time, though, a solid layer of frost quickly formed. I even took off my gloves and used my fingernails, which only gave me frostbite. That windshield was lousy anyway, configured to the bullet-shaped fuselage and affording limited visibility. It was hard to see out during landings, but I had never expected to fly the X-1 on instruments. I radioed Dick Frost, flying low chase, and told him the problem. "Okay, pard," he said. "I'll talk you in. You must've done a lot of sweating in that cockpit to ice the damned windshield." I told him, "Not as much as I'm gonna do having you talk me in. You better talk good, Frost." He laughed. "I know. A dumb bastard like you probably can't read instruments."

The X-1 wasn't the Space Shuttle. There were no on-board computers to line you up and bring you down. The pilot was the computer. Under normal flight conditions, I'd descend to 5,000 feet above the lakebed and fly over the point where I wanted to touch down, then turn and line up downwind, lowering my landing gear at around 250 mph. The X-1 stalled around 190 mph, so I held my glide speed to around 220 and touched down at around 190. The ship rolled out about three miles if I didn't apply the brakes. Rogers Dry Lake gave me an eight-mile runway, but that didn't make the landing untricky. Coming in nose-high, you couldn't see the ground at all. You had to feel for it. I was sensitive to ground effect, and felt the differences as we lowered down. There was also that depth perception problem, and a lot of pilots bent airplanes porpoising in, or flaring high then cracking off their landing gears. My advantage was that I had landed on these lakebeds hundreds of times. Even so, the X-1 was not an easy-landing airplane. At the point of touchdown, you had to discipline yourself to do nothing but allow the ship to settle in by itself. Otherwise you'd slam it on its weak landing gear.

So, landing blind was not something you'd ever want to be forced to do. I had survived the Ughknown only to be kicked in the butt by the Unexpected. But that was a test pilot's life, one damned thing after another. Frost was a superb pilot, who knew the X-1's systems and characteristics even better than I did. I had plenty of experience flying on instruments, and in a hairy deal like this, experience really counted. Between the two of us we made it look deceptively easy, although we both knew that it wasn't exactly a routine procedure. Frost told me to turn left ten degrees, and I followed by using my magnetic compass, monitoring my rate of turn by the needle and ball. I watched the airspeed and rate of descent, so I knew how fast I was coming down from that and the feel of the ground effect. I followed his directions moving left or right to line up on the lakebed, which was also five miles wide, allowing him to fly right on my wing and touch down with me.

He greased me right in, but my body sweat added another layer of frost to the windshield. "Pard," Dick teased, "that's the only time you haven't bounced her down. Better let me hold your hand from now on."

Before my next flight, Jack Russell, my crew chief, applied a coating of Drene Shampoo to the windshield. For some unknown reason it worked as an effective antifrost device, and we continued using it even after the government purchased a special chemical that cost eighteen bucks a bottle.

Despite the frosted windshield, I now had renewed confidence in the X-1. We had licked the elevator problem, and Ridley and I phoned Colonel Boyd and told him we thought we could safely continue the flights. He told us to press on. This was on Thursday afternoon. The next scheduled flight would be on Tuesday. So we sat down with the NACA team to discuss a flight plan. I had gone up to .955 Mach, and they suggested a speed of .97 Mach for the next mission. What we didn't know until the flight data was reduced several days later, was that I had actually flown .988 Mach while testing the stabilizer. In fact, there was a fairly good possibility that I had attained supersonic speed.

Instrumentation revealed that a shock wave was interfering with the airspeed gauge on the wing. But we wouldn't learn about this until after my next flight.

All I cared about was that the stabilizer was still in one piece and so was I. We were all exhausted from a long, draining week, and quit early on Friday to start the weekend. I had promised Glennis that I would take her to Elly Anderson's, in Auburn, for a change of scene and to get her away from the kids. As cautiously as we were proceeding on these X-1 flights, I figured that my attempt to break the barrier was a week or two away. So I looked forward to a relaxed few days off. But when I got home, I found Glennis lying down, feeling sick. We canceled the babysitter and called Elly. By Sunday she was feeling better, so we went over to Pancho's place for dinner. On the way over, I said to Glennis, "Hey, how about riding horses after we eat?" She was raised around horses and was a beautiful rider.

Pancho's place was a dude ranch, so after dinner we walked over to the corral and had them saddle up a couple of horses. It was a pretty night and we rode for about an hour through the Joshua trees. We decided to race back. Unfortunately there was no moon, otherwise I would have seen that the gate we had gone out of was now closed. I only saw the gate when I was practically on top of it. I was slightly in the lead, and I tried to veer my horse and miss it, but it was too late. We hit the gate and I tumbled through the air. The horse got cut and I was knocked silly. The next thing I remember was Glennis kneeling over me, asking me if I was okay. I was woozy, and she helped me stand up. It took a lot to straighten up, feeling like I had a spear in my side.

Glennis knew immediately. "You broke a rib," she said. She was all for driving straight to the base hospital. I said, no, the flight surgeon will ground me. "Well, you can't fly with broken ribs," she argued. I told her, "If I can't, I won't. If I can, I will."

Monday morning, I struggled out of bed. My shoulder was sore, and I ached generally from bumps and bruises, but my ribs near to killed me. The pain took my breath away. Glennis drove me over to Rosemond, where a local doctor confirmed I had two cracked ribs, and taped me up. He told me to take it easy. The tape job really helped. The pain was at least manageable and I was able to drive myself to the base that afternoon.

I was really low. I felt we were on top of these flights now, and I wanted to get them over with. And as much as I was hurting, I could only imagine what the old man would say if I was grounded for falling off a horse. So, I sat down with Jack Ridley and told him my troubles. I said, "If this were the first flight, I wouldn't even think about trying it with these busted sumbitches. But, hell, I know every move I've got to make, and most of the major switches are right on the control wheel column."

He said, "True, but how in hell are you gonna be able to lock the cockpit door? That takes some lifting and shoving." So we walked into the hangar to see what we were up against.

We looked at the door and talked it over. Jack said, "Let's see if we can get a stick or something that you can use in your left hand to raise the handle up on the door to lock it. Get it up at least far enough where you get both hands on it and get a grip on it." We looked around the hangar and found a broom. Jack sawed off a ten-inch piece of broomstick, and it fit right into the door handle. Then I crawled into the X-1 and we tried it out. He held the door against the frame, and by using that broomstick to raise the door handle, I found I could manage to lock it. We tried it two or three times, and it worked. But finally, Ridley said, "Jesus, son, how are you gonna get down that ladder?"

I said, "One rung at a time. Either that or you can piggyback me."

Jack respected my judgment. "As long as you really think you can hack it," he said. We left that piece of broomstick in the X-1 cockpit.

Ninth Powered Flight: October 14, 1947

Glennis drove me to the base at six in the morning. She wasn't happy with my decision to fly, but she knew that Jack would never let me take off if he felt I would get into trouble. Hoover and Jack Russell, the X-1 crew chief, heard I was dumped off a horse at Pancho's, but thought the only damage was to my ego, and hit me with some "Hi-Ho Silver" crap, as well as a carrot, a pair of glasses, and a rope in a brown paper bag—my bucking bronco survival kit.

Around eight, I climbed aboard the mother ship. The flight plan called for me to reach .97 Mach. The way I felt that day, .97 would be enough. On that first rocket ride I had a tiger by the tail; but by this ninth flight, I felt I was in the driver's seat. I knew that airplane inside and out. I didn't think it would turn against me. Hell, there wasn't much I could do to hurt it; it was built to withstand three times as much stress as I could survive. I didn't think the sound barrier would destroy her, either. But the only way to prove it was to do it.

That moving tail really bolstered my morale, and I wanted to get to that sound barrier. I suppose there were advantages in creeping up on Mach 1, but my vote was to stop screwing around before we had some stupid accident that could cost us not only a mission, but the entire project. If this mission was successful, I was planning to really push for a sound barrier attempt on the very next flight.

Going down that damned ladder hurt. Jack was right behind me. As usual, I slid feet-first into the cabin. I picked up the broom handle and waited while Ridley pushed the door against the frame, then I slipped it into the door handle and raised it up into lock position. It worked perfectly. Then I settled in to go over my checklist. Bob Cardenas, the B-29 driver, asked if I was ready.

"Hell, yes," I said. "Let's get it over with."

He dropped the X-1 at 20,000 feet, but his dive speed was once again too slow and the X-1 started to stall. I fought it with the control wheel for about five hundred feet, and finally got her nose down. The moment we picked up speed I fired all four rocket chambers in rapid sequence. We climbed at .88 Mach and began to buffet, so I flipped the stabilizer switch and changed the setting two degrees. We smoothed right out, and at 36,000 feet, I turned off two rocket chambers. At 40,000 feet, we were still climbing at a speed of .92 Mach. Leveling off at 42,000 feet, I had thirty percent of my fuel, so I turned on rocket chamber three and immediately reached .96 Mach. I noticed that the faster I got, the smoother the ride.

Suddenly the Mach needle began to fluctuate. It went up to .965 Mach—then tipped right off the scale. I thought I was seeing things! We were flying supersonic! And it was as smooth as a baby's bottom: Grandma could be sitting up there sipping lemonade. I kept the speed off the scale for about twenty seconds, then raised the nose to slow down.

I was thunderstruck. After all the anxiety, breaking the sound barrier turned out to be a perfectly paved speedway. I radioed Jack in the B-29. "Hey, Ridley, that Machmeter is acting screwy. It just went off the scale on me."

"Fluctuated off?"

"Yeah, at point nine-six-five."

"Son, you is imagining things."

"Must be. I'm still wearing my ears and nothing else fell off, neither."

The guys in the NACA tracking van interrupted to report that they heard what sounded like a distant rumble of thunder: my sonic boom! The first one by an airplane ever heard on earth. The X-1 was supposedly capable of reaching nearly twice the speed of sound, but the Machmeter aboard only registered to 1.0 Mach, which showed how much confidence they had; I estimated I had reached 1.05 Mach. (Later data showed it was 1.07 Mach—700 mph.)

And that was it. I sat up there feeling kind of numb, but elated. After all the anticipation to achieve this moment, it really was a let-down. It took a damned instrument meter to tell me what I'd done. There should've been a bump on the road, something to let you know you had just punched a nice clean hole through that sonic barrier. The Ughknown was a poke through Jello. Later on, I realized that this mission had to end in a let-down, because the real barrier wasn't in the sky, but in our knowledge and experience of supersonic flight.

I landed tired, but relieved to have hacked the program. There is always strain in research flying. It's the same as flying in combat, where you never can be sure of the outcome. You try not to think about possible disasters, but fear is churning around inside whether you think of it consciously or not. I thought now that I'd reached the top of the mountain, the remainder of these X-1 experimental flights would be downhill. But having sailed me safely through the sonic barrier, the X-1 had plenty of white-knuckle flights in store over the next year. The real hero in the flight test business is a pilot who manages to survive.

And so I was a hero this day. As usual, the fire trucks raced out to where the ship had rolled to a stop on the lakebed. As usual, I hitched a ride back to the hangar with the fire chief. That warm desert sun really felt wonderful. My ribs ached.

Orbit

BY TOM WOLFE

(Excerpted from *The Right Stuff*)

Although America was desperate to catch and pass the Russians in the so-called "Space Race" in the late nineteen fifties and early sixties, the first two Americans into space did not orbit the earth. Alan Shepard and Gus Grissom rode their Project Mercury capsules into space for only a few minutes respectively, then splashed down in the Atlantic off the coast of Florida where their missions had been launched. For the United States, it was a beginning, but still far short of the Russian accomplishments of launching Cosmonauts Yuri Gagarin and Gherman Titov into earth orbit.

For most Americans, the curtain on the biggest space show on earth went up on February 20, 1962, when Astronaut John Glenn commanded his Friendship 7 Mercury space capsule on three orbits of the earth. By this time, people had come to understand that there was nothing routine about going into space, and particularly about going into earth orbit. Glenn's flight was fraught with high suspense and danger—all witnessed live by a worldwide television audience, for America did not cloak its space activities in secret as the Russians had done. Few Americans who were alive at that time will ever forget the drama of John Glenn's flight and his return to earth.

Today you can visit Glenn's Friendship 7 capsule in the National Aeronautical and Space Museum in Washington, D.C. And right here, thanks to the vivid prose of Tom Wolfe, you can experience what it must have been like in the capsule during that historic flight.

On February 20 Glenn was once again squeezed inside the Mercury capsule on top of the Atlas rocket, lying on his back, whiling away the holds in the countdown by going over his checklist and looking at the scenery through the periscope. If he closed his eyes it felt as if he were lying on his back on the deck of an old ship. The rocket kept creaking and twisting, shaking the capsule this way and that. The Atlas had 4.3 times as much fuel as the Redstone, including 80 tons of liquid oxygen. The liquid oxygen, the "lox," had a temperature of 293 degrees below zero, so that the shell and tubing of the rocket, which were thin, kept contracting and twisting and creaking. Glenn was at the equivalent of nine stories up in the air. The enormous rocket seemed curiously fragile, the way it moved and creaked and whined. The contractions created high-frequency vibrations and the lox hissed in the pipes, and it all ran up through the capsule like a metallic wail. It was the same rocket lox wail they used to hear at dawn at Edwards when they fueled the D-558–2 many years before.

Through the periscope Glenn could see for miles down the Banana River and the Indian River. He could just barely make out the thousands of people along the beaches. Some of them had been camping out along there in trailers since January 23, when the flight was first scheduled. They had elected camp mayors. They were having a terrific time. A month in a Banana River trailer camp was not too long to wait to make sure you were here when an event of this magnitude occurred.

There they were, thousands of them, off on the periphery as Glenn looked out. He could only see them through the periscope. They looked very small and far away and far below. And they were all wondering with a delicious shudder what it must be like to be in his place now. How frightened is he! *Tell us! That's all we want to know!* The fear and the gamble. Never mind the rest. Lying on his back like this, with his legs jackknifed up above him, stuffed blind into the holster, with the hatch closed, he couldn't help but be aware of his own heartbeat from time to time. Glenn could tell that his pulse was slow. Out loud, if the subject ever came up, everyone said that pulse rates didn't matter; it was a very subjective thing; many variables; and so on. It had only been within the past five years that biosensors had ever been put on pilots. They resented them and didn't care to attach any importance to them. Nevertheless, without saying so, everyone knew that they provided a rough gauge of a man's emotional state. Without saying so—not a word!—everyone knew that Gus Grissom's pulse rate had been *somewhat panicky*. It kept jumping over 100 during the countdown and then spurted up to 150 during the lift-off and stayed that high throughout his weightless flight, then jumped again, all the way to 171, just before the retrorockets went off. No one—certainly not out loud—no one was going to draw any conclusions from it, but . . . it was not a sign of the right stuff. Add to that his performance in the water . . . In his statement about people who get panicky over the flight test business, Glenn had said you had to know how to control your emotions. Well, he was as good as his word. Did any yogi ever control his heartbeat and perspiration better! (And, as the biomedical panels in the Mission Control room showed, his pulse never went over 80 and was holding around 70, no more than that of any normal healthy bored man having breakfast in the kitchen.) Occasionally he could feel his heart skip a beat or beat with an odd electrical sensation, and he knew that he was feeling the tension. (And at the biomedical panels the young doctors looked at each other in consternation—and then shrugged.) Nevertheless, he was aware that he was feeling no fear. He truly was not. He was more like an actor who is going out to perform in the same play yet once again—the only difference being that the audience this time is enormous and highly prestigious. He knew every sensation he would feel once the event began. The main thing was not to . . . "foul up." Please, dear God, don't let me foul up. In fact, there was little chance that he would forget so much as a word or a single move. Glenn had been the backup pilot—everyone said *pilot* now—for both Shepard and Grissom. During the charade before the first flight, he had gone through all of Shepard's simulations, and he had repeated most of Grissom's. And the simulations he had gone through as prime pilot for the first orbital flight had surpassed any simulations ever done before. They had even put him in the capsule on top of the rocket and *moved the gantry* away from the rocket, because Grissom had reported the odd sensation of perceiving the gantry as *falling over,* as he witnessed the event through his periscope, just before lift-off. Therefore, this feeling would be *adapted out* of Glenn. They put him in the capsule on top of the rocket and instructed him to watch the gantry move away through his periscope. *Nothing* must be novel about the experience! On top of all that, he had Shepard's and Grissom's descriptions of variations from the simulations. "On the centrifuge you feel thus-and-such. Well, during the actual flight it feels like that but with this-and-that difference." No man had ever lived an event so completely ahead of time. He was socketed into the capsule, lying on his back, getting ready to do precisely what his enormous Presbyterian Pilot self-esteem had been dying to do for fifteen years: demonstrate to the world his righteous stuff.

Exactly that! The Presbyterian Pilot! Here he is!—within twenty seconds of lift-off, and the only strange thing is how little adrenaline is pumping when the moment comes . . . He can hear the rumble of the Atlas engines building up down there below his back. All the same, it isn't terribly loud. The huge squat rocket shakes a bit and struggles to overcome its own weight. It all happens very slowly in the first few seconds, like an extremely heavy elevator rising. They've lit the candle and there's no turning back, and yet there's no surge inside him. His pulse rises only to 110, no more than the minimum rate you should have if you have to deal with a sudden emergency. How strange that it should be this way! He has been more wound up for a takeoff in an F-102.

"The clock is operating," he said. "We're underway."

It was all very smooth, much smoother than the centrifuge . . . just as Shepard and Grissom said it would be. He had gone through the same g-forces so many times . . . he hardly noticed them as they built up. It would have bothered him much more if they had been less. Nothing novel! No excitement, please! It took thirteen seconds for the huge rocket to reach transonic speed. The vibrations started. It was just as Shepard and Grissom said: it was much gentler than the centrifuge. He was still lying flat on his back, and the g-forces drove him deeper and deeper into the seat, but it all felt so familiar. He barely noticed it. He kept his eyes on the instrument panel the whole time . . . All quite normal, every little needle and switch in the right place . . . No malevolent instructor feeding *Abort* problems into the loop . . . As the rocket entered the transonic zone, the vibration became intense. The vibrations all but obliterated the roar of the engines. He was entering the area of "max q," maximum aerodynamic pressure, in which the pressure of the shaft of the Atlas forcing its way through the atmosphere at supersonic speed would reach almost a thousand pounds per square foot. Through the cockpit window he could see the sky turning black. Almost 5 g's were driving him back into his seat. And yet . . . *easier than the centrifuge* . . . All at once he was through *max q*, as if through a turbulent strait, and the trajectory was smooth and he was supersonic and the rumble of the rocket engines was more muffled than ever and he could hear all the little fans and recorders and the busy little kitchen, the humming little shop . . . The pressure on his chest reached 6 g's. The rocket pitched down. For the first time he could see clouds and the horizon. In a moment—*there it was*—the Atlas rocket's two booster engines shut down and were jettisoned from the side of the shaft and his body was slammed forward, as if he were screeching to a halt, and the g-forces suddenly dropped to 1.25, almost as if he were on earth and not accelerating at all, but the central sustainer engine and two smaller engines were still driving him up through the atmosphere . . . A flash of white smoke went up past the window . . . *No! The escape tower was firing early—but the* JETTISON TOWER *light wasn't on!* . . . He didn't see the tower go . . . Wait a minute . . . There went the tower, on schedule . . .

The JETTISON TOWER light came on green . . . The smoke must have been from the booster rockets as they left the shaft . . . The rocket pitched back up . . . going straight up . . . The sky was very black now . . . The g-forces began pushing him back into his seat again . . . 3 g's . . . 4 g's . . . 5 g's . . . Soon he would be forty miles up . . . the last critical moment of powered flight, as the capsule separated from the rocket and went into its orbital trajectory . . . or didn't . . . *Hey!* . . . All at once the whole capsule was whipping up and down, as if it were tied to the end of a diving board, a springboard. The g-forces built up and the capsule whipped up and down. Yet no sooner had it begun than Glenn knew what it was. The weight of the rocket on the launch pad had been 260,000 pounds, practically all of it rocket fuel, the liquid oxygen. This was being consumed at such a furious rate, about one ton per second, that the rocket was becoming merely a skeleton with a thin skin of metal stretched over it, a tube so long and light that it was flexing. The g-forces reached six and then he was weightless, just like that. The sudden release made him feel as if he were tumbling head over heels, as if he had been catapulted off the end of that same springboard and was falling through the air doing forward rolls. But he had felt this same thing on the centrifuge when they ran the g-forces up to seven and then suddenly cut the speed. At the same moment, right on schedule . . . a loud report . . . the posigrade rockets fired, throwing the capsule free of the rocket shaft . . . the capsule began its automatic turnabout, and all the proper green lights went on in front of him, and he knew he was "through the gate," as they said.

"Zero-g and I feel fine," he said. "Capsule is turning around . . ."

Glenn knew he was weightless. From the instrument readings and through sheer logic he knew it, but he couldn't feel it, just as Shepard and Grissom had never felt it. The turnaround brought him up to a sitting position, vertical to the earth, and that was the way he felt. He was sitting in a chair, upright, in a very tiny cramped quiet little cubicle 125 miles above the earth, a little metal closet, silent except for the humming of its electrical system, the inverters, the gyros, the cameras, the radio . . . *the radio* . . . He had been specifically instructed to violate the Fighter Jock code of No Chatter. He was supposed to radio back every sight, every sensation, and otherwise give the taxpayers the juicy stuff they wanted to hear. Glenn, more than any of the others, was fully capable of doing the job. Yet it was an awkward thing. It seemed unnatural.

"Oh!" he said. "That view is tremendous!"

Well, it was a start. In fact, the view was not particularly extraordinary. It was extraordinary that he was up here in orbit about the earth. He could see the exhausted Atlas rocket following him. It was tumbling end over end from the force of the small rockets throwing the capsule free of it.

He could hear Alan Shepard, who was serving as capcom in the Mercury Control Center at the Cape. His voice came in very clearly. He was saying, "You have a go, at least seven orbits."

"Roger," said Glenn. "Understand Go for at least seven orbits . . . This is *Friendship 7*. Can see clear back, a big cloud pattern way back across toward the Cape. Beautiful sight."

He was riding backward, looking back toward the Cape. It must be tremendous, it must be beautiful—what else could it be? And yet it didn't look terribly different from what he had seen at 50,000 feet in fighter planes. He had no greater sense of having left the bonds of earth. The earth was not just a little ball beneath him. It still filled his field of consciousness. It slid by slowly underneath him, just the way it did when you were in an airplane at forty or fifty thousand feet. He had no sense of being a *star voyager.* He couldn't see any stars at all. He could see the Atlas booster tumbling behind him and beginning to grow smaller, because it was in a slightly lower orbit. It just kept tumbling. There was nothing to stop it. Somehow the sight of this colossal great tumbling cylinder, which had weighed more than the average freighter while it was on the ground and which now weighed nothing and had been discarded like a candy wrapper—somehow it was more extraordinary than the view of earth. It shouldn't have been, but it was. The earth looked the way it had looked to Gus Grissom. Shepard had seen a low-grade black-and-white movie. Through his window Glenn could see what Grissom saw, the brilliant blue band at the horizon, a somewhat wider band of deeper blues leading into the absolutely black dome of the sky. Most of the earth was covered in clouds. The clouds looked very bright, set against the blackness of the sky. The capsule was heading east, over Africa. But, because he was riding backward, he was looking west. He saw everything after he had passed over it. He could make out the Canary Islands, but they were partly obscured by clouds. He could see a long stretch of the African coast . . . huge dust storms over the African desert . . . but there was no sense of taking in the whole earth at a glance. The earth was eight thousand miles in diameter and he was only a hundred miles above it. He knew what it was going to look like in any case. He had seen it all in photographs taken from the satellites. It had all been flashed on the screens for him. Even the view had been simulated. *Yes . . . that's the way they said it would look . . .* Awe seemed to be demanded, but how could he express awe honestly? He had lived it all before the event. How could he explain that to anybody? The view wasn't the main thing, in any case. The main thing . . . was *the check-list!* And just try explaining that! He had to report all his switch and dial readings. He had to put a special blood-pressure rig on the arm of his pressure suit and pump it up. (His blood pressure was absolutely normal, 120 over 80—*perfect stuff!*) He had to check the manual attitude-control system, swing the capsule up and down, side to side, roll to the right, roll to the left . . . and there was nothing novel about it, not even in orbit, a hundred miles above the earth. *How could you explain that!* When he swung the capsule, it felt the same as it did in a one-g state on earth. He still didn't feel weightless. He merely felt less cramped, because there were no longer any pressure points on his body. He was sitting straight up in a chair drifting slowly and quietly around the earth. Just the hum of his little shop, the background noises in his headset, and the occasional spurt of the hydrogen-peroxide jets.

"This is *Friendship 7,*" he said. "Working just like clockwork on the control check, and it went through just about like the procedures-trainer runs."

Well, that was it. The procedures trainer and the ALFA trainer and the centrifuge . . . He noticed that, in fact, he seemed to be moving a little faster than he had been on the ALFA trainer. When you sat in the trainer, cranking your simulated hydrogen-peroxide thrusters, they ran films on the screen of the earth rolling by below you, just the way it would be in orbital flight. "They didn't roll it by fast enough," he said to himself. Not that it mattered particularly . . . The sensation of speed was no more than that of being in an airliner and watching a cloud bank slide by far below . . . The world demanded awe, because this was a voyage through the stars. But he couldn't feel it. The backdrop of the event, the stage, the environment, the true orbit . . . was not the vast reaches of the universe. It was the simulators. *Who could possibly understand this?* Weightless he was, in the vacuum of space, humming around the earth . . . but his center of gravity was still back in that Baptist hardtack Low Rent stretch of sand and palmetto grass in Florida.

Ahhhh—but now this was truly something. Forty minutes into the flight, as he neared the Indian Ocean, off the east coast of Africa, he began sailing into the night. Since he was traveling east, he was going away from the sun at a speed of 17,500 miles an hour. But because he was riding backward, he could see the sun out the window. It was sinking the way the moon sinks out of sight as seen on earth. The edge of the sun began to touch the edge of the horizon. He couldn't tell what part of the earth it was. There were clouds everywhere. They created a haze at the horizon. The brilliant light over the earth began to dim. It was like turning down a rheostat. It took five or six minutes. Very slowly the lights were dimming. Then he couldn't see the sun at all, but there was a tremendous band of orange light that stretched from one side of the horizon to the other, as if the sun were a molten liquid that had emptied into a tube along the horizon. Where there had been a bright-blue band before, there was now the orange band; and above it a wider dimmer band of oranges and reds shading off into the blackness of the sky. Then all the reds and oranges disappeared, and he was on the night side of the earth. The bright-blue band reappeared at the horizon. Above it, stretching up about eight degrees, was what looked like a band of haze, created by the earth's atmosphere. And above that . . . for the first time he could make out the stars. Down below, the clouds picked up a faint light from the moon, which was coming up behind him. Now he was over Australia. He could hear Gordon Cooper's voice. Cooper was serving as the capcom at the tracking station in the town of Muchea, out in the kangaroo boondocks of western Australia. He could hear Cooper's Oklahoma drawl.

"That sure was a short day," said Glenn.

"Say again, *Friendship 7*," said Cooper.

"That was about the shortest day I've ever run into," said Glenn. Somehow that was the sort of thing to say to old Oklahoma Gordo sitting down there in the middle of nowhere.

"Kinda passes rapidly, huh?" said Gordo.

"Yessir," said Glenn.

The clouds began to break up over Australia. He could make out nothing in the darkness except for electric lights. Off to one side he could make out the lights of an entire city, just as you could at 40,000 feet in an airplane, but the concentration of lights was terrific. It was an absolute mass of electric lights, and south of it there was another one, a smaller one. The big mass was the city of Perth and the smaller one was a town called Rockingham. It was midnight in Perth and Rockingham, but practically every living soul in both places had stayed up to turn on every light they had for the American sailing over in the satellite.

"The lights show up very well," said Glenn, "and thank everybody for turning them on, will you?"

"We sure will, John," said Gordo.

And he went sailing on past Australia with the lights of Perth and Rockingham sliding into the distance.

He was over the middle of the Pacific, about halfway between Australia and Mexico, when the sun began to come up behind him. This was just thirty-five minutes after the sun went down. Since he was traveling backward, he couldn't see the sunrise through the window. He had to use the periscope. First he could see the blue band at the horizon becoming brighter and brighter. Then the sun itself began to slide up over the edge. It was a brilliant red—not terribly different from what he had seen at sunrise on earth, except that it was rising faster and its outlines were sharper.

"It's blinding through the scope on clear," said Glenn. "I'm going to the dark filter to watch it come on up."

And then—*needles!* A tremendous layer of them—Air Force communications experiment that went amok . . . Thousands of tiny needles gleaming in the sun outside the capsule . . . But they couldn't be needles, because they were luminescent—they were like snowflakes—

"This is *Friendship 7,*" he said. "I'll try to describe what I'm in here. I am in a big mass of some very small particles that are brilliantly lit up like they're luminescent. I never saw anything like it. They're round, a little. They're coming by the capsule, and they look like little stars. A whole shower of them coming by. They swirl around the capsule and go in front of the window and they're all brilliantly lighted. They probably average maybe seven or eight feet apart, but I can see them all down below me also."

"Roger, *Friendship 7.*" This was the capcom on Canton Island out in the Pacific. "Can you hear any impact with the capsule? Over."

"Negative, negative. They're very slow. They're not going away from me more than maybe three or four miles per hour."

They swirled about his capsule like tiny weightless diamonds, little bijoux—no, they were more like fireflies. They had that lazy but erratic motion, and when he focused on one it would seem to be lit up, but the light would go out and he would lose track of it, and then it would light up again. That was like fireflies, too. There used to be thousands of fireflies in the summers, when he was growing up. These things were like fireflies, but they obviously couldn't be any sort of organism . . . unless all the astronomers and all the satellite recording mechanisms had been fundamentally wrong . . . They were undoubtedly particles of some sort, particles that caught the sunlight at a certain angle. They were beautiful, but were they coming from the capsule? That could mean trouble. They must have been coming from the capsule, because they traveled along with him, in the same trajectory, at the same speed. But wait a minute. Some of them were far off, far below . . . there might be an entire field of them . . . a minute cosmos . . . something never seen before! And yet the capcom on Canton Island didn't seem particularly interested. And then he sailed out of range of Canton and would have to wait to be picked up by the capcom at Guaymas, on the west coast of Mexico. And when the Guaymas capcom picked him up, he didn't seem to know what he was talking about.

"This is *Friendship 7,*" said Glenn. "Just as the sun came up, there were some brilliantly lighted particles that looked luminous that were swirling around the capsule. I don't have any in sight right now. I did have a couple just a moment ago, when I made the transmission over to you. Over."

"Roger, *Friendship 7.*"

And that was it. "Roger, *Friendship 7.*" Silence. They didn't particularly care.

Glenn kept talking about his fireflies. He was fascinated. It was the first true unknown anyone had encountered out here in the cosmos. At the same time he was faintly apprehensive. *Roger, Friendship 7.* The capcom finally asked a polite question or two, about the size of the particles and so on. They obviously were not carried away by this celestial discovery.

All of a sudden the capsule swung out to the right in a yaw, out about twenty degrees. Then it was as if it hit a little wall. It bounced back. Then it swung out again in the yaw and hit the little wall and bounced back. Something had gone out in the automatic attitude control. Never mind the celestial fireflies. He was sailing over California, heading for Florida. Now all the capcoms were coming alive, all right.

President Kennedy was supposed to come on the radio as Glenn came over the United States. He was going to bless his single-combat warrior as he came over the continental U.S.A. He was going to tell him the hearts of all his fellow citizens were with him. But that all went by the boards in view of the problem with the automatic controls.

Glenn went sailing over Florida, over the Cape, starting his second orbit. He couldn't see much of anything down below, because of the clouds. He no longer cared particularly. The attitude control was the main thing. One of the small thrusters seemed to have gone out, so that the capsule would drift to the right, like a car slowly skidding on ice. Then a bigger thruster would correct the motion and bounce it back. That was only the start. Pretty soon other thrusters began acting up when he was on automatic. Then the gyros started going. The dials that showed the angle of the capsule with respect to the earth and the horizon were giving obviously wrong readings. He had to line it up visually with the horizon. Fly by wire! Manual control! It was no emergency, however, at least not yet. As long as he was in orbit, the attitude control of the capsule didn't particularly matter, so far as his safety was concerned. He could be going forward or backward or could have his head pointed straight at the earth or could be drifting around in circles or pitching head over heels, for that matter, and it wouldn't change his altitude or trajectory in the slightest. The only critical point was the re-entry. If the capsule were not lined up at the correct angle, with the blunt end and the heat shield down, it might burn up. To line it up correctly, fuel was required, the hydrogen peroxide, no matter whether it was lined up automatically or by the astronaut. If too much fuel was used keeping the capsule stable while it sailed around in orbit, there might not be enough left to line it up before the re-entry. That had been the problem in the ape's flight. The automatic attitude control had started malfunctioning and was using up so much fuel they brought him down after two orbits.

Every five minutes he had to shift his radio communications to a new capcom. You couldn't receive and send at the same time, either. It wasn't like a telephone hookup. So you spent half the time just making sure you could hear each other.

"*Friendship 7, Friendship 7,* this is CYI." That was the Canary Islands capcom. "The time is now 16:32:26. We are reading you loud and clear; we are reading you loud and clear. CYI."

Glenn said: "This is *Friendship 7* on UHF. As I went over recovery area that time, I could see a wake, what appeared to be a long wake in the water. I imagine that's the ships in our recovery area."

"*Friendship 7* . . . We do not read you, do not read you. Over."

"*Friendship 7,* this is Kano. At G.M.T. 16:33:00. We do not . . . This is Kano. Out."

"*Friendship 7, Friendship 7,* this is CYI Com Tech. Over."

Glenn said, "Hello, Canary. *Friendship 7.* Receive you loud and a little garbled. Do you receive me? Over."

"*Friendship 7, Friendship 7,* this is CYI Com Tech. Over."

"Hello, Canary, *Friendship 7.* I read you loud and clear. How me? Over."

"*Friendship 7, Friendship 7,* this is CYI Com Tech. Over."

"Hello, CYI Com Tech. *Friendship 7.* How do you read me? Over."

"*Friendship 7, Friendship 7,* this is CYI, CYI Com Tech. Do you read? Over."

"Roger. This is *Friendship 7,* CYI. I read you loud and clear. Over."

"*Friendship 7, Friendship 7,* this is CYI Com Tech, CYI Com Tech. Do you read? Over."

"Hello, CYI Com Tech. Roger, read you loud and clear."

"*Friendship 7,* this is CYI Com Tech. Read you loud and clear also, on UHF, on UHF. Standby."

"Roger, *Friendship 7.*"

"*Friendship 7, Friendship 7, Friendship 7,* this is Canary capcom. How do you read? Over."

"Hello, Canary capcom. *Friendship 7.* I read you loud and clear. How me?

Finally, the Canary Islands capcom said: "I read you loud and clear. I am instructed to ask you to correlate the actions of the particles surrounding your spacecraft with the actions of your control jets. Do you read? Over."

"This is *Friendship 7*. I did not read you clear. I read you loud but very garbled. Over."

"Roger. Cap asks you to correlate the actions of the particles surrounding the vehicle with the reaction of one of your control jets. Do you understand? Over."

"This is *Friendship 7*. I do not think they were from my control jets, negative. Over."

There—exactly five minutes to get one question out and one answer. Well, at least they finally showed an interest in the fireflies. They wondered if they might have something to do with the malfunctioning thrusters. Oh, but it was a struggle.

In any case, he was not particularly worried. He could control the attitude manually if he had to. The fuel seemed to be holding out. Everything hummed and whined and buzzed as usual inside the capsule. The same high background tones came over the radio. He could hear the oxygen coursing through his pressure suit and his helmet. There was no "sensation" of motion speed at all, unless he looked down at the earth. Even then it slid by very slowly. When the thrusters spurted hydrogen peroxide, he could feel the capsule swing this way and that. But it was like the ALFA trainer on earth. He still didn't feel weightless. He was still sitting straight up in his chair. On the other hand, the camera—when he wanted to reload it, he just parked it in the empty space in front of his eyes. It just floated there in front of him. Way down there were little flashes all over the place. It was lightning in the clouds over the Atlantic. Somehow it was more fascinating than the sunset. Sometimes the lightning was inside the clouds and looked like flashlights going on and off underneath a blanket. Sometimes it was on top of the clouds, and it looked like firecrackers going off. It was extraordinary, and yet there was nothing new about the sight. An Air Force colonel, David Simons, had gone up in a balloon, alone, to 102,000 feet, for thirty-two hours and had seen the same thing.

Glenn was now over Africa, riding over the dark side of the earth, sailing backward toward Australia. The Indian Ocean capcom said: "We have message from MCC for you to keep your landing-bag switch in off position. Landing-bag switch in off position. Over."

"Roger," said Glenn. "This is *Friendship 7*."

He wanted to ask why. But that was against the code, except in an emergency situation. That fell under the heading of nervous chatter.

Over Australia old Gordo, Gordo Cooper, got on the same subject: "Will you confirm the landing-bag switch is in the off position? Over."

"That is affirmative," said Glenn. "Landing-bag switch is in the center off position."

"You haven't had any banging noises or anything of this type at higher rates?"

"Negative."

"They wanted this answer."

They still didn't say why, and Glenn entered into no nervous chatter. He now had two red lights on the panel. One was the warning light for the automatic fuel supply. All the little amok action of the yaw thrusters had used it up. Well, it was up to the Pilot now . . . to aim the capsule correctly for re-entry . . . The other was a warning about excess cabin water. It built up as a by-product of the oxygen system. Nevertheless, he pressed on with the checklist. He was supposed to exercise by pulling on the bungee cord and then take his blood pressure. The Presbyterian Pilot! He did it without a peep. He was pulling on the bungee and watching the red lights when he began sailing backward into the sunrise again. Two hours and forty-three minutes into the flight, his second sunrise over the Pacific . . . seen from behind through a periscope. But he hardly watched it. He was looking for the fireflies to light up again. The great rheostat came up, the earth lit up, and now there were thousands of them swirling about the capsule. Some of them seemed to be miles away. A huge field of them, a galaxy, a microuniverse. No question about it, they weren't coming from the capsule, they were part of the cosmos. He took out the camera again. He had to photograph them while the light was just right.

"Friendship 7." The Canton Island capcom was coming in. "This is Canton. We also have no indication that your landing bag might be deployed. Over."

Glenn's first reaction was that this must have something to do with the fireflies. He's telling them about the fireflies and they come in with something about the landing bag. But who said anything about the landing bag being deployed?

"Roger," he said. "Did someone report landing bag could be down? Over."

"Negative," said the capcom. "We had a request to monitor this and to ask you if you heard any flapping, when you had high capsule rates."

"Well," said Glenn, "I think they probably thought these particles I saw might have come from that, but these are . . . there are thousands of these things, and they go out for it looks like miles in each direction from me, and they move by here very slowly. I saw them at the same spot on the first orbit. Over."

And so he thought that explained all the business about the landing bag.

They gave him the go-ahead for his third and final orbit as he sailed over the United States. He couldn't see a thing for the clouds. He pitched the capsule down sixty degrees, so he could look straight down. All he could see was the cloud deck. It was just like flying at high altitudes in an airplane. He was really no longer in the mood for sightseeing. He was starting to think about the sequence of events that would lead to the retrofiring over the Atlantic after he had been around the world one more time. He had to fight both the thrusters and the gyros now. He kept releasing and resetting the gyros to see if the automatic attitude control would start functioning again. It was all out of whack. He would have to position the capsule by using the horizon as a reference. He was sailing backward over America. The clouds began to break. He began to see the Mississippi delta. It was like looking at the world from the tail-gun perch of the bombers they used in the Second World War. Then Florida started to slide by. Suddenly he realized he could see the whole state. It was laid out just like it is on a map. He had been around the world twice in three hours and eleven minutes and this was the first sense he had had of how high up he was. He was about 550,000 feet up. He could make out the Cape. By the time he could see the Cape he was already over Bermuda.

"This is *Friendship 7*," he said. "I have the Cape in sight down there. It looks real fine from up here."

"Rog. Rog." That was Gus Grissom on Bermuda.

"As you know," said Glenn.

"Yea, verily, sonny," said Grissom.

Oh, it all sounded very fraternal. Glenn was modestly acknowledging that his loyal comrade Grissom was one of the only three Americans ever to see such a sight . . . and Grissom was calling him "sonny."

Twenty minutes later he was sailing backward over Africa again and the sun was going down again, for the third time, and the rheostat was dimming and he . . . saw *blood*. It was all over one of the windows. He knew it couldn't be blood, and yet it was blood. He had never noticed it before. At this particular angle of the setting rheostat sun he could see it. Blood and dirt, a real mess. The dirt must have come from the firing of the escape tower. And the blood . . . *bugs,* perhaps . . . The capsule must have smashed into bugs as it rose from the launch pad . . . or *birds* . . . but he would have heard the thump. It must have been bugs, but bugs didn't have blood. Or the blood red of the sun going down in front of him diffusing . . . And then he refused to think about it any more. He just turned the subject off. Another sunset, another orange band streaking across the rim of the horizon, more yellow bands, blue bands, blackness, thunderstorms, lightning making little sparkles under the blanket. It hardly mattered any more. The whole thing of lining the capsule up for retrofire kept building up in his mind. In slightly less than an hour the retro-rockets would go off. The capsule kept slipping its angles, swinging this way and that way, drifting. The gyros didn't seem to mean a thing anymore.

And he went sailing backward through the night over the Pacific. When he reached the Canton Island tracking point, he swung the capsule around again so that he could see his last sunrise while riding forward, out the window, with his own eyes. The first two he had watched through the periscope because he was going backward. The fireflies were all over the place as the sun came up. It was like watching the sunrise from inside a storm of the things. He began expounding upon them again, about how they couldn't possibly come from the capsule, because some of them seemed to be miles away. Once again nobody on the ground was interested. They weren't interested on Canton Island, and pretty soon he was in range of the station on Hawaii, and they weren't interested, either. They were all wrapped up in something else. They had a little surprise for him. They backed into it, however. It took him a while to catch on.

He was now four hours and twenty-one minutes into the flight. In twelve minutes the retro-rockets were supposed to fire, to slow him down for re-entry. It took him another minute and forty-five seconds to go through all the "do you reads" and "how me's" and "overs" and establish contact with the capcom on Hawaii. Then they sprang their surprise.

"*Friendship 7,*" said the capcom. "We have been reading an indication on the ground of segment 5–1, which is Landing Bag Deploy. We suspect this is an erroneous signal. However, Cape would like you to check this by putting the landing-bag switch in auto position, and seeing if you get a light. Do you concur with this? Over."

It slowly dawned on him . . . *Have been reading* . . . For how long? . . . Quite a little surprise. And they hadn't told him! They'd held it back! *I am a pilot and they refuse to tell me things they know about the condition of the craft!* The insult was worse than the danger! If the landing bag had deployed—and there was no way he could look out and see it, not even with the periscope, because it would be directly behind him—if it had deployed, then the heat shield must be loose and might come off during the re-entry. If the heat shield came off, he would burn up inside the capsule like a steak. If he put the landing-bag switch in the automatic control position, then a green light should come on if the bag was deployed. Then he would know. Slowly it dawned! . . . That was why they kept asking him if the switch were in the off position!—they didn't want him to learn the awful truth too quickly! Might as well let him complete his three orbits—then we'll let him find out about the bad news!

On top of that, they now wanted him to fool around with the switch. *That's stupid!* It might very well be that the bag had not deployed but there was an electrical malfunction somewhere in the circuit and fooling with the automatic switch might then cause it to deploy. But he stopped short of saying anything. Presumably they had taken all that into account. There was no way he could say it without falling into the dread nervous chatter.

"Okay," said Glenn. "If that's what they recommend, we'll go ahead and try it. Are you ready for it now?"

"Yes, when you're ready."

"Roger."

He reached forward and flipped the switch. Well . . . this was it—

No light. He immediately switched it back to off.

"Negative," he said. "In automatic position did not get a light and I'm back in off position now. Over."

"Roger, that's fine. In this case, we'll go ahead, and the re-entry sequence will be normal."

The retro-rockets would be fired over California, and by the time the retro-rockets brought him down out of his orbit and through the atmosphere, he would be over the Atlantic near Bermuda. That was the plan. Wally Schirra was the capcom in California. Less than a minute before he was supposed to fire the retro-rockets, by pushing a switch, he heard Wally saying: "John, leave your retropack on through your pass over Texas. Do you read?"

"Roger."

But why? The retropack wrapped around the edges of the heat shield and held the retro-rockets. Once the rockets were fired, the retropack was supposed to be jettisoned. They were back to the heat shield again, with no explanation. But he had to concentrate on firing the retro-rockets.

Next to the launch this was the most dangerous part of the flight. If the capsule's angle of attack was too shallow, you might skip off the top of the earth's atmosphere and stay in orbit for days, until long after your oxygen had run out. You wouldn't have any more rockets to slow you down. If the angle were too steep, the heat from the friction of going through the atmosphere would be so intense you would burn up inside the capsule, and a couple of minutes later the whole thing would disintegrate, heat shield or no heat shield. But the main thing was not to think about it in quite those terms. The field of consciousness is very small, said Saint-Exupéry. *What do I do next?* It was the moment of the test pilot at last. Oh, yes! *I've been here before! And I am immune! I don't get into corners I can't get out of!* One thing at a time! He could be a true flight test hero and try to line the capsule up all by himself by using the manual controls with the horizon as his reference—or he could make one more attempt to use the automatic controls. Please, dear God . . . don't let me foul up! What would the Lord answer? (Try the automatic, you ninny.) He released and reset the gyros. He put the controls on automatic. The answer to your prayers, John! Now the dials gibed with what he saw out the window and through the periscope. The automatic controls worked perfectly in pitch and roll. The yaw was still off, so he corrected that with the manual controls. The capsule kept pivoting to the right and he kept nudging it back. The ALFA trainer! One thing at a time! It was just like the ALFA trainer. . . no sense of forward motion at all . . . As long as he concentrated on the instrument panel and didn't look at the earth sliding by beneath him, he had no sense at all of going 17,500 miles an hour . . . or even five miles an hour . . . The humming little kitchen . . . He sat up in his chair squirting his hand thruster, with his eyes pinned on the dials . . . Real life, a crucial moment—against the eternal good beige setting of the simulation. One thing at a time!

Schirra began giving him the countdown for firing the rockets. "Five, four—"

He nudged it back one more with the yaw thruster.

"—three, two, one, fire."

He pushed the retro-rocket switch with his hand.

The rockets started firing in sequence, the first one, the second one, the third one. The sound seemed terribly muffled—but in that very moment, the jolt! Pure gold! One instant, as Schirra counted down, he felt absolutely motionless. The next . . . *thud thud thud* . . . the jolt in his back. He felt as if the capsule had been knocked backward. He felt as if he were sailing back toward Hawaii. All as it should be! Pure gold! The retro-light was lit up green. It was all going perfectly. He was merely slowing down. In eleven minutes he would be entering the earth's atmosphere.

He could hear Schirra saying: "Keep your retropack on until you pass Texas."

Still no reason given! He couldn't see the pattern yet. There was only the dim sense that in some fashion they were jerking him around. But all he said was: "That's affirmative."

"It looked like your attitude held pretty well," said Schirra. "Did you have to back it up at all?"

"Oh, yes, quite a bit. Yeah, I had a lot of trouble with it."

"Good enough for government work from down here," said Schirra. That was one of Schirra's favorite lines.

"Do you have a time for going to Jettison Retro?" said Glenn. This was an indirect way of asking for some explanation for the mystery of keeping the retropack on.

"Texas will give you that message," said Schirra. "Over."

They weren't going to tell him! Not so much the thought . . . as the *feeling* . . . of the insult began to build up.

Three minutes later the Texas capcom tracking station came in: "This is Texas capcom, *Friendship 7*. We are recommending that you leave the retropackage on through the entire re-entry. This means that you will have to override the zero-point-oh-five-g switch, which is expected to occur at 04:43:53. This also means that you will have to manually retract the scope. Do you read?"

That did it.

"This is *Friendship 7*," said Glenn. "What is the reason for this? Do you have any reason? Over."

"Not at this time," said the Texas capcom. "This is the judgment of Cape Flight . . . Cape Flight will give you the reason for this action when you are in view."

"Roger. Roger. *Friendship 7*."

It was really unbelievable. It was beginning to fit—

Twenty-seven seconds later he was over the Cape itself and the Cape capcom, with the voice of Alan Shepard on the radio, was telling him to retract his periscope manually and to get ready for re-entry into the atmosphere.

It was beginning to fit together, he could see the pattern, the whole business of the landing bag and the retropack. This had been going on for a couple of hours now—and they were telling him nothing! Merely giving him the bits and pieces! But if he was going to re-enter with the retropack on, then they wanted the straps in place for some reason. And there was only one possible reason—something was wrong with the heat shield. And this they would not tell him! *Him!*—the pilot! It was quite unbelievable! It was—

He could hear Shepard's voice.

He was winding in the periscope, and he could hear Shepard's voice: "While you're doing that . . . we are not sure whether or not your landing bag has deployed. We feel it is possible to re-enter with the retropackage on. We see no difficulty at this time in that type of re-entry."

Glenn said, "Roger, understand."

Oh, yes, he understood now! If the landing bag was deployed, that meant the heat shield was loose. If the heat shield was loose, then it might come off during the re-entry, unless the retropack straps held it in place long enough for the capsule to establish its angle of re-entry. And the straps would soon burn off. If the heat shield came off, then he would fry. If they didn't want him—*the pilot!*—to know all this, then it meant they were afraid he might panic. And if he didn't even *need* to know the whole pattern—just the pieces, so he could follow orders—*then he wasn't really a pilot!* The whole sequence of logic clicked through Glenn's mind faster than he could have put it into words, even if he had dared utter it all at that moment. He was being treated like a passenger—a redundant component, a backup engineer, a boiler-room atten- dant—in an automatic system!—like someone who did not have that rare and unutterably righteous stuff!—as if the right stuff itself did not even matter! It was a transgression against all that was holy—all this in a single limbic flash of righteous indignation as John Glenn re-entered the earth's atmosphere.

"*Seven,* this is Cape," said Al Shepard. "Over."

"Go ahead, Cape," said Glenn. "You're ground . . . you are going out."

"We recommend that you . . ."

That was the last he could hear from the ground. He had entered the atmosphere. He couldn't feel the g-forces yet, but the friction and the ioniza- tion had built up, and the radios were now useless. The capsule was beginning to buffet and he was fighting it with the controls. The fuel for the automatic system, the hydrogen peroxide, was so low he could no longer be sure which system worked. He was descending backward. The heat shield was on the out- side of the capsule, directly behind his back. If he glanced out the window he could see only the blackness of the sky. The periscope was retracted, so he saw nothing on the scope screen. He heard a *thump* above him, on the outside of the capsule. He looked up. Through the window he could see a strap. *From the retropack. The straps broke! And now what!* Next the heat shield! The black sky out the window began to turn a pale orange. The strap flat against the window started burning—and then it was gone. The universe turned a flaming orange. That was the heat shield beginning to burn up from the tremendous speed of the re-entry. This was something Shepard and Grissom had not seen. They had not re-entered the atmosphere at such speed. Nevertheless, Glenn knew it was coming. Five hundred, a thousand times he had been told how the heat shield would *ablate,* burn off layer by layer, vaporize, dissipate the heat into the atmo- sphere, send off a corona of flames. All he could see now through the window were the flames. He was inside of a ball of fire. But!—a huge flaming chunk went by the window, a great chunk of something burning. Then another . . .

another . . . The capsule started buffeting . . . The heat shield was breaking up! It was crumbling—flying away in huge flaming chunks . . . He fought to steady the capsule with the hand controller. *Fly-by-wire!* But the rolls and yaws were too fast for him . . . The ALFA trainer gone amok, inside a fireball . . . The heat! . . . It was as if his entire central nervous system were now centered in his back. If the capsule was disintegrating and he was about to burn up, the heat pulse would reach his back first. His backbone would become like a length of red-hot metal. He already knew what the feeling would be like . . . and when . . . *Now!* . . . But it didn't come. There was no tremendous heat and no more flaming debris . . . Not the heat shield, after all. The burning chunks had come from what remained of the retropack. First the straps had gone and then the rest of it. The capsule kept rocking, and the g-forces built up. He knew the g-forces by heart. A thousand times he had felt them on the centrifuge. They drove him back into the seat. It was harder and harder to move the hand controller. He kept trying to damp out the rocking motion by firing the yaw thrusters and the roll thrusters, but it was all too fast for him. They didn't seem to do much good, at any rate.

No more red glow . . . he must be out of the fireball . . . seven g's were driving him back into the seat . . . He could hear the Cape capcom:

". . . How do you read? Over."

That meant he had passed through the ionosphere and was entering the lower atmosphere.

"Loud and clear; how me?"

"Roger, reading you loud and clear. How are you doing?"

"Oh, pretty good."

"Roger. Your impact point is within one mile of the up-range destroyer."

Oh, pretty good. It wasn't Yeager, but it wasn't bad. He was inside of one and a half tons of non-aerodynamic metal. He was a hundred thousand feet up, dropping toward the ocean like an enormous cannonball. The capsule had no aerodynamic qualities whatsoever at this altitude. It was rocking terribly. Out the window he could see a wild white contrail snaked out against the blackness of the sky. He was dropping at a thousand feet per second. The last critical moment of the flight was coming up. Either the parachute deployed and took hold or it didn't. The rocking had intensified. The retropack! Part of the retropack must still be attached and the drag of it is trying to flip the capsule. . . He couldn't wait any longer. The parachute was supposed to deploy automatically, but he couldn't wait any longer. Rocking . . . He reached up to fire the parachute manually—but it fired on its own, automatically, first the drogue and

then the main parachute. He swung under it in a huge arc. The heat was ferocious, but the chute held. It snapped him back into the seat. Through the window the sky was blue. It was the same day all over again. It was early in the afternoon on a sunny day out in the Atlantic near Bermuda. Even the
landing-bag light was green. There was nothing even wrong with the landing
bag. There had been nothing wrong with the heat shield. There was nothing
wrong with his rate of descent, forty feet per second. He could hear the rescue
ship chattering away over the radio. They were only twenty minutes away from
where he would hit, only six miles. He was once again lying on his back in the
human holster. Out the window the sky was no longer black. The capsule
swayed under the parachute, and over this way he looked up and saw clouds
and over that way blue sky. He was very, very hot. But he knew the feeling. All
those endless hours in the heat chambers—it wouldn't kill you. He was coming down into the water only 300 miles from where he started. It was the same
day, merely five hours later. A balmy day out in the Atlantic near Bermuda. The
sun had moved just seventy-five degrees in the sky. It was 2:45 in the afternoon. Nothing to do but get all these wires and hoses disconnected. *He had
done it.* He began to let the thought loose in his mind. He must be very close
to the water. The capsule hit the water. It drove him down into his seat again,
on his back. It was quite a jolt. It was hot in here. Even with the suit fans still
running, the heat was terrific. Over the radio they kept telling him not to try
to leave the capsule. The rescue ship was almost there. They weren't going to
try the helicopter deal again, except in an emergency. He wasn't about to attempt a water egress. He wasn't about to hit the hatch detonator. The Presbyterian Pilot was not about to foul up. His pipeline to the dear Lord could not
be clearer. He had done it.

PART TWO

Pilots in Command

In the fell clutch of circumstance
I have not winced nor cried aloud.
Under the bludgeonings of chance
My head is bloody but unbowed.

It matters not how strait the gate
How charged with punishments the scroll,
I am the master of my fate;
I am the captain of my soul.

—Excerpt from "Invictus" by W. E. Henley

"Dooley"

BY ERNEST K. GANN

(Excerpted from the novel *Island In the Sky*)

During his lifetime (1910–1991), Ernest K. Gann flew a lot and he wrote a lot. His numerous books are not all devoted to his passion for aviation, for he was a sea-going man who sometimes turned his prose to the oceans that shared his love with the sky. But because he was arguably the most skilled writer of flying stories of them all, his name will always be an icon in aviation literature. Novels such as *Blaze of Noon, Island In the Sky,* and *The High and the Mighty*—along with the classic *Fate Is the Hunter,* a nonfiction account of his years as a commercial airline pilot—bring readers as close to cockpit reality as any pages ever written. In Gann's prose, his metaphorical "Island In the Sky," the cockpit, is a place of intrigue and challenges, of exhilaration and terror, felt by the reader with a life force few writers have successfully re-created so well. Were there not many other splendid aviation writers, I would not have the privilege of editing this book. But there was only one Ernest K. Gann.

Traditional publishing restraints have been known to dictate that any anthology should bear only one selection from the works of a particular author. Because I am here to serve readers—not publishing attitudes—I have tossed such notions overboard and have included two Ernest Gann tales in this anthology. An overdose of Gann? Not in my eyes. [To learn a great deal more about Ernest K. Gann—his life, books, and films—click his name into the Google.com search engine and check out the several fine sites listed.]

"Dooley" is the lead character of Gann's novel of civilian military air transport pilots in the early stages of World War Two. The pages excerpted are from the opening of the novel and were chosen for the powerful portrait they paint of life in the cockpit during a time of crisis.

The film version of *Island In the Sky,* with The Duke, John Wayne, playing the role of Dooley, was pretty darn good, in my opinion. I used to watch it on television years ago. Now, for some strange reason (no doubt involving some kind of legal squabble), *Island In the Sky* has disappeared from television and has never been released on video or DVD as of this writing. I really wish The Duke could come back here and kick some butt over this issue.

Dooley rubbed the gray stubble on his chin and stared into the evening. His heavy, sharp-cut face, Chinese red from the glow behind the overcast, made no secret of the disgust that filled his being.

"I don't like," he mumbled in a leathery voice, "the look of that there goddamned cloud bank." Dooley always referred to things as "that there"—it pleased him immensely. It was an exact impression, something you could depend upon, something everyone in Dooley's vast experience had always been able to understand. He looked down 10,000 feet to the placid waters of Davis Strait. For a moment, the deep-cut lines about his mouth ceased to look like spokes around a wagon wheel. "I don't like the look of it," he repeated to the world aloft in general, and then his attention returned to the cloud bank straight ahead. His eyes shifted and sought out the temperature gauge on the crowded instrument panel. The narrow pointer on the needle indicated precisely 20 degrees above zero.

Dooley lolled back in his seat and alternately rubbed his miserably chapped lips and the shock of gray hair that stood defiantly forward from the top of his head. He was tired. His eyes burned and his legs ached. His stomach, a recent host to a quantity of canned meat, felt as if it belonged to some other person. His tongue was dry. He placed an oxygen mask over his face for a moment, breathed deeply, and stared again at the cloud bank ahead. His lungs temporarily satisfied, he placed the mask carefully upon his knee and spoke to the young man at his side.

"We are going to pick up some ice." He said it thoughtfully and very quietly. He seemed not to care. There was no twinkle in his blue eyes. Dooley had to think—and think swiftly too, as swiftly as his passage through the upper atmosphere. He had to balance things, many things, one against the other. He had to work into nice juxtaposition a series of related facts that were constantly at odds with each other. In addition, they were facts that grew and changed with rapidity. Failure on Dooley's part to compromise with them would result in disaster.

Most of his preliminary thinking was automatic, based on long-established conditions. He was the captain of a four-engined transport airplane, presently somewhere between Greenland and North America. The logistics of modern warfare had brought Dooley to this forsaken portion of the sky. The Army Air Transport Command had provided the airplane, which someone had named the *Corsair*. The name was prominently lettered in yellow paint along its nose. A commercial airline had provided Dooley and his crew. It was a satisfactory arrangement for everyone concerned, including Dooley, whose forty-three years would otherwise have bound him to an Army desk. Now he was proceeding toward the Labrador Coast at an estimated speed of 200 miles per hour. Beneath him, far below in the twilight vastness, were numerous icebergs, strangely linked together by vagrant wisps of sea mist. Above his head, the plexi-glass window revealed the Arctic evening sky. Just ahead, a scant few feet, the instrument panel offered assorted information on the airspeed, altitude, course, engine temperatures, manifold pressures, and revolutions per minute. They were pleasantly old, familiar, hard facts—easy to comprehend and easy to observe. Dooley's twenty years of flying had taught him over and over again that he must not be content with them alone.

It was the things you couldn't see that counted—the hidden, never tangible series of upper world plots and fancies. They invariably joined company with the unwary, always with an air of deceptive innocence. They frequently killed the unwary.

Dooley twisted about in his seat and faced the radio operator.

"D'Annunzia," he said in a voice pitched just loud enough to compete with the four engines, "see if you can get a cross-bearing from Consolation Island." The black-haired young man removed his earphones and bent forward in his leather jacket.

"Been trying for the past hour. I can't raise 'em."

"How about Chapel Inlet?"

"Nothing doing." He threw up his hands.

"Who can you raise? Are they all asleep?"

"It's the damn Northern Lights . . . we gotta wait."

Dooley looked at the floor, then swung completely around in his seat and eyed the young man who sat directly behind him. He had the air of someone peering into a goldfish bowl. It was difficult for him to approach this youngster with the close-cropped hair and the apple cheeks. Like all his comrades, Dooley had been accustomed to doing his own navigation through space. Through the years it had become a deep-seated habit, an almost unconscious mental manipulation that seldom failed. Now the groundings in power had seen fit to place a navigator beneath his wing. The compass in Dooley's head, they maintained, was not sufficiently accurate. Dooley was open-minded but skeptical. He could not for the life of him decide whether the youngster, who lent the cockpit rather an air of the college dormitory, was a nuisance or a help.

"Murray," he said patiently, "where do *you* think we are?" Murray snapped his head up. His eyes were alive with enthusiasm. With a sudden jerky movement he retrieved the pencil from behind his ear and placed its point upon the chart before him.

"Right hear, Captain," he answered confidently. "I got a sun line an hour ago and just advanced it."

"Where's here?" Dooley twisted his neck around so he could see better. There was a small cross near the middle of the chart.

"How far to the coast?"

"Two hundred . . . perhaps two twenty . . ."

Dooley breathed deeply and thought for a moment. He continued to stare at the blank white representation of a thousand miles of open ocean.

"How high was the sun when you shot it?"

"Thirteen degrees and ten minutes."

"Do you think that was high enough to trust?" Dooley raised an eyebrow.

"I had perfect shooting—steady as a rock. Very small area with the marks."

"I'd be inclined to doubt it. My guess is we're coming up on the coast right now." He looked over his shoulder at the now fast approaching cloud bank. It was a cinch they wouldn't be able to top it. He turned back to Murray.

"Figure up what kind of a ground speed we'd have to make to be over that there Martin River peninsula about ten minutes from right now."

Murray picked up a small round celluloid disk, made a few gestures with his dividers on the chart, and frowned.

"We'd have to make 260 miles per hour, Captain . . . that could just hardly be possible!"

Dooley squinted at the sky above and noted the high wraiths of cirrus clouds. A tolerant smile played across his lips. He must remember when he got back on the ground to tell those bastards in the snug, warm offices so far over the horizon, that he was not running an aerial kindergarten, and that furthermore he, Dooley, wanted a say in the training of navigators—among many other things. He rubbed his eyes and sighed.

"Murray," he said in the manner of a high priest delivering a dictum, "anything . . . anything, mind you, Murray . . . can happen up here."

"Something wrong?" For the first time Murray's easy confidence seemed shaken.

"No . . . that is, not yet. You'd better split the difference between your assumed position and the Martin River peninsula. That'll put us a hundred or so offshore. Keep careful track of your dead reckoning from now on."

"Yes, sir." The answer was barely audible above the engine drone. Dooley turned around and faced the task ahead.

There was a jagged cut in the gray shapeless cloud mass. Dooley turned the *Corsair* so as to split the exact center of the cut. He wanted to slip in easily. He half hoped there might be layers within the great mass itself, layers that would enable him to change altitude from time to time and thus remain in contact flight. Although the big ship would then become even more obviously a fly speck in creation, Dooley was not concerned with futile comparisons. To remain in contact meant there would be no ice. The ice crystals and snow which swept hand in hand through the upper world would not leave his radios a pile of useless metal boxes.

For a time, the false horizon moved with the *Corsair*. A few low-hanging tendrils of mist from the upper cloud deck whipped past the wings. Then suddenly, as if crushed within a vise, the horizon merged into nothing. The bulbous snout of the *Corsair* was enveloped in dank, heavy vapor. Dooley flipped on the cockpit lights and glanced at the outside temperature gauge— that was the important thing, it must not get any warmer. He knew that now there was no postponing the encounter and he wanted as many things on his side as he could muster. As they penetrated more deeply into the overcast, certain elements, most of them invisible, would oppose the *Corsair's* swift passage. Working together, they might stop it altogether. They were old and familiar enemies to Dooley. They were merciless and unforgiving enemies. They were tricky and powerful.

Dooley was neither helpless nor afraid. He had a few tricks of his own and a certain number of weapons at his disposal, which when called upon in emergencies were invariably able to perform greater miracles than anyone ever had a right to expect. It was, however, a characteristic of Dooley's trade to keep the point of expectancy low, thus preventing disappointment, particularly when disappointment was but a prelude to instant death.

The most valuable of Dooley's weapons was his experience in the air, a quality matched only by his comrade captains. It dated back to the real beginning of flying, to the barnstorming days of the Twenties, when Dooley, a red-faced irrepressible youth, managed a precarious living by walking the wings of a wood and fabric biplane. He learned then, although it failed to turn him to other pursuits, the number one rule of flight—that human beings were completely out of their element when attempting it. He learned the law of gravity, not in school but at first hand, when he assisted in separating the various remains of his fellows from the torn earth and the smoking, twisted metal parts of an OX-5 engine. Some five thousand air hours later, and it was in air hours that all of Dooley's kind marked the passage of time, he learned that although his engines were a little more powerful and his airplanes faster, the fundamental penalty for aerial carelessness had not changed. When the air mail came along, he saw that ignorance could kill as quickly as anything else, and that unless he gained some scientific knowledge as to the causes and movements within the upper atmosphere, his new engine and new airplane must remain only a transport to destruction.

At ten thousand hours, Dooley was suspicious, cynical, and most of all humble. Personal modesty had become almost a fetish in his profession. The old-timers, the tough guys, the goggled heroes of derring-do, had almost without exception made their last take-off. Dooley stood now on a pinnacle of fifteen thousand hours. Behind him were the days and nights of airline flying— the simple formula that if he took good care of his own neck, his passengers would likewise remain in good health. In war, there were problems to be solved that reached into the very foundations of Dooley's experience.

Dooley's second aid was the airplane in which he sat. With certain major limitations it was a remarkable contrivance, almost as versatile as his own knowledge. The *Corsair* had been conceived in a sprawling California plant. Its 1200 horsepower engines swung four heavy black propellers usually at 2000 revolutions per minute. When properly synchronized, and it was Dooley's art and pleasure to see that they were, a smooth tremor spaced at even periods ran the length of the ship. The four propellers blended perfectly together in an even rhythm. Dooley knew that unless his luck had run out altogether, he could depend upon their muffled harmony until the fuel supply was exhausted. Fuel was one more very important factor Dooley had to think about. He pointed a stumpy finger at the two quivering flow meter gauges on the instrument panel and leaned over to the broad-nosed young man who occupied the other pilot's seat.

"Frank," he said, without taking his eyes from the flight instruments before him, "I got a hunch we're in for a long ride. We can't get a bearing from Consolation Island. The damn static has knocked out the Chapel Inlet range. I think I know where we are but I can't be sure until we get something on the radio."

Frank Lovatt's pale face remained expressionless as he listened to the older man. It was his duty to aid Dooley in the actual flying of the *Corsair.* Like Dooley, he was a pilot. The unpleasant did not surprise him since he had learned to keep a low point of expectancy. Between him and Dooley rested a peculiar respect, a mysterious bond that ties all pilots together. Frank's four years of itinerary flying had brought him to the threshold of a professional career. He knew that someday he too would be a captain. It was Dooley's obligation to see that he would be prepared.

"We may not get a thing until we're somewhere around Victoria," Dooley went on. "We can't let down through this stuff without the Chapel Inlet range."

"Maybe we'll find a hole."

"I doubt it." Dooley glanced at the swirling mist outside the cockpit windows. "We're going right in to a front. That there low-pressure area they showed us on the map as being around Montreal must be a lot further east than they thought it was."

"How about going back to Sparkle Ten? We've got enough gas . . . I think." Frank hunched his big shoulders and ran a hand through his kinky brown hair.

Dooley thought of the long Greenland fjord twisting between jagged iron mountains. He thought of the field, of Sparkle Ten, half hidden at the base of the brooding ice cap.

"It would be dark by the time we got back there. Do you want to tangle with that fjord at night, young fella?" A light danced in Dooley's blue eyes.

"Not me, Daddy." Frank shook his head solemnly. "It was just an idea I had."

"You can bust your ass with ideas like that. If we can't get in to Chapel Inlet we'll keep right on for Victoria. This front can't stretch much further than Maine. We're going to need all the gas we've got to do it. Watch your head temperatures and lean the mixtures out all you can. Our consumption so far has been around 200 gallons per hour. Let me know what you can get 'em down to."

The factors were beginning to arrange themselves. They were maneuvering for position. Dooley knew that the balance of power was about even with perhaps a slight edge on his side. He could not permit himself to believe otherwise. Before him was a mass of weather, unlike the surrounding atmosphere, known to airmen as a "front." It could, and probably did, extend for a thousand miles or more.

In the air mass there would be turbulence. The *Corsair* would be tossed up and down, its body twisted and wracked. Every rivet, every sheet of thin aluminum would be subjected to ten times the normal strain. The engines would shudder in their mountings. The pressure on Dooley's ears would increase and decrease. The whole instrument panel would joggle in its sponge-rubber mounting until it was difficult to read. Dooley's rump would be alternately lifted into the air and squashed against the seat. Only his safety belt would keep him in place. The flush of the air stream past the windows would change from a roar to a whisper and back to a roar. The wing tips, now still cutting evenly through the cloud vapor, would bend alarmingly. But Dooley was not worried about turbulence.

There would be static in the front. Several billion tiny electrical discharges would each be faithfully recorded in the *Corsair*'s radios—unfortunately to the exclusion of everything else. The headphones would crackle and sizzle as if some hand had tossed a side of bacon into hot grease. Occasionally, there might be moments of comparative peace when faint signals from a ground station could be heard. Dooley was counting on that, but he was worried about static.

There would be snow in the front. It would enter the vital carburetors of the engines through the hungry air scoops. It would brush against the cockpit windows and form a magnificent corona if the landing lights were turned on. Some would find its way through the cracks in the windows. Dooley was not worried about snow.

There would be wind in the front, contrary invisible forces. Dooley had no means with which to tell either the direction from which they came or their strength. He could only look back upon his experience and guess. The winds could hold him back or speed him on his way. They could push him far to one side of the course or to the other, as the tide does a mariner. A cross wind of 40 miles an hour could put the *Corsair* 200 miles either side of Victoria—200 endless miles which would require the last drop of gasoline to recover. Dooley had considered this likely chance. The *Corsair* was proceeding on dead reckoning, a simple record of time and compass course that took no accounting of outside forces. Dooley was worried about the winds aloft.

Yet within the front could lurk the one thing that Dooley feared the most. The *Corsair*, like every other thing that flew, was poorly equipped to fight it. Dooley and crew, the engines, cargo, and the heavy gasoline, were supported in the atmosphere only by the *Corsair*'s wings. They were shaped with the utmost care. Nothing else kept the *Corsair* from the surface of the earth. Dooley was not flying in a magic metal construction but on an exact mathematical formula. Even a little ice would force Dooley to increase his gasoline consumption. The *Corsair* would grow heavier. It would vibrate viciously—but that was unimportant. The shape of the wings would change, the formula upon which Dooley depended for his very existence in the air would be altered. It would depend, in part, on how much ice there was. The rubber de-icing boots along the leading edges of the wings were a feeble defense. The factors meshed in together like a pair of clenched fists. Dooley was very worried about ice.

Hardly an hour had passed before Dooley knew again, as he had so many times before, that feeling close akin to hunger—an emptiness in his digestive parts that became almost nausea. Dooley knew the feeling, knew it every time there was reason to fear. In his twenty years of flying it had come to him regularly, sometimes once, sometimes twice each year. It was useless to rationalize, to explain the thing away as a mean, an average, for every so many hundreds of air hours. It came faithfully to all men who flew to live. The face became hot. The temples pounded, not painfully but with noticeable acceleration. The hands became wet. There was a recurrent surge to the feeling. It would mount within the body to an almost unbearable degree, then fade, only to return again with increasing strength. It came when something was wrong—seen or unseen. Sometimes it would flash through the system quickly and be gone for good. Sometimes it would arrive stealthily and linger long after the body had sought rest upon the ground. It came when an engine sputtered and died. It came in a thunderstorm when the instruments went mad. It came at the end of a blind approach when the ground failed to appear. It came, as now, to Dooley, when doubt crept through his mind. For a pilot flies on a series of certainties, else he could not long withstand the strain. He is certain of his airplane. He is certain where he has come from. He must be certain where he is, and where he is going. Only so can he operate with continued efficiency. Doubt came to Dooley when D'Annunzia touched his arm.

"I just worked Chapel Inlet, Captain."

"Good. Why don't they turn on the range?"

"He says it's on, Captain."

"The hell he says . . . I can't hear a thing!" Dooley reached above his head and turned the small radio crank back and forth. He pressed the headphones against his ears until the static crackle stabbed his brain. ". . . not one goddamned thing."

D'Annunzia brushed a stray lock of hair from his forehead. "I tried to get him to take a bearing on us but he said our signals were too weak."

"Did you try to work Sparkle Ten?"

"Yes. No go."

"How about Victoria?"

"I heard him working somebody else a while ago but I guess he can't hear us."

"Did you try Montreal?" Dooley sensed the feeling in his stomach and hated it.

"Yes . . . on that special frequency."

"Nothing?"

"Nothing."

"Son of a bitch." Dooley pronounced the words very slowly with even emphasis. Could it be that the winds aloft had carried the *Corsair* so far to one side or the other of Chapel Inlet that no radio signals could be heard? Of course it could be. Any damn thing could be. But no—Chapel Inlet was too powerful. He had frequently heard it 600 miles out. The operator on the ground was a liar. The range wasn't on. The man on the ground was in a crap game and didn't want to be disturbed. He was too lazy to walk out through the snow and turn the range on. It was much too powerful. And yet tonight . . .

"D'Annunzia."

"Yes."

"What's the frequency of that little broadcasting station—Ste. Beaupré, on the St. Lawrence?"

"Seven ninety I think. Wait a second. I'll look it up."

D'Annunzia went back to his radio table and flipped through a book. Dooley dimmed the cockpit lights to a ghostly glow, rubbed the condensation from the windows, and peered into the blackness. He flipped on the landing lights. It was snowing. The flakes spewed at the lights in long phosphorescent lines, perfectly horizontal. The red wingtip light seemed to explode vaporously as it sliced through the murk. Dooley carefully examined the engine cowlings and propeller hubs jutting out from the wing a few feet behind him. No ice. That was something. No sign of it even. Dooley thought about the top of the overcast. It would probably reach 30,000 feet. It would be nice up there with lots of stars. Murray could take shots on a few and determine where in space the *Corsair* flew. There might be radio signals too—but the *Corsair* would never make 30,000 feet with her load. Why not dump the load then? Dooley was thankful that for once he had no passengers. But gasoline—it would take at least 150 gallons to reach 30,000 feet. It was futile thinking, he decided. The *Corsair* would ice up anyway. Then how about going down, slip through underneath maybe? No, not down into the Labrador wilderness, not down—not until he knew where they were, not until things got a lot worse.

The four engines were purring a heavy song. There was not even a rough magneto. The compass course was the same—hadn't varied a degree. Frank was seeing to that. Good boy, Frank. He'd make a good captain some day. But the wind, the damn wind. What was it doing? Where was it coming from—for certain? Yet it would be impossible to miss the continent of North America and fly this course. And Chapel Inlet? Why in hell couldn't the range be heard? Dooley flipped off the landing lights.

"Seven ninety is right." D'Annunzia stood at his side. The lights from the instruments made a solemn purple mask of his face.

"Good. We'll try it on the compass." Dooley reached above his head and turned another crank beside the first one. Out of the corner of his eye he watched a radium-faced pointer swing lazily about a calibrated dial. It crept up and down as if animated by some supernatural force. He pressed the head-phones to his ears again. His face lit up.

"I can hear some guy talking." He watched the pointer move slowly upward. That was it. That must be it. Wonderful things, radio compasses. They point right where you want to go. Ste. Beaupré—a few hundred miles ahead perhaps and just a few miles to the right of the course.

D'Annunzia pressed his own phones to his ears. His eyes were tight shut as if to hear better. "That must be it, Captain. Kinda weak, but it's there."

Dooley's face had grown quiet again. "He's talking English. That's wrong. It should be French. I can't quite make out what he's saying."

"There'll be a station announcement in a minute. It's fifteen past the hour right now. They always do it then."

"If he doesn't make the station announcement in French, it's not Ste. Beaupré." Dooley fumed and rubbed the back of his neck. The announcement came in English. Pittsburgh, Pennsylvania. Dooley removed his headphones slowly.

"Skip," said D'Annunzia as if personally offended. "Funny how it does that so much up here."

Corporal Angeles in the warm, bright radio room at Victoria put down his cigarette. His bony fingers tapped the key before him. Like a stone dropped in quiet water, his fingers sent concentric wavelets of energy into the night. The wavelets spread out in all directions, bounced against the iono-sphere, and came down to earth again in a million places. But Corporal Angeles sought only one answer from one place. His log was incomplete. The sergeant would be mad about that. It would eventually get to the lieutenant and he would be madder still. There would be tedious explanations and little use to say that Chapel Inlet was always hard to read. Everybody knew it including the sergeant and he was tired of hearing about it. His fingers continued their staccato movement, calling and recalling, begging an answer through the stars.

"What's on your mind?" There he was, faint and clear. Corporal Angeles's fingers ticked off official mumbo-jumbo. Strictly regulation. All right for those guys to play around. They couldn't get leave anyway. Poor suckers; 50 below most of the time and not a woman within 500 miles. Nothing but the frigging army.

"Need position reports. Dooley. 17895. From 22:30. Zones. Fox. Baker. Zed."

"None received."

"Anything from Sparkle Ten?"

"No."

"What goes?" He hoped no one else was listening.

"Dunno. He asked for the range to be turned on. It's been on all the time. Will request position and advise."

"OK." Corporal Angeles leaned back and picked up a funny paper.

Dooley's mind was at rest again. The feeling, the empty sensation, had gone. He had done some careful figuring. Let it snow, to hell with it. To hell with the radio too. The *Corsair* had gasoline, plenty of it—enough to reach down the coast to Boston if need be. He'd wake Stankowski, the engineer, after a while and have him transfer gas from the outboard wing tanks. With that 400 extra gallons, the *Corsair* would be sailing free just so long as he, Dooley, could stay on course. A compass was a compass, a lovely, indisputable thing—light and sight to the blind. It could point the way to Dooley who had lost his vision. If he kept faith with it, the *Corsair* would cut a straight path through the atmosphere. Dooley intended to keep faith with his compass.

When the *Corsair* first slipped into the overcast, Dooley automatically abandoned his natural instincts. He purposely ignored the messages that traveled upward from the base of his spine—gentle suggestions that the *Corsair* was going up or down. Dooley responded only to the altimeter. It was right. When relation with the outside world had ceased to be, his spine was wrong. Dooley was flying in an aluminum cocoon, as separate from earthly things as if he had yet to be born. Robbed of sight, suspended in air, Dooley knew also that he could not depend on his other human instincts. The liquid in his ears that brushed against a million tiny hairs and in so doing transmitted certain sensations to his brain was useless now. On earth, it permitted him to walk without falling down, to sit or stand without reeling. It told him when to turn and when to lean. In the *Corsair,* Dooley could see and feel, yet he was blind and numb. Without some relation to the earth, his sense of inner balance was faulty. The birds had known of this since the beginning of time. Dooley knew it. It was second nature for him to switch from his human senses to the man-made things before him. He watched the artificial horizon. It would tell him when a wing was down. Each instrument, the compass, the gyro, the altimeter and horizon, airspeed, turn and bank, whispered information Dooley's mind must know.

"Captain. Chapel Inlet wants to know our position." D'Annunzia stood at his side again. Dooley did not reply at once. The correct answer was one he very much wanted himself. The only correct answer he knew was that he did not know. Nor could anyone know unless the strength of D'Annunzia's signals soon increased. A conundrum rolled suddenly, foolishly, through Dooley's head—He wants to know what I want to know; we both want to know what I can't know. It repeated itself over and over again in exact rhythm with the beat of the engines. Dooley flung it from his mind.

"Murray! Come here and bring your chart."

The youth scrambled forward.

"Where do you think we are now?"

"According to my dead reckoning, we're just coming up on the St. Lawrence . . . maybe 50 miles this side of it. It's pretty hard to be sure."

"Good boy," Dooley smiled. "When you talk like that, I believe you. I think you're right this time. We must have missed Chapel Inlet altogether. Went way to the south."

"We couldn't have gone north. Not on this course."

"Right. We're east of the course then, over the Kimball Mountains, maybe still over the sea. So, we'll take a new course, two seventy degrees. Frank, fly two seventy."

"Two seventy."

"D'Annunzia. Transmit blind and keep transmitting. Ask for bearings from anyone you can raise—bad or good we can't be too particular now—and be careful you're not getting a bearing from a submarine. Try Windsor and Kinkaid first. They're wide awake down there. Murray, if D'Annunzia gets anything besides the German Navy, plot them on your map, then show them to me right away."

"Yes, sir."

"Remember, right away—no matter what's going on. Stankowski!" Dooley snapped the name and searched the gloom behind him. A sharp-faced young man rose from his seat at the rear of the cockpit. He moved forward through the half-light, rubbing his eyes. Dooley's speech was rapid now. The forces had run up their colors. It would depend now on the kind and the amount. At least he was ready. In a way he was glad the thing had come.

"Stankowski. Go back and transfer everything you can to the main tanks."

"Okay."

"Let me know when you've finished."

"Okay."

"Did you have a good snooze?"

"Pretty good. It's cold back there."

"Did you fill the alcohol tanks before we left Sparkle Ten?"

"Sure. Aren't we going to land at Chapel Inlet, Captain?"

"Not tonight."

"Straight through to Victoria . . . hell, I got some laundry for a guy at Chapel. Carried it right over his head the last time. He's gonna be sore."

"Forget it. Get going on that fuel."

"OK." Stankowski sighed and vanished through a small door at the rear of the cockpit.

Dooley kept his face turned forward; the thing he feared most had come at last. His eyes never left the growing fringe of white along the windshield.

Corporal Angeles put down the funny paper and looked out the window. Past the green lights at the end of the runway, in the distance, he could see the fitful street lights of Victoria swinging in the wind. The view was familiar to him. There was the Victoria Hotel. In it, you could drink whiskey. The hotel where people didn't care whether you were an officer or an enlisted man, and that was nice. Across the street in the drugstore there were girls, both good and bad. They'd be lined up along the soda fountain now, eating ice cream although it was below zero outside. Corporal Angeles knew most of them. They were the regulars. They arrived about this time every night in a flurry of laughter and snow. They were pink-cheeked and breathless. Most of them wore tight-fitting sweaters beneath their heavy coats, like the girls at home. It made their breasts stand out. He glanced down at his radio log and observed that there were still some vacant spaces. There would be no girls and no laughter until those were filled. His fingers ticked off another call to Chapel Inlet. Would those guys, please to God, stay on the job once in a while?

"Still need Dooley 17895 position reports."

"No got." The answer was surprisingly prompt.

"Who has?"

"Dunno. He's asking Kinkaid for bearings now."

For a moment Corporal Angeles sat motionless. His mind left the girls in Victoria and searched the radio log. Maybe this was something he should call the sergeant about or even the lieutenant. No. They'd both be mad as hell. They wouldn't say so because it was their duty, but they'd be mad as hell anyway. Besides, that Dooley—he was one of those commercial pilots. They always came in sooner or later.

"Is he in trouble?" he asked, more to satisfy himself than anything else.

"Dunno."

"Is he landing Chapel?"

"Dunno. Am standing by."

"Same here." He glanced up at the clock and watched the sweep second hand swing around to exactly twenty-three hours Greenwich time. In thirty minutes he would be with the girls.

The *Corsair* staggered beneath her load of ice. From a few indefinite traceries of white across the windshield, the enemy had gathered strength with deliberate progression. The outsides of the windshields were opaque—a dead white wall of coagulated crystals. The *Corsair* heaved and swerved from side to side. Sickening tremors ran the length of the fuselage, beginning as a hardly noticeable vibration, then building to a mighty shudder, then ceasing again as chunks of ice from the propellers, too heavy to maintain their grasp on the whirling blades, struck the metal sides like pistol shots. The altitude had dropped from 10,000 to 8000—nothing, not even Dooley's knowing hands, now white about the wheel, could hold the *Corsair* higher.

"Frank!"

He leaned to Dooley quickly. There was sweat across his brow. "Yes."

"Run 'em up to 2500 r.p.m. and 35 inches of boost. Make it 5 inches of super-charger instead of 3. To hell with the gas. She's sinking again." Frank's heavy hands sought the controls. The engines howled in agony. The *Corsair* lunged forward.

"D'Annunzia! Murray! Come here!" They rose quickly out of the darkness and came to Dooley's side. Their faces were solemn and pale. They leaned close to Dooley. It was hard to hear above the din and they wanted desperately to hear every word that fell from his lips. Dooley fought off the feeling that twisted his stomach, and spoke to them slowly, thoughtfully.

"Boys, we're in trouble. The bearing from Kinkaid was good. We're somewhere over the St. Lawrence all right, but even if the weather at Victoria was perfect we can't hold this course any longer. We've got to get out of this ice somehow—right now. D'Annunzia, does Chapel Inlet still want that position report?"

"Yes."

"OK. Tell them we're in a jam. Our bearing from Kinkaid was 120 degrees. Tell them we're icing up and can't hold our altitude. We're taking a course of three thirty degrees until our gas runs out. Got that?"

"I got it."

"Keep your key down as much as you can. Go to 500 KC and get everyone who'll listen to take a bearing on you."

"If we hold three thirty degrees, Captain, we'll fly right off the map. It says uncharted, up that way." Murray held the map as if it had suddenly betrayed him.

"I know it. But up there it's flat, they say. We may get a break down through. If we can stay in the air for three hours more, we may be able to get out of this front. It's moving east. I think, we can get out of it to the northwest."

Dooley wiped his lips. "Here, Frank, take this thing. Just hold her level if you can. Turn very slowly, not over a 5-degree bank now, to 330 degrees." He sighed and loosened his belt. "Do you know what that means, boys?"

"Jump?" ventured Murray.

Dooley smiled in spite of himself. But the eagerness with which the boy was willing to plummet himself into the unknown made him feel sad.

"No, I wouldn't do that, although you can if you want. It's awful cold down there and a long walk home. It's lonesome too—nobody lives down there but the beavers. Stick around. We'll get out of this yet. I haven't jumped out of an airplane in sixteen years and I'm not going to start tonight. Now listen, Murray—"

"Yes, Captain."

"I want to tell you now because I may not have time later on. Get Stankowski to help you. Get out everybody's heavy flying suits and lay them beside each man along with a Mae West. You and D'Annunzia put yours on as soon as you can. When I tell you, go back to the cabin, turn your back to the nose and brace it against a strong support. Wrap yourself in something, blankets, sleeping bags, anything you can find. Go way back, almost to the trail. Put your hands behind our head and stay there until we stop. Then get out as quick as you can. Don't wait for us."

"What about you and Frank?"

Dooley grunted. As well ask the doctor to cut the patient's throat, the priest to smash the chalice. "We'll ride the bastard down."

Corporal Angeles twisted a dial ever so slightly, cocked his head to one side, and wrote rapidly:

> . . . 17895 *Dooley advises quote* AM IN A JAM. ICING UP. BELIEVE CAN HOLD *8000.* PROCEEDING *330* DEGREES TILL GAS RUNS OUT. ALL STATIONS TAKE BEARINGS. *Unquote. Can you read him Victoria? He's on* 500 KC *now* . . .

"No." His fingers barely touched the key.

"Advise operations officer. Instruct your DF station take bearings if possible. Stand by."

"OK." There would be no girls or laughter for Corporal Angeles this night. He snatched at the telephone.

"Lieutenant Cord?"

"He isn't here."

"Well get him."

"He'll be back in about an hour. Gone up to the mess."

"Get him, damn it. 17895's in trouble."

"Roger." The receiver clicked.

"Officers' mess."

"Is Lieutenant Cord there?"

"Just a minute."

Corporal Angeles tapped the end of his pencil against the radio panel.

"Lieutenant Cord speaking."

"This is Corporal Angeles, sir. 17895's in trouble."

"What's the matter?"

"I couldn't read him but Chapel informs he's icing up and proceeding three thirty degrees until his gas runs out."

"What was his last position?"

"I don't know, sir."

"Find out if you can. Have operations inform all stations to maintain silence. All messages in the clear until further notice. I'll be there in five minutes."

The *Corsair* had ceased to fly—it wallowed through the sky. The altimeter read 6000 feet. Dooley could not gain another inch. Like an animal in the final, violent throes of death, not one thing within the ship was still. The instrument panel shook until it was almost unreadable. The airspeed read zero—frozen, useless, gone. The tachometers swung wildly. All unity had left. The compass danced and swung, chased its tail around and around. The white crustations along the leading edges of the wings were so heavy that the rubber de-icing boots expanded and deflated helplessly beneath the ice. And Dooley knew that there was doubtless even more that he could not see along the trailing edges of the wings—on the underside where the formula on which he flew would be even more radically changed. There would be ice—heavy, increasing, and building with dreadful certainty—far back on the *Corsair's* big rudders. Ice on the horizontal stabilizer, on the fuselage, on the propellers, on the engine cowlings—each bit adding not only its deathly weight to the *Corsair* but destroying a certain section of the precious formula. The engines fought with all their tremendous power, fiercely, unheeding the terrific strain. But the engines demanded fuel for power, fuel and more fuel, now 5 gallons for every single minute. Dooley knew that the fuel was almost gone. Without power to ease it down, the *Corsair* would be almost instantly done.

Only the clock, the *Corsair's* one remaining tie with life on earth, remained the same—or so it seemed to Dooley who watched the hands stand still when they should have been creeping. Time, rigid as steel yet variable on demand, was unyielding to Dooley now, determined to hide the outcome of his battle until the last intolerable moment. It seemed he had been flying for many hours, yet the clock claimed only two. The clock was obviously a liar. It must have stopped, reversed itself in fact, set back the passage of time against all the laws of God. Dooley wanted to smash it, to yell to all the heavens he knew the time had come. It said ten o'clock. That was ridiculous. The children would be in bed—his wife yawning over a book. It still said ten o'clock, a manifest impossibility since he had been flying one long, full, interminable night. He could fly on forever thus, and that dogmatic deadline when but a few remaining gallons sloshed within the *Corsair's* tanks would never come. And yet at last it had.

"Frank. Open your window. Smash it if you have to." Frank pounded at the frozen hinges. Dooley did the same. Inch by inch the windows parted. Numbing cold, the outside air swept through the openings with a hollow roar. Dooley allowed the *Corsair's* nose to settle slightly.

"D'Annunzia! Murray! Stankowski! Go back to the cabin!" He shouted now. His words were almost drowned in the angry bellow of the night.

Dooley took a last look at Frank. "Let me know if you see anything!" He switched the cockpit lights off completely. The cockpit became a cold black cell. He cut the engines back a trifle. The wind eased to a deep rumble. The altimeter slipped steadily down to 4000 feet. He flipped on the landing lights.

"See anything?"

"Not yet." Frank wiped the moisture from his eyes.

Dooley pushed the nose down. He had to keep plenty of speed. The ice had changed the formula—he must allow for that. She'd land like a grand piano thrown off a mountain. Three thousand feet.

"I think she's breaking, Dooley!" Frank strained forward against his safety belt.

"Did you see a hole?"

"No, but we passed through a layer!"

Two thousand feet. Dooley held his breath. Time again. Time, evenly divided now, composed of seconds and minutes. Capricious. Now it rushed with unbelievable speed. During any one of the next few hundred seconds he, Dooley, would cease to live. That was all it took, less than a second. The *Corsair* would meet with the planet earth and earth would resist. The *Corsair* was too frail, too small a thing to emulate a meteor. One thousand feet. Could the ceiling reach to the very ground? His legs were taut as if to catch the impact by themselves.

"It better be quick!" The altimeter moved down to 800 feet. Dooley braced himself for death. He'd always wanted it to come this way, or so he'd said.

"Dooley! A hole! Jesus Christ—a hole!" Frank was beating his fists against the windows.

Dooley swerved the *Corsair*, then suddenly as if shot from a catapult, they were skimming along beneath the overcast a few feet above the ground. Pine trees, almost covered with snow, outlined the low hilly terrain.

"I see a lake . . . I think!" yelled Dooley.

"There's one over here too!"

"Wheels down! Gimme half flaps quick!"

"You going to land with the wheels down?" Frank's voice was incredulous.

"Hell, yes. We'll walk away from this yet!"

"OK!" Frank tripped a lever on the control pedestal. The roar of wind increased suddenly as the nose wheel went down, then died again as Dooley pulled back on the throttles. The *Corsair* hung as though suspended on an invisible wire.

"Full flaps, Frank . . . hold your hat!" The *Corsair* swooped down over the treetops and plumped into a shower of snow.

"Cut the switches, Frank!" He snapped the metal bar at his side.

"Thank you, God Almighty," whispered Dooley, for suddenly there was blessed silence.

Flight Of Passage

BY RINKER BUCK

(Excerpted from the book *Flight Of Passage*)

There is in the world of publishing and entertainment alike a certain almost indefinable and elusive phenomenon known as "The Buzz." Aviation books don't sneak into this exalted listing very often, but one that did in 1997 was *Flight Of Passage* by Rinker Buck.

With your permission, I will quote the jacket copy of the book in order to quickly set the stage for what I trust you will find a profound reading experience, albeit a brief one considering the excellence of the entire book.

"In the summer of 1966, Rinker and Kernahan Buck—two teenaged schoolboys from New Jersey—bought a dilapidated Piper Cub airplane for $300, rebuilt it, and piloted it on a record-breaking flight across America—navigating all the way to California without a radio because they couldn't afford one. Their trip retraced a mythical route flown by their father, Tom Buck, a brash, colorful ex-barnstormer who had lost a leg in a tragic air crash before his sons were born—but who so loved the adventure of flight that he taught his boys to fly before they could drive."

As is evident, Rinker Buck waited quite a spell to write the story of this great flight, and in that time he had developed and honed prose skills that would serve him well. A distinguished journalist with bylines in such publications as *New York, Life,* and *Adweek* magazines, Rinker Buck made his first book a powerful reading experience.

This excerpt concerns only the launch of the flight. But for the ultimate reading and flying experience, you really ought to get hold of the book and go along for the entire ride.

I t was madness out at the airport. My father had invited a number of his friends to watch us take off, and some of the Basking Ridge pilots brought their wives out to see us off too, but they were the nonflying wives and none of them knew squat about aviation. While Kern and I were trying to ready the plane, there was a crowd milling around the Cub, and people kept banging their heads into the wing struts, pestering us with idiotic questions and changing babies' diapers back on the tail section. Everybody was astounded by the simplicity of 71-Hotel. They kept sticking their heads into the cockpit, wiggling the stick for a second or two and then staring at us in disbelief, asking one of the pilots if this really was the plane the Buck boys were flying to California, or just a toy. On top of this, my father had arranged to have the soda machine opened up for everyone's use, and my little brothers and sisters kept spilling root beer and Coke on Kern's new paint job.

While Kern preflighted and gassed the plane, I stowed our gear. The only way to wedge everything into the baggage compartment, I discovered, was to cram the pillowcases with our clothes at the very bottom, the shopping bag full of our maps in the middle, then the sleeping bags on top. Still, there wasn't enough space. The sleeping bags on top overflowed up past the windows, which would block my view out the back, and we were concerned that the protruding gear would bang up against the fabric headliner and cause it to break apart in heavy turbulence. At the last minute we took everything out again and started jettisoning things from the pillowcases—duplicate tubes of toothpaste, paperback books, extra pants and sneakers. I handed the discarded items over to my mother in a big, messy armful, and stuffed everything back into the baggage compartment. More or less, all we had to wear now were the penny loafers and the Levi's that we already had on.

I jammed the sectional maps we would need that day, and my clipboard for keeping flight notations, into the leather pocket behind Kern's front seat.

The nonexistent waterbag was another royal pain in the ass. My father had told everyone about the waterbag, but of course he forgot to update them when we couldn't find one. Curious onlookers kept coming over to the plane, dipping their head down between the wheels, where the waterbag was supposed to be lashed, and coming back up disappointed. I was civil to the first seven or eight people who asked about the waterbag—maybe it was even a dozen. Then I blew my stack.

"Hey, Rinker, Kern!" my little brother Nicky howled. "Where's the waterbag? Daddy said you can't take off without the waterbag."

"Shut up, Nicky," I said. "It's none of your business."

"It is too! Daddy says you have to have a waterbag!"

"Nicky, we don't have it yet. But I'll tell you what. If I do find a wa-terbag, I'm going to shove it straight up your butt. Now scram!"

Nicky ran off to inform my father that I was threatening to shove a waterbag up his butt.

"Screw this Kern," I said. "Let's fly."

"Yeah. This is a train wreck. Where's Dad?"

Finally, my father bounded over to the plane with an exasperated look on his face, as if this was all our fault.

"Hey, boys, c'mon now! You can't get to California sitting here on the ramp. Hop in. I'll prop you."

Kern strapped himself into the front seat, I took the rear. With a raised hand and a growl, my father cleared the area of kids. Then he leaned into the cockpit for a final chat.

"All right boys," my father said. "Now I'm not going to give you the big lecture or anything. Just pace yourselves, that's all. Six or seven hours of fly-ing a day is plenty. Nobody cares how long it takes you to get out to the coast."

Kern was impatient to go.

"Got it Dad."

"Now another thing," my father said. "We've got a nice crowd here. Everybody came out to watch you take off. Once you get in the air, circle the pattern once, and then come back down the runway for a flyby. A flyby, I said, not a buzz job. There's a difference. Don't get too low. Then just wiggle the wings a little for Mother. Okay?"

Kern wasn't even listening.

"Dad, I think I can get this airplane off this strip, okay? My switches are off, and I'm priming."

I always liked the way my father turned a prop. He had this graceful, muscular way with a propeller in his hands, a jaunty confidence that bespoke years of flying. Embarrassed as I was, sitting there with a crowd watching us, I enjoyed watching him throw the blades.

"All right boys," my father called out. "Make us proud now, and have a great trip. Brakes, Throttle, Contact!"

"Contact!"

The cylinders fired on my father's first throw, blurt-blurting and coughing through the stack, blowing back an aromatic puff of smoke. We waved goodbye to my mother, taxied down to the end of Runway 28, ran up the engine and cleared the controls.

As soon as Kern ruddered onto the strip and firewalled the throttle, I loved that Cub. Despite two passengers, baggage, and a full load of fuel, we only traveled a hundred feet or so down the runway before 71-Hotel popped off the ground and clawed for air. Kern immediately trimmed for some down elevator, to get the nose lower.

He and Lee had rigged the plane well. As we passed the windsock I wiggled my stick in the rear to signal Kern that I wanted to feel the controls, and he wiggled back and gave me the plane. I did a gentle aileron bank to either side, and tapped the rudders. The controls were firm and responsive, not at all like the other old taildraggers we flew, where you moved the controls and then waited a second or two for a response. There was no doubt about it. 71-Hotel was our best restoration yet.

We leveled off downwind of the runway and then Kern banked and drove the Cub to come back in over the crowd. He yelled back to me over the roar of the engine.

"Hey Rink. Did he say a flyby, or a buzz job?"

"Buzz job!"

It was a lie, but I'd had enough of my father at that moment. We were up here, he was down there, and for the next few weeks there would be 2,400 miles of open country between us. Fuck 'im. A buzz job it would be.

Kern gave the runway a close shave. He dove for the grass from 800 feet, pushed the throttle to the stop, and roared past the gas-pump crowd at 120 miles per hour. As we were passing the windsock I looked out the side. My father was kangarooing across the ramp to the edge of the runway in his crazy wooden-legged gait.

Going by the gas pumps, Kern put the right wing over, pulled the stick back into our laps, and turned up and over the crowd, Eddie Mahler style. Below us, everyone's necks were craned straight up. There was a decent quartering wind on our nose when Kern leveled the wings and hung 71-Hotel on her prop.

We climbed almost vertically over the crowd. As we passed through 700 feet I opened the right window and leaned out into the slipstream, looking back over the tail to my father. When he saw me hanging out the side of the plane my father started waving both of his hands over his head. I took the stick and wiggled the wings in reply and then I washed the rudders back and forth with my feet to fishtail the plane and my father waved some more. He kept waving, waving both arms over his head, growing smaller and smaller as we climbed the plane, and he was waving still as we disappeared over the hills.

I looked back several times at my father as he waved, wiggling the wings for him a couple of more times. Behind and below me, he was framed by the tail section of the plane, as if in a picture. I remember the way the sunlight turned the grass around him a hard green, and the way the image of him was blurred and kept going double from the slipstream beating my hair into my face and whipping up tears in the corners of my eyes. I was filled with an immense sadness and happiness for him at once, and afterward I couldn't understand why that particular vision of him moved me so much, or why it returned so often in my dreams. After a while I just accepted it as a portrait of contentment between us. Maybe we would never say it that way but the truth was that we were happiest watching each other recede in the distance.

There wasn't a lot of time to dwell on that right away. As we climbed through 3,000 feet I could just barely make out through the haze our first navigation checkpoint, the big man-made reservoir at Clinton. The air was choppy already, a bad sign so early in the day, because the turbulence would only build as the sun rose higher. Kern climbed and descended several times, up above 5,000 feet and then back down to 2,500, trying to find a scud-level altitude where the haze thinned out. But it was useless. At any height we were assaulted by the same blinding white glare and low visibility, with the amber sun on our nose burning through the windshield.

Momentarily I lapsed into an old habit, my first navigation chore when we flew west, and decided to tune in the radio frequency for the Flight Service Station at Allentown, Pennsylvania, to check the weather ahead of us. Then I had to laugh at myself. There was no Allentown FSS for us, and there would be no FSS all the way across the country. We didn't have a radio in this plane.

I pulled my seatbelt tight against the turbulence and squinted ahead into the glare. The old Continental roared, the cockpit smelled of burnt oil, and the slipstream rushing in through the open window rifled my shirt and my hair. I was glad to be launched at last. In the seat ahead of me Kern looked back, smiled, and gave me a thumbs-up. I knew that he would push hard to beat the storms ahead, which frightened me, but we were still several hours away from the worst of the weather and there wasn't much point in worrying about what I couldn't yet see. Probably we could find a way through. I knew all along that we'd have to fly like the blazes to make Indiana by nightfall.

As soon as we got out over the Delaware River, barely fifty miles from home, we could see what kind of trouble we were in. Menacing and black anvil-head clouds, their tops silver-bright in the sun, towered up on our right, blocking our planned route to the northwest. To the south, vapory sheets of rain fell to the green fields. In between there was still an open patch of sky.

Kern hunched forward over the instrument panel and peered intently through the windshield. Pushing over the rudder and the stick, he steered southwest over the picturesque farmlands of Bucks County, Pennsylvania.

We hadn't expected the weather to develop this quickly. The flight-briefer at Teterboro had forecast these conditions for later in the day, and much further west. But weather is weather and obviously the warring masses of air were pushing trouble eastward faster than predicted.

From the backseat, I leaned over sideways and looked at my brother's face. His mouth was turned up into a grim half-smile. I had never flown alone with him in bad weather but I knew what this expression meant. We were now in a race. He was determined to outfox the advancing edges of the storm and beat the front to Pittsburgh.

I was surprised by Kern's decision to push ahead in the face of such early signs of adverse weather, and I had to deliberately work at resisting the urge to panic. Kern and I still didn't know each other very well in the air. His dauntless, supremely confident attitude in a plane, so different from his uptight personality on the ground, was a mystery to me. I couldn't understand where it came from or how one person could be transformed so dramatically by environment. Kern, likewise, had no idea of just how frightened I was by turbulence and poor visibility. Fear in the air was something my father simply wouldn't tolerate, and over the years I had devised a variety of physical and mental stratagems to hide this weakness—closing my eyes during spins, breaking up cross-country trips in marginal weather into ten-minute segments, which made them easier to bear. This was the price I had paid for inclusion in our weekend flying excursions, and now I would have to do the same with Kern. I didn't want to disappoint him this early in the trip, and I vowed not to complain or reveal my fears no matter how much this leg across Pennsylvania rattled me.

Hitching my seatbelt up another notch, I stuck my head out into the slipstream for some fresh air, and closed my eyes for a while and pretended that we weren't surrounded by angry, jagged walls of clouds and rain falling in several directions.

After Quakertown the air turned rough and the visibility was constantly changing. Suddenly the clouds on either side of us would drop low and wedge in sideways, forcing us down against smudgy barn silos and power-line pylons, and then we'd come around the gauzy corners of the cloud into open skies dazzling with rainbows, sparkling fields, and tidy, white-washed Pennsylvania Dutch farms. We were "scud-running," trying to get below and between the clouds, and I couldn't believe were violating all the lessons of our training during literally the first hour of our trip.

Kern anxiously looked from side to side as he flew the plane, jabbing his finger against the map on the magenta symbol of an airport. I could see what he was doing, flying us from airport to airport in case the weather forced us down. But after Pottstown there weren't any strips for a long stretch. As the turbulence became stronger, the compass spun crazily and the plane plunged up and down like a cork bobber. Kern threw back the map and yelled.

"Rink! I've got to concentrate on flying this airplane. Get me to Harrisburg."

Harrisburg. I focused hard on the map, determined to deliver for Kern. I was actually quite prepared for this assignment, but didn't appreciate that about myself yet. The summer before my father had spent nearly two months giving me cross-country flying lessons. I was only fourteen that year and hadn't spent much time in planes equipped with sophisticated radios and instruments, and he could see that I was only comfortable with straight "pilotage" navigation, flying point-to-point by reference to landmarks. But that was fine with an old barnstormer like him, in fact it was preferable. Anybody could scream around in an expensive Bonanza stuffed with all the latest radio-navigation equipment. But what happened when the radio broke or the electrical system failed? Too many pilots couldn't navigate merely by direct reference to the ground. So I had spent several Saturdays and Sundays in a row navigating by rivers, roads, and gravel quarries marked on sectional maps, learning how to pick up a compass course and correct for wind drift when the landmarks ran out. Mapreading and pilotage the old-fashioned way were my father's gift to me, a very simple gift. But that's all that Kern and I needed in 71-Hotel. Simplicity was the only asset we had.

Our original plan called for skirting the north edge of the Allegheny Mountains, where the peaks are lower and there would be less turbulence, following the gentle rolling valleys of central Pennsylvania into Youngstown, Ohio. But that course was now obstructed by clouds and the weather had formed a narrow, irregular chute forcing us to divert south through Harrisburg and the lower Susquehanna Valley. This would mean facing the Alleghenies head-on in rough air, but we would have to worry about that when we got there.

On the map I found a rail line just south of Pottstown that meandered west to the Susquehanna, up through Reading and Hershey. If clouds blocked our way there were several intersecting power lines that we could pick up. Wiggling the stick to signal Kern, I ruddered over for the tracks and pointed them out on the map. Throwing open the side windows, I kept my head out in the slipstream to look for landmarks. The ceiling was dropping again and we didn't have a lot of airports in front of us. I didn't have the luxury of guessing at our position and I focused like a gnome on the land and then back to the map.

That's how we flew the first leg, like a pair of old airmailers. While Kern manned the plane and kept us straight and level from the front seat, I hung out the side from the rear, battered by the rain and the slipstream as I concentrated on the terrain. Kern and I seamlessly adjusted to flying this way and barely exchanged a word about it. When the Cub strayed from the rail line I would look ahead to find the tracks again, and then step on the rudder to steer us over and hold our course on the rails while Kern held the plane steady with the stick. In the rougher air we flew like that for three or four minutes at a stretch, sharing the plane and speaking to each other through the feel of the controls. This was immensely comforting to me because I'd always found that my turbulence jitters eased when I could control the plane. Occasionally, the turbulence blowing in through the windows kited the map off my lap, and the map was hard to hang on to because it was slippery with sweat from my hands. Still, I could almost enjoy this adventure now, hanging out the side and peering forward as I ruddered us along the rails.

As we bobbed over the Pennsylvania farmlands, sometimes we could see ahead of us for a good seven or eight miles and sometimes we could barely see a mile or two. But the Continental roared and the floorboards throbbed and the cockpit was filled with the soothing ether of burnt oil. I liked the thunderous, intense propinquity of the two-seat Cub, flying through this black purgatory with my brother. Right there, on the first leg of the trip, I discovered something important about us. Kern's determination and self-confidence were contagious. All I had to do was look forward at his face. He was grinning, enjoying the chase against the weather. He handled the controls maturely, gently managing the stick with just a couple of fingers and softly cupping his left palm over the throttle, not overcontrolling, as most pilots do in rough air. Up here, I was very committed to him; I yearned to please him. Kern was preternaturally gifted as a pilot and so intent on outwitting those clouds that I could almost feel his skill as a physical sensation, and it would have shattered me to disappoint him.

And so, under increasingly inclement skies, we followed the rails into Hershey, crossed the Susquehanna south of Harrisburg, and flew on to the small grass strip at Carlisle, Pennsylvania. It was the last airport where we could refuel before we faced the Alleghenies. The weather was always closing in behind us. We would get past a nice stretch of farms, or over the big Bethlehem Steel blast furnaces along the Susquehanna, and a few minutes later I'd look back over the tail. It was raining where we just had been.

But we'd beat the weather so far, barely. As we entered the traffic pattern at Carlisle, clouds closed in on us from all sides, swirling around and narrowing the clear space of air in front of us like a funnel of water churning down a tub drain. To dodge the clouds, Kern swung the Cub around in a dive and cut off the edge of the pattern, and then quickly cross-controlled, throwing us into a steep, shuddering sideslip to make the field. As we were rolling out on the spongy grass runway, a light rain began pattering onto the windshield and the wing fabric.

We called them the "geezers," the airport geezers. Every little airport in America had one or two, and still does. They're the old-timers in the Dickie pants, the matching Dickie shirt, and the broad leather belt, sitting on the gas pump bench. They might be seventy or seventy-five now, not flying much anymore, but geezers aren't envious of the younger pilots, just solicitous. Geezers pour a lot of oil into hot engines and know how to squeak bugs off the windshield without wasting any cleaner. Student pilots on their first solo cross-countrys get to know the geezers quite well. It's the geezer who tells them that they have just landed at the wrong airport, and then talks to them for a while about how easy a mistake that is—all of these little airports look the same from the air—and he steers the young pilot to the right strip. Then the geezer takes the student's logbook, makes an entry for the flight, and signs his name. According to him, the kid landed at the correct field.

The airport geezer at Carlisle was a jowl-faced, big-bellied fellow named Wilbur—you never get more than a first name. When he saw us taxi in with the Cub he pulled on a rain slicker and a ball-cap and jogged out to meet us. There was an empty T-hangar by the fuel pumps. We shut down the engine and Wilbur helped us push 71-Hotel out of the rain.

Wilbur was surprised that we were flying around in this kind of weather, in a Cub without a radio no less. But geezers become geezers because they've survived many mistakes themselves and they are not dogmatic in their old age. As we stood in the T-hangar, in gentle tones, Wilbur was slowly backing into the standard lecture on the perils of "Get-There-Itis" when the rain stopped and the skies all around Carlisle cleared. He asked us where we were headed. Kern and I weren't ready yet to tell a stranger that we were flying all the way to California in a Piper Cub so we told him Pittsburgh instead.

The airport manager strolled out and said he was making a run down the road for hamburgers. I was suddenly hungry and I rode off with him. Kern and Wilbur went into the pilots' shack to phone the FAA weather station at Allentown.

When I got back, they were bent over a table in the shack, looking at the Detroit sectional map. The weather situation was still complicated and unfavorable for us. The front we were worried about wasn't due to arrive in Pittsburgh until three-thirty or four in the afternoon, and we could probably beat it there. But ahead of the storm, directly along our route, there were reports of moderate to severe turbulence and scattered rain showers. We had just flown through the advance squall line of that system. It was clearing now around Carlisle, but that didn't mean much. The mountains were only a few miles away and there would be plenty of stratocumulus bangers behind them.

Wilbur had us cased out pretty well. He could see that Kern was determined to get to Pittsburgh, but that I was more tentative, maybe even afraid. A good geezer never tells a transient pilot what to do, especially young ones, because for sure they'll take right off and do the opposite. Instead, geezers anticipate the inevitable fuckup and provide a backup plan. Wilbur suggested that we take off, fly west for fifteen minutes, and see what it was like. In a firm voice he told Kern to turn back if the visibility was poor, the air too rough. Carlisle wasn't going anywhere.

Considering the conditions—a lot of low clouds and poor visibility—following a compass course was unrealistic. Wilbur suggested that we follow the Pennsylvania Turnpike, which ran right by the airport. It wasn't the most direct route over the mountains, but we couldn't get lost if we stayed over the Pike. The windward side turbulence, on the west face of the peaks, would be quite bad. After Shippensburg we would begin passing over the tunnels in the mountains and we'd lose sight of the road for a while, but he told us to just pick a landmark or rock formation up on the high ridge, fly for it, and then we could pick up the Pike again as we came down the other side.

Wilbur advised us to turn southwest after we passed Latrobe. In this kind of weather, with all the pollution from the steel mills ringing Pittsburgh, we wouldn't be able to see anything anyway. Just past the town of Mount Pleasant, the Turnpike would split off for Route 70. Wilbur recommended that we avoid the smog and air traffic around Pittsburgh by following Route 70 into the airport in Washington, Pennsylvania.

It was sparkling and clear outside when we stepped out of the shack. The mowed grass runway was a watery, bright green. The airport sat on a high, broad plateau, commanding a breathtaking prospect of Cumberland County. But it was a deceptive, killer beauty. As I polished off my second hamburger, I felt like the condemned with his last meal. It was going to be a hell of a ride over the mountains.

Wilbur helped us fuel the Cub, checked the oil, and gave us a prop. He acted as if he expected to see us back in Carlisle in twenty minutes, and even offered us a room that night at his house. "The Mrs." would make us a real Pennsylvania Dutch dinner. As we climbed out past the pumps Kern rocked the wings and Wilbur waved.

It was the most murderous corridor of turbulence I have ever experienced. For the next hour and a half I detested my brother. I hated him for catapulting so hard over the mountains, hated my father for letting us make this trip, hated that society of hard, cynical pilots into which we were born and which now obligated me, more or less, to earn my manhood by proving that I could take this abuse. After an hour my knees and my shins, hammered by the turbulence and the shuddering stick against the cockpit walls, ached like scavenged meat.

Bang, bang, bang, bang, across the wretched, washerboard peaks of the Alleghenies we hauled. Kern peered out through the windshield with his chin just above the instrument panel and then looked back, smiling, actually smiling back at me, as our butts and stomachs got walloped by the updrafts and downdrafts. It was maddening, the way he kept flashing me that earnest *Leave It to Beaver* grin of his. It was his way of communicating his apologies for making me sit through all this turbulence, and also showing me that there wasn't anything to worry about, a kind of aerial twenty-third Psalm. Yea though we were flying through the shadow of death, that frigging smile on my brother's face was supposed to comfort me.

He was too good a pilot to pick up a downed wing every time we keeled over in the turbulence. Three seconds later, the back side of the buffet would hit us again and knock us down the other way. I had nothing to do, no task to perform or to distract me, because all we were doing was following the Pennsylvania Turnpike. We couldn't climb and get away from the worst ground effects, because an overcast was starting to drop low. We couldn't descend, and fly the valleys to get some relief from the ridge effect near the peaks, because then we would lose contact with the Turnpike. As it was, we were skirting over the tops of the ridges with just a few hundred feet of clearance, and every one of those damn ridge lines packed a ferocious wallop for us.

Turnpike, tunnel, Turnpike, tunnel, bounce-jolt-bounce. The Alleghenies are bleak, featureless terrain, and there wasn't a single farm or a town to break the monotony of the range. The mountains were flinty and hard, gray and black as they ran to the narrow horizons, with endless hardwood stands and pine barrens on top. Perhaps seven or eight seconds passed between turbulent buffets, and it was very rough turbulence, enough to jiggle the throttle and force Kern to constantly adjust it. I shook all over, trembling from the cold blasts of air coming in through the windows, which made my perspiring hands and chest shiver. But Kern wouldn't throttle back and slow the plane down to make the ride a little easier. He was determined to make it to Washington County Airport and refuel before the front came through. Indiana. If this was the price I had to pay to get there, I wanted out. It was the worst spell of flying in my life.

After the town of Ligonier the mountains thinned out and the turbulence was milder. But the storm seemed to resent our progress over the mountains and pushed forward new artillery. Mean, snarling black clouds were popping up everywhere, and we were just flying in a maze, poking around for white patches of sky ahead. The headwinds were even stronger than before, a good sign, in a way. We were getting close to the center of the front and it was obvious that it was blowing through quite hard and wasn't going to get stalled. Behind it there would be clear skies and beautiful conditions across Ohio and Indiana.

But where to go? We couldn't turn southwest at Latrobe as Wilbur advised, because the city was socked in by rain. Diverting north instead, we lost the Pike for a while, then picked it back up later. But we had missed the turnoff for Route 70. Now we were almost upon Pittsburgh itself. In this weather, there was no way, legally or safely, to proceed due west over the city. Our destination at Washington was still fifty miles to the southwest, behind a solid wall of gray and black clouds.

Then I just had this lucky little run of memory and map reading.

Barnstorming blarney. I had reached the age of weariness concerning my father's fabulous talk. He was an Olympic-class bullshitter and everyone loved him for it, but not me. By the age of ten I knew every one of his yarns by heart and after that they were maddening, starting to come around for the fourth or fifth time. It never occurred to me, however, that barnstorming blarney might someday prove useful to me, that it was in fact an education, a ground-school course in geography and aerial escape routes, another one of my father's cryptic gifts to me.

But it occurred to me right there, because now we were in trouble over Pittsburgh. My father had always spoken affectionately of this city on the western frontier of Pennsylvania, which he regarded as an industrial utopia. It was his kind of town, big-shouldered and loud, full of beefy, fun-loving steelworkers and insane millionaires. Toward the end of World War II, before his big

crash, my father had a friend who flew military cargo out of the North Phila-
delphia Airport in twin-engine Lockheeds. My father was working for *Life*
then, but he liked the Lockheed and he often flew along on weekends as a
copilot, to get some free stick time. They landed at Pittsburgh a lot, at the old
Allegheny County Airport on the western side of the Monongahela River, and
loved to carouse at night in the waterfront bars. Pittsburgh had always been
legendary among pilots for its miserable weather. The big weather systems
pouring out of the Great Lakes condensed near the junction of the Ohio and
the Monongahela rivers with the smoke from the giant U.S. Steel mills, smoth-
ering the terrain all the way to West Virginia with a dense blanket of smog.

"Pittsburgh," my father used to say to us. "You know what they call it?
'Hell with the lid taken off.' It's awful there, but we always got through. What
we did, see, was stay out over the rivers, between the mills, and then we'd fly
the steel plants stack-to-stack."

Stack-to-stack. Ahead of us, flashing pinkish red in the haze, was the
beacon on a tall obstruction. I stared at the map. We were quite lost at the moment but I quickly scrolled through all the possibilities, eliminating other
nearby obstructions by poking my head out into the air and squinting down at
the obvious landmarks, the rail junctions and roads. The tower ahead of us was
the first tall one east of the Monongahela, which had to be the big U.S. Steel
blast furnace at Braddock, clearly marked on the sectional map as a "Stack"
with a flashing red beacon.

I wanted to be positive—we couldn't risk a navigation error at this
point, because the weather behind us was closed off. I gave myself an extra fifteen seconds, peering around for other beacons, and then checking the location of the stack against the city below. I was as certain as I could be. That stack
was Braddock.

"Kern!"

"Yo."

"I want this airplane over that stack."

"Done. Rink, are you sure you know what you're doing?"

"I know what I'm doing."

Whoa! Elevator going up. There was quite a strong updraft rocketing
out of that stack. It hadn't occurred to either of us that the blast furnace exhaust churning up through a U.S. Steel stack was just about the strongest thermal you could find. We launched straight up toward the clouds, with the altimeter winding around like a second hand.

As the smell of sulfur and molten steel swirled through the cockpit,
Kern closed the throttle, downed the left wing and cross-controlled to slip

sideways and spill off lift, and pointed the nose down. It was awesome and harshly beautiful, slipping sideways at a 45-degree bank in the effluent of a steel mill.

But I was jubilant and swelling with pride as we boomeranged through that blast furnace exhaust. There, on the immense roof of the mill, in white block letters, I could read: BRADDOCK.

Stack-to-stack, that's how we flew it. From Braddock we crossed the Monongahela to Duquesne, from Duquesne we flew to McKeesport, from McKeesport to Clairton, and then we flew down over the glorious, giant oxbow in the mighty Mon and south to Monessen.

It was a lovely stretch of flying down through the American Ruhr. Now and then the sky opened up and the sun poked through, shimmering off the river and the ceramic-tile exterior of the stacks. Below us spread a busy industrial setting. The steel mills on either side of the river belched out smoke, the rail lines were full of cars, and barges eased under picturesque steel bridges, which glared softly in the haze. Down along the waterfront the buildings and mills were made of brick and hard granite. Up on the ridgelines, row upon row of white wooden houses perched on the slopes, with terraced gardens thrown out in every direction from the back yards. And the churches. There were so many churches in the Mon Valley. Every town hugging the river—McKeesport and Duquesne, Clairton, Dravosburg, Donora, and Charleroi—had a bright cluster of cathedrals and brick parishes in its midst, immaculate and brimming with wide green lawns in the back. Shiny onion domes and Cyrillic crosses from the church steeples reached up to us in the plane, jewels in the necklace formed by the course of the broad river.

I was very confident of our position now, and I could have flown those industrial landmarks forever. After the bleak, turbulent agony of the mountains, the Mon Valley was a joy to fly. Now we were approaching the last oxbow on the river and our first full state—Pennsylvania—would be behind us. My heart had always belonged to Pennsylvania anyway, because that is where my father's family came from and those were the stories I loved best. Now Kern and I had flown it all and been delivered to this magical western edge.

But there wasn't a lot of time for appreciating the beauty of the steel country after we picked up Route 70 at Monessen. The storm raging out of Ohio still meant to defeat us. The stretch of country west of the Mon Valley is wooded and lonely, and we flew on for fifteen minutes in nasty turbulence and spitting rain. Finally we could make out the hangars of the Washington County airport. The sky above them was twisting up a dragon's tail of angry gray and black clouds. As we firewalled for the field everything ahead of us turned a solid wall of black. We weren't beating the front any longer. We had met it.

Kern kept the Cub high to avoid any obstructions beneath us and punched through some yellowish, bumpy clouds. The runway ahead of us was obscured by rain. When I threw open the side door to stare down and see where we were, the propwash from the propeller sprayed my face and shoulders with rain. But I could see our position on the highway now and then I pulled back into the cockpit to look at the map. The runway ran due east and west, at a slight angle to the highway.

"Kern! Fly 270 degrees, now! Start your descent."

"Rink, where's the wind? We don't know the wind."

"Fuck the wind, Kern! We've got plenty of room. It's a 5,000-foot, paved runway."

The wind wasn't consistent anyway. By looking straight down at the tree branches, I could see that it was kiting all over the place, gusting hard and presenting us with a 25-degree crosswind from either side. But that was the great thing about Kern—I knew that he could sideslip or crab through anything, a freaking gale if he had to. We were weathervaning all over the place as we descended on final approach, but Kern was holding my 270 degrees pretty well.

Finally, through the vortices of rain and cloud ahead of us, we could see the white centerline of the runway. Kern closed the throttle and slipped hard to make the pavement.

We were fast and pushed both sideways and from behind, by gusts that couldn't decide whether they were a crosswind or a tailwind. The Cub took up a lot of runway, jackassing down the wet tar. Kern would think he had the wings stalled, then a gust would balloon us up again, he'd slip and stall us again, and then we bounced some more. The wheels skittered all over the place, hydroplaning on the wet runway, and Kern didn't dare touch the brakes. But finally we were stalled hard and the wheels felt solid in the puddles, and the tail would stay down.

As he turned off for a taxiway, Kern reached back and squeezed my knee, which I could never forget because my knee was tender to the touch from banging against the cockpit walls in the turbulence.

"Jesus, Rink. Way to go. That was aces—all day."

I was happy about that compliment from my brother, and relieved to be on the ground. The raging stratocumulus couldn't hurt us now. The front was right on top of the airport and all we had to do was wait an hour or two for it to cleanse the skies ahead of us, and then we'd crop the bejesus out of 71-Hotel that evening to make Indiana.

But it was more than beating the weather that thrilled me. I was enchanted with this journey now, in love with my brother and myself. Everything about us had led to this moment. My brother's fortitude and skill with a plane had been outstanding. I had listened inside me and found a use for barnstorming blarney. Understanding navigation, enjoying it, seemed a divine gift now. And I had to admit that even my father's kick-ass style of flying, my long education in being afraid in planes, had helped me across the Alleghenies. These forces had merged as one and delivered us through the storms. River to river we'd flown Pennsylvania in the most awful weather imaginable, and I was confident that the rest of the country couldn't throw anything worse at us.

The wings echoed like timpani drums as we taxied in. Torrential sheets of rain were falling now. Water came through the seals on the windshield, beaded up on the ceiling, and fell on our heads. Drenched, our loafers and socks filling up to our ankles, we tied down the Cub and ran across the tarmac in the rain to the pilots' shack. We were laughing and elated as we pushed through the door.

I was exhausted. Collapsing onto the couch in the pilots' shack, I slept soundly for two hours. Kern woke me after the front passed through. I could see as soon as I rose off the couch that it was beautiful outside and I felt refreshed, expectant. We had earned this—clear, open skies across Ohio and into Indiana. Slumber had cleared my head. Already, we'd made the longest cross-country flight of our lives, and that hateful ride across the Alleghenies seemed like it happened a year ago.

I don't recall what I dreamed about on that couch and I don't remember thinking about it much, but not again on that trip, not for the rest of my life, did I fear flying with my brother.

Approaches

BY LAURENCE GONZALES

(Excerpted from the book *One Zero Charlie: Adventures in Grass Roots Aviation*)

The universe of general aviation—what Laurence Gonzales so aptly calls "grass roots aviation"—is made up of smallish airports and planes like Cessnas, Pipers, Beechcrafts, and a plethora of others with less-familiar names. Even though general aviation facilities are usually not far outside of towns and villages across America, many people have about as much understanding about what makes these facilities tick as they do when regarding the behemoth super airline terminals, like JFK in New York or O'Hare in Chicago. Airplanes come and go. That's about all the average citizen knows or wants to know.

As experienced aviators and wannabes alike realize, however, general aviation facilities are veritable theaters where both great dramas and all sorts of tiny adventures unfold, and high emotions are set loose. Flying lessons, cross-country trips, air taxi operations, and overnight light-plane cargo hauls—"freight dogs" they're called, ferrying bundles of stuff like bank transactions and newspapers—these are the events unfolding every day at any successful small airport.

For the average pilot, the initial pilot's license granted by the FAA allows the operator to fly in VFR conditions only, which means Visual Flight Rules, which in turn means just what they say: You have to be able to see things. Especially Mother Earth, which, as aviators like to joke, "may rise up and smite thee" when you venture into clouds.

Flying in low-visibility weather means Instrument Flight Rules, IFR, requiring an instrument license, which means a great deal of training and experience. Almost every VFR pilot at some time or another longs to have an

117

instrument rating, to be set free to fly whenever and wherever they wish, regardless of weather. No more hanging out in motels or at the airport while they wait for the weather to clear.

Alas, things aren't quite that simple out there. Dangers can still lurk in the clouds, turbulence and thunderstorms in particular, both of which are usually quite fatal no matter what kind of flying license is in your wallet. And then there's that other little three-letter word which strikes pure dread into all but the most foolhardy fliers: Ice.

Laurence Gonzales has written novels, short stories, essays, and plays, and has worked on the editorial staffs of several magazines. He also has found time to fly a lot of general aviation airplanes, a passion that led to his wonderful book *One Zero Charlie,* detailing life at one facility in the subculture of general aviation—Galt airport in Northern Illinois—known to pilots as "One Zero Charlie." This excerpt from the book, which was published in 1992, takes us flying when weather is very much a factor, and an old aviation bromide flashes through one's mind: "It's better to be on the ground and wishing you were flying, than to be flying and wishing you were on the ground."

The coming of dawn was almost imperceptible. A faint gray light spread upon the low hills that surround Galt Airport. The cold had returned, and now the corn stubble poked through snow, which had fallen during the night. A turbulent wind blew from the northwest, as dark and vagrant clouds hunched over the hills munching on a stand of naked trees. A wood frame building, which resembled a control tower with an orange wind sock on top, stood watch over the flight line where single-engine Cessnas and Pipers waited, tied down with muddy nylon ropes in rows out on the white and frozen grass. As gusts of wind hit them, they shook like horses left too long in the fields. They were not shiny airplanes. Most of them had a weathered appearance, their once-brilliant blues and reds and greens now scoured and faded by the ceaseless buffing of the wind.

Parallel to the flight line were ranks of rusty corrugated steel hangars as long as city blocks. Their sliding doors—great sheets of metal fixed on wooden frames—sounded like stage thunder as they rumbled in the wind against their rails.

I stood in the second-floor office in the square frame building sipping coffee and waiting for the weather to lift. The room had windows all around, as a control tower might have. The office was cramped by the clutter of desks and file cabinets and cardboard boxes. Facing north, I could see the first students of the morning taxiing out in their Cessna trainers. They would motor slowly down the runway toward the east; I could hear them on the radio speaker across the room, declaring their position so that no one would land on top of them. At the far end of the field, more than half a mile away, they would turn around and rush toward us, still silent in the snowscape, and wobble into un-certain flight, the fragile wings kicked this way and that by the wind. At last they would clamor past the office window, their tiny and pitiful racket as tri-umphant as a child's string of firecrackers tossed into the air.

They wouldn't leave the landing pattern this morning for fear of los-ing the airport. But since the instructors weren't paid unless they flew, some of them would drag their students up in nearly any kind of weather, saying, "This is good practice for you. If you can land in this wind, you can land in any-thing." Or: "Now, don't ever do this by yourself, but . . ."

Out the west windows I could see the maintenance hangar, a great gray metal building called the shop (the new shop), large enough for five air-planes, and beyond that, in a field overgrown with dried thistles and dusted with snow, a kind of hillbilly heaven of antique rusted trucks and tractors, old airplane wings and tails and motors in various stages of disintegration. Farther still beyond that, closer to the main road, was an old yellow barn with a con-crete silo attached, the gray shaft of concrete gripped by the black veins of naked vines. In the middle distance: three rusted wire corncribs which had once stood in a neat row. Although two had been crushed and tossed aside in a storm, one still stood mysteriously intact, as if to prove the defiantly illogical nature of Nature and its random, unaccountable calamities. Between the three corncribs and the barn was the hog shed, peeling yellow paint, abandoned now for more than a decade. On one side of the driveway that ran past the farrow-ing pens, a bulldozer was frozen in rust and snow, its scoop slightly raised, as if it were the victim of a spell cast upon the whole place, the yellow house, the yellow barn, the yellow shed, even the shattered gray-green carcass of a World War II B-17 Flying Fortress with weeds growing up through its ribs and noth-ing but spiders dropping on silken bungees from its bomb-bay doors. For someone who just happened by, it would be difficult to tell if this was an air-port that had once been a farm or a farm that had once been an airport.

Behind me at her desk, Carla talked to pilots on the radio. "Six Fox Sierra, wind at Galt is north eighteen gusting thirty-five, and it looks like they're using either end today." Either way there would be an uncomfortable wind across the runway.

Carla hung the microphone on its hook beside a speaker on the wall between the photographs of her son, Chance, and her husband, Steve, and another photograph, equally prominent, of their yellow airplane in the act of landing over the pines at Galt Airport. It was an old Cessna 170, which she referred to as "the family car."

The airplane whose pilot Carla had been addressing as 6FS[1] on the radio appeared out of the mists. It was a miracle that would have been easy to miss: something being created out of nothing. The airplane simply materialized with no fanfare or ceremony, and what was not there before all at once existed, as divine and inconspicuous as an amoeba being born in the sea.

The plane landed, and the two Cessna 152 trainer planes took off again, dipping left and right in the stiff wind, and began grinding their way around the pattern like toys on a wire.

Directly across the east-west runway were the ruins of the original maintenance hangar, or shop, which had been left exactly as it was after the fire, its corrugated metal roof and walls curled up where they had melted. Certainly human beings have had a hand in shaping this place, but there has been a curious fatalism in the way that Art has allowed Fate and Nature to co-produce. He has always seemed to know that there were other forces at work here, and he's made no attempt to deny them. His overt, almost defiant, statements in leaving the battered corncribs where they fell, the frozen bulldozer gaping at the wind, were more than mere acts of neglect. Art has never shied away from work—he literally built the farm and the airport with his own hands. No, what Art left undone was left carefully, like the silence between notes in a symphony.

Carla sighed at the pile of invoices on the desk before her. Galt Flying Service, Inc., was a flight school and a repair shop. It sold gas and oil and assorted pilot supplies. More than 200,000 gallons of gasoline were pumped in 1991. GFS provided parking for airplanes, either transient or permanent, either outside or in a hangar. Galt had seven airplanes and rented them for a total of 7,800 hours in 1991. In other words, speaking in averages, a Galt airplane was flying not quite every hour of every day of the year. For such a small airport, Carla had a lot of invoices. "So. Are you going to fly or not?" she asked me.

"Nobody's paying me to go up in this weather," I said. And then I thought, well, actually, it might be a good day to practice blind flying. And then I thought better of it. "Where's Steve?"

---

[1] An aircraft's call sign is always pronounced digit by digit, and the letters are given according to the official phonic alphabet. Thus, 1237U is "one two three seven Uniform" and 6FS is "six Foxtrot Sierra."

"He's out in the shop," she said. "Where else?"

Bill Tate, an instructor in his early twenties, came in to check the schedule with Carla. Bill was tall and thin and had dark hair. He wore prefaded jeans and Reeboks and a white and gray sweater. Bill looked as if he belonged in a movie about a fighter squadron. His walk was cool and loose, and he flipped a pencil as he sauntered in and sat down to look at the weather computer. The report looked something like this: RFD RS COR 1251 M8 OVC 11/2SF 927/29/25/3619G30.

"You going up?" I asked him.

"Not in this," Bill said. "Ground school today. Mucho ground school."

"What about those guys?" I asked, pointing to the 152s.

"That's Jay and Brett. They're coming in. It's too low. What about you? You going up?"

"I don't think so," I said. I could feel myself being drawn toward flight. There was something—definitely something—in the idea of going into those clouds with a little airplane strapped onto me. But I gave it up. "Nah," I said. "It's too low."

I descended the stairs from the office to the flight desk and said hello to Jay Pettigrove, the curly-headed chief flight instructor, who had just landed. Jay now sat at the scheduling desk behind two glass display cases placed in an L-shape to provide a demarcation between the public and employee areas—in front of the flight desk or behind the flight desk—though people walked behind the desk and helped themselves to what they needed, and the cash register and merchandise cases were never locked. One morning I found the register drawer sitting out. It was full of cash. No one was around. All was quiet. I called out for Carla, and she answered from the office upstairs, "Good morning, Laurence."

"Carla, did you know there was a drawer full of cash sitting out here?" I asked.

"Oh, yeah, why don't you put it in the cash register?" she called.

In the glass display cases were fanfold charts, fuel testers, flashlights, logbooks, textbooks, plotter-computers, and sheet metal screws for replacing the ones that fell out of airplanes as they rattled through the sky.

I could hear the wind humming against the picture windows that dominated two walls facing north and west, giving a view of the ramp and runway and parking lot. The lot was nearly empty, and only a few students straggled in.

Jay glanced up from the schedule on the desk before him. He had a disheveled appearance and always looked as if he needed a new wardrobe and a haircut, but he was meticulous and organized when it came to instructing. He was one of the few flight instructors I knew who took notes during flight.

"You going up?" Jay asked me, thumbing through a gray steel box of three-by-five cards, which were scribbled with the names and telephone numbers of the members of the Galt Flying Club. On the front of the box was the number of the Hebron Crash-Fire-Rescue Squad, which had only been used twice, once for Lloyd and Scotty, once for the shop fire.

"Probably not," I said. "I don't know. I got a briefing for Kenosha. There aren't any pilot reports of icing."

"You're right on the ILS for Runway Six," Jay said.

"Yeah," I said without enthusiasm. "So what are you going to do?" I asked.

"Sim rides today," he said, referring to the instrument flight simulator, a device in the back room for practicing flight by reference to the instruments. IFR means instrument flight rules. The world is crisscrossed by an invisible spider web of electronic signals that connect points of interest, say, Indianapolis and Champaign. Instruments on board the aircraft receive those signals and guide us from point to point so that we can zigzag from anywhere to anywhere.

Like chess players, we sit for hours studying dials that barely move, while we barely move the controls. But after what seems an eternity of sweaty, trembling concentration, we appear at the end of a runway at some distant place. It's a game of intellect, and if we don't play well, we can be sublimated into the very clouds themselves and never reappear. On a single-engine airplane, a single vacuum pump runs the critical instruments that keep us right side up in the clouds. One day I saw Bob Russell, the airport manger and chief mechanic, carrying a new vacuum pump in a plastic wrapper on his way out to the shop, and I asked, "What's that?"

"A time bomb," he said.

"What?" I asked, taken unaware.

"Yeah," he said. "Because just when you get into the clouds, they quit." And he went off to install it on an airplane.

I pulled my coat around me and hurried across the parking lot to the shop on no particular errand. It was a gray metal cube of a building with a greasy concrete floor and a colossal door that curtsied across a horizontal crease and lifted away like aluminum skirts gathered up by steel fingers. As I entered, I was greeted by the whine of a drill, which ricocheted around the big open space and came screaming down from the rafters like an electric hawk. A faint and oily light filtered in from somewhere, illuminating three small Cessnas in different stages of disassembly. An aircraft engine on a hoist sat in one corner, and the airplane from which the engine had been removed had scrap iron chained to its nose to keep it from tipping backward onto its tail. Some of the walls of the hangar had been covered with a kind of chicken-wire mesh, and here and there scraps of plastic sheeting kept the wind out. High up by the roof, rectangular holes had been cut in the steel siding and fitted with translucent yellowed green plastic material to admit a kind of dingy curdled light, which mixed with the harsh fluorescence of industrial fixtures. The room smelled like fuel and urethane.

It was there, in the dark glare of the shop, amid the seeming chaos of mysterious mechanical undertakings, that I found Carla's husband, Steve Nusbaum, a bearded man in a leather vest and greasy jeans, wearing battered blue Velcro sneakers and a threadbare seed cap that read, obscurely, "39395." Steve was tall and rangy with a salt-and-pepper beard that grew completely untrimmed, mountain-man style. When I entered, he looked up from his work at a table fashioned from plywood and sawhorses and glared suspiciously at me. He looked like a moonshiner who had just been discovered in the most secluded part of the woods.

"What's this?" I asked, indicating a fifteen-foot-long skinless fuselage, which sported a shiny blond wooden propeller and sat back comfortably on new tires but had as yet no wings.

"A *Stork*," Steve said and continued to work on the tube-steel frame with hands so rough they looked as if the fingers had been welded together at the knuckles. *Stork* was the name that he and his partner, Don Ericson, were going to use for their miniature single-place aircraft. Like a work of art, it was all original hand work. When I first saw it, the *Stork* was barely a skeleton. But in its strange incipient qualities, I could see the promise of flight, as in a newly hatched dragonfly. The original Fieseler *Storch* had little in common with its namesake, except high lift and slow flight. "In 1943," Don told me, "one of these flown by a German colonel picked Mussolini off of a hilltop and got him away to Northern Italy."

Just behind the pilot's seat an ominous-looking canister was fixed to the frame. "What's that?" I asked Steve.

"A parachute," he said. He was not a man who gave himself over to voluble discourse, but I stayed and pestered him until he showed me how, in an emergency, the pilot could pull a red knob, and a ballistic device would fire the parachute right through the Dacron skin, bringing the airplane and all its contents, living or otherwise, gently back to earth.

Don Ericson, a local pilot and industrialist businessman from an airfield ten miles south of Galt, had lost his pilot's medical certificate to a heart condition. He could no longer fly commercially produced airplanes—not legally. Anyone can fly an ultralight aircraft without a pilot's license or a medical certificate. So Don, who was a smiling pixie of a man with an elfish grin and a snap-brim cap concealing a slightly bald head, a man who spoke quickly and confidentially, as if he were letting us in on a secret, and always with a smirk and a friendly laugh, had begun to design the *Stork* ultralight aircraft as an attempt to get back into the air after being grounded. Now with Steve's help and tight-lipped encouragement, Don was refining the design, and they were actually fabricating the prototype. Don had the engineering expertise, the equations, but Steve had the practical know-how and the skill of having built or rebuilt many airplanes by hand. One day I watched Don compare notes on model airplanes with Steve's 14-year-old son Chance. Chance had all the models and Don had all the history. As Chance brought out more and more of them, Don became more and more excited. "Oh!" he would say. Or: "*Well, now!*" And: "When I was a kid, I built one of these." He narrated how one particular model went up and up and caught a thermal and crossed the river, "And I never saw it again." He had wanted to build a real one ever since.

Yet no one asked why Don would bother. All conversation about it began from the assumption that he had to get back into the air. And most of the people involved in the effort looked at it as a necessary test, as in a mythical story, a rite of passage; and they were only too glad to help the hero of the tale.

But viewed from the outside, Don's course of action seemed much less clear. He did not need to go anywhere, so the aircraft could not rightly be viewed as a mode of transportation. He did not fly for a living, so it would represent no income. On the contrary, it represented quite a substantial expense. Why did he need to fly then? To entertain himself? He could have bought a boat. He could have golfed. I have a friend who in his sixties races cars, a sport for which no medical exam is necessary. And my friend has only one kidney. Don could have raced.

But even granting that Don must fly, why didn't he buy an ultralight airplane? No. This went deeper. For Don was being pursued by something, no more or less than Lloyd had been.

Don had told me that store-bought ultralights would not do for him. They simply did not have the qualities of a real airplane. His would be a real airplane, but it would also fit the FAA's definition of ultralight.[2] Even after all

---

[2] Ultralight is defined by Federal Aviation Regulation (FAR) Section 103.7. The craft must not weigh more than 254 pounds empty. It can carry no more than 5 gallons of fuel. Its top speed cannot exceed 55 knots, while landing speed must be 24 knots or less. Once a vehicle meets those requirements, it is not considered an airplane, and no license is required to fly it.

this, Don's explanation seemed flimsy. I felt that his quest went beyond what logic could support. It as evident in his unbridled enthusiasm, in his gleeful concentration on the task of figuring out the plan for his airplane, that Don was setting out to shape and craft a fitting image of himself, like Queequeg building his own coffin on the deck of the *Pequod*. And in the act he would renew and redefine his life. His heart had been taken away, and this would be his new heart, with rudder pedals and a silver parachute. This *Stork,* like the German original, would pluck him from a hilltop in a daring escape.

When I came upon the *Stork,* most remarkable to me was not the airplane itself but the childlike temerity of the men who were making it with their hands. I had always thought of airplanes as arising from sophisticated research projects at some secret headquarters out west (or east). Real airplanes were created somewhere else by someone else. And yet here, among the cow pies and cornfields, was an airplane as real as any, being born in this unlikely manger scene, the big airplanes like cows breathing warm life onto the baby. How appropriate that they had called it the *Stork.* And there were Steve's hands, with their stigmata of grease and the scabbed-over scrapes and cuts on the scarred knuckles; those very human hands, which trembled when he tried to quit smoking Camels, were raising—out of the elements of the earth by a power just as mysterious as rain—an apparatus that would fly. Their innocence struck me more than their boldness. It was the same innocence, I reckoned, with which someone first hollowed out a log and set forth across the infinite sea, casting adrift without the hint of a question about where to find land again or the tremor of apprehension for a safe return. What we celebrate as heroic vision is more often a deadly naiveté for those who don't make it to the celebration.

"When will it fly?" asked.

"Maybe April," Steve said.

"Who will fly it first?"

"I suppose I will," he said, not looking up from his work, a Camel dangling from his lips, smoke curling up across his frown, seemingly unimpressed with the notion of sending his one and only body up in the contraption. Certainly greater minds than his and Don's had created machines that failed to fly. Some of them were fairly recent, too. The first P-51 Mustang fighter plane ever built had crashed. The first B-17 bomber ever built had also crashed. The first P-38 Lightning twin-engine fighter ever built had crashed. The first X-15 had crashed, too.

I stood there with the growing sense that I needed to fly. The craving for food or sex, for music or travel or a swim in the sea, is no different. Steve asked me the same question: Would I fly today? But I found myself inexplicably giving him a different answer.

"No. Yeah. Maybe so. I don't know," I said, trying to understand the process that was taking place inside me. Surely I believed that I had a choice in the matter, and surely, somewhat earlier, I thought I had made the choice not to fly in this weather. Perhaps I shared with Don that peculiar compulsion. I found myself telling Steve, "I was thinking about going over to Kenosha and just shooting a couple of approaches."

"Pretty scummy out there today," Steve said, buried in his work within the wing.

"Yeah," I said. "Just right for approaches."

My ten-year-old daughter, Elena, had refused to fly with me until I earned my instrument rating. Since she was five or six, she had been telling me that airplane pilots, the kind she flew with on her way to Grandma's house, had instrument ratings, and that if I wanted her to fly with me, I had to get one, too. Consequently, I had earned my instrument rating the previous year, and discovered that I had inherited from my father a flair for flying on the gauges. But it took constant practice to remain sharp, and it was not an endeavor to be taken lightly. Once I got myself up in the clouds, I would have to make a successful approach—a rather complicated technical procedure that led to a glorious moment when I'd burst out of the clouds over the runway. But going into the clouds represents a big commitment: Once up there, I can't stop to close the baggage door if I forgot. I can't take a coffee break or go to the bathroom. If it's bumpy, I can barely stop to wipe the sweat out of my eyes. Sometimes, without explanation, pilots find themselves upside down in the clouds and simply come spinning back to earth. Those crashes are usually written off by the government body that investigates them, the National Transportation Safety Board, as disorientation accidents, although a lot of pilots believe they are the result of that idiotic vacuum pump, the Time Bomb, which fails in flight, taking away the primary reference instruments with no bells or whistles to tell you there's anything wrong. Whatever the case, thoughts of those accidents always give me pause before I blast off into the clouds in just any old kind of weather, in just any old mood, in just any old bucket of bolts. One instructor told me, "Whenever someone comes to me and says, 'I want to learn instrument flying,' I put them in a big galvanized garbage can, and I give them a flashlight and about a dozen needles to thread, and then I close the lid and start throwing rocks at the garbage can. Because that's what single-engine IFR is like, and anybody who wants to do that on purpose is crazy."

Even so, I found myself drifting back from the shop and standing behind the flight desk thumbing through the schedule to see what was available in the way of IFR airplanes.

I believe that certain people—Lloyd and Steve and Don and I among them—fly once and are never the same again. The question is whether the effect is something particular to the flying or to the compulsion of the person and whether that same person would have, like Toad of Toad Hall, fallen into the thrall of something else had the circumstances been different. Kenneth Grahame in *The Wind in the Willows* described how Toad, overcome by his enchantment for a motorcar, actually stole one.

> The next moment, hardly knowing how it came about, he found he had hold of the handle and was turning it. As the familiar sound broke forth, the old passion seized on Toad and completely mastered him, body and soul. As if in a dream he found himself, somehow, seated in the driver's seat; as if in a dream, he pulled the lever and swung the car round the yard and out through the archway; and, as if in a dream, all sense of right and wrong, all fear of obvious consequences, seemed temporarily suspended. He increased his pace and as the car devoured the street and leaped forth on the high road through the open country, he was only conscious that he was Toad once more, Toad at his best and highest, Toad the terror, the traffic-queller, the Lord of the lone trail, before whom all must give way or be smitten into nothingness and everlasting night. He chanted as he flew, and the car responded with sonorous drone; the miles were eaten up under him as he sped he knew not whither, fulfilling his instincts, living his hour, reckless of what might come to him.

I do not mean to say that anyone who learns to fly an airplane is an addict. For many people it is merely a pastime or a profession, and an airplane is simply a tool, like a halberd or a hypodermic needle. But there are those of us who fly—I mean who really fly a lot—and with a mad insatiable passion and all the time thinking and dreaming of it with that tingling in our fingers and toes, yearning for the feel of the stick and rudder and with a deep socket in our bellies that can only be filled by G-force and swirling mists, and I believe that we perform our feats of daring in the service of altered brain chemistry. It is precisely this *Impulse* that has been the cornerstone of aviation since the very beginning. Why else would someone throw himself off a cliff with a pair of homemade wings strapped to his back, while the certainty of death and the hope of flight struggle for dominion over his leaping heart? The Impulse. A mystical disease of the mind which overwhelms our will and propels us against all good sense not simply into flight but into more and more dangerous and audacious methods of rocketeering, controlled falling, death leaps, and fiery plummets to earth. The Impulse: It's the only explanation.

So it was that I stood on the ramp that snowy foggy day looking longingly at the sky, seriously considering going up "just to see what it's like." I had no destination. Yet I craved flight so desperately that I was on the point of risking all just for the pluperfect sensation of having done it.

I believe it's the same impulse described in Yeats's poem "An Irish Airman Foresees His Death."

> *Nor law, nor duty bade me fight,*
> *Nor public men, nor cheering crowds.*
> *A lonely impulse of delight*
> *Drove to this tumult in the clouds.*

In essence, he's saying, "Hey, all this stuff about politics and war and love of country is all bullshit: I'm doing this because it's fun." Or as the Great Santini said, "It's better than dying of piles." Yeats again, saying the same thing:

> *I balanced all, brought all to mind,*
> *The years to come seemed waste of breath,*
> *A waste of breath the years behind*
> *In balance with this life, this death.*

There are people who become legends by dying and there are people who become legends by living. Amos Buettell, owner of Crown Industries in Hebron, Illinois, did both, and Russell, especially, was one of the keepers of his legend. Russell liked to talk about Amos's Pitts special, which had such a big engine that the exhaust flames burned the Lexan window out of the floor between Amos's feet as he flew in aerobatics competition. Better than any historical figure, Amos conformed to Russell's ideal of the perfect aviation story, and often at lunch or on break, Russell would tilt his head back and look at the ceiling and say, "Boy, Amos could fly." Amos also was the perfect embodiment of the obsessed pilot, perhaps even more so than Lloyd had been. There was a time in the early 1980s when Amos Buettell's presence dominated Galt Airport. There used to be a big framed color photograph of Amos on the wall in the briefing room, but his legend grew so large that someone actually walked away with it.

I found myself nosing around the blue-and-white Cessna 172 Cutlass, taking a fuel sample, tightening little screws that were always coming loose, preparing it for flight. One of the line boys came out and said, "Are you going to take the Cutlass?"

"Yeah, I'm just going over to Kenosha to shoot a couple of approaches." There: I had said it. I was going after all. How could I do it? Why would I do it?

Even as I checked the rudder, the wheel wells, the Pitot tube, I could not walk my way mentally through the logic that had taken me this far, that I was freezing out on the snowy grass, trying to take a collection of bolts into the air on a day such as today. And yet I plunged heedlessly on, swept away, bewitched.

"Okay," Greg said. He was one of the line guys. He began to walk away, calling back across the ramp, "Want me to put you down for two hours?"

"That sounds fine," I said.

I filed my flight plan and rolled down to the end of the runway, still thinking that this might not be such a great idea. As I sat at the end of the runway running up the engine, I half hoped that something would go wrong. If I had a mechanical problem, then I could turn back and not have to go. I'd have something to talk about. There was always apprehension, a longing to go back, a sudden sense of loneliness in the desperation of it all, which laid bare our deeper motives. At the beginning of many flights, I was very eager to take to the air, but even then there was always something held in reserve, some occult sense of peril, well hidden from myself and from those around me.

I think my greatest fear was ice. Dr. Boone Brackett, my instrument instructor, liked to fly to New York to attend the opera. Dr. Brackett was an orthopedic surgeon, a lawyer, and an airline transport pilot, a man well read in science, literature, and the arts, compact, somewhat bald, forceful, quick-witted, sometimes wearing a beard and always with a gruff Texas accent. He had been a combat surgeon in Vietnam and wanted to go to the Gulf War when it started. On a moment's notice, he and his daughter had flown to Germany to see the Berlin Wall come down. His favorite routine involved sewing someone up on Friday night, rushing out to the airport on his motorcycle (which he had once driven to Alaska with his sons), flying his twin-engine Baron to New York, taking in a weekend of opera, and then rushing back to Chicago just in time to see patients on Monday morning.

When he first began his shuttle from Chicago to New York, Dr. Brackett had a single-engine Cessna Skylane 182RG (for retractable gear), not very different from the aircraft I was about to fly, and one night he was coming back late on a Sunday, in pretty solid weather with his pretty solid girlfriend, when they entered an area of ice that had not been forecast. Up in the higher levels of the atmosphere, where the temperature is well below freezing, sometimes moisture exists that has not yet come out of the vapor stage, but it is supercooled. The presence of almost any object will cause it to form droplets—a dust particle, for example, or smoke. An airplane wing works nicely, too. Two problems develop rapidly when water, being supercooled, comes into solid existence as ice just as it touches the airplane wing. Water weighs a lot, for one thing. And ice works like magic clay, molding and shaping a new wing, changing its aerodynamic characteristics. In fact, a nicely curved wing that flies can turn into a square wing that doesn't fly in a matter of one minute in heavy icing conditions.

So it was that Dr. Brackett flew into some deadly place over Pennsylvania. He had begun to notice that he needed more power just to stay level. The airplane was slowing down. He added power and nose-up trim to compensate. Sometimes ice can form in the carburetor and close off the throat like diphtheria, so he pulled the carburetor heat to keep the engine from coughing. He looked out the window at the wing struts and felt his heart cease and then begin again as he realized that his airplane was being transformed into a gay and decorative confection: Ice had formed a frosting that winked and sparkled every time his wingtip strobes popped off in the murky nighttime clouds. All at once a chunk let loose from the spinning propeller and exploded on his windshield. The unbalanced propeller shook the airplane so hard Dr. Brackett thought the engine would come loose from its mounts, and then another chunk hit the fuselage like a cannon shell. "It was like we were taking flak," he told me.

In the end, he had the throttle pushed all the way to the fire wall, and they were still going down. That was when he realized that he had no approach plates for Pennsylvania. An approach plate shows how to line up on the electronic beam that leads us down to the runway. ("Well, I didn't intend to land in Pennsylvania. I intended to fly *over* it to Midway and go home!") He called the air traffic controller on the radio and told him that he was in trouble and that he needed an airport with an ILS, an instrument landing system. The weather was low, and he had to shoot an instrument approach blind if he was going to have any hope of landing on concrete and not on the side of a hill in the forest. One thing was certain: He was coming down.

"That air traffic controller saved my life," Dr. Brackett told me. "Do you know what he said that saved my life?"

I said I could not imagine.

"He read me the frequency of the ILS and turned me onto a course that would intercept it." Then Dr. Brackett's radio antennas iced up so that he couldn't hear the controller. The ILS is a radio beam, like an extension of the runway in the air, and in order to fly down that beam, the on-board receiver must be tuned to the correct frequency. "I wrote the frequency on my leg, because I had already dropped my clipboard I was so nervous. I dialed in that ILS and rode it down with the stall warning horn blasting and the power on full, and I honest-to-God didn't think I was going to make it. But when we broke out, there we were, right over the numbers, and when I landed about eight hundred pounds of ice fell off onto the runway. I mean, we were *loaded.*" After that, they taxied to the ramp and took a look at the plane. It was encased in ice. "I couldn't believe the thing flew at all," he told me. "The ramp guy came out and took one look at the airplane, and he said, 'Did you land like that?' I said, 'Yeah, but we left half the ice on the runway.'"

"Then what did you do?" I asked.

"Oh, I had 'em haul it into a hangar and melt off the ice. Then we took off and finished the trip."

Dr. Brackett figured that the worst was over. "How bad could one trip get?" he asked me. After that, sometimes when I'd call him in the operating room, he'd answer the phone by saying, "Lone Star Airlines."

So it was as I sat on the end of Runway 27 at Galt, preparing to cast myself into the clouds that wintry day, that I thought, Well, if Dr. Brackett can do it, then I can do it.

I checked the magnetos, and they were normal. I pulled the carburetor heat to make sure the RPM's dropped, indicating that heated air was being fed to the venturi. I turned on the heater for the Pitot tube. I checked the propeller; its pitch is controlled by a knob on the instrument panel. I checked the vacuum pump, the Time Bomb, the only source of power for the gyroscopic instruments that are meant to keep us level in the clouds. The attitude indicator, for example, displays an airplane that moves in reference to an artificial horizon line. Whatever the make-believe airplane in the round glass does, the real airplane does, too. It's so simple that a child could do it. So why do people get killed trying to fly the artificial airplane?

Sometimes the Time Bomb goes off and the vacuum system fails. Then the gyroscope winds down and the little airplane tips slowly and sadly over, and we don't notice, because we've become hypnotized by the drone and the darkness and the swirling mist outside. We innocently follow the little airplane over, over, tipping, tipping, down and down, failing to notice that our compass is going around and around as our airspeed increases. Then it's too late; we're in a graveyard spiral. Suddenly we get all excited and pull back to stop the spiral, but we're going so fast that we pull the wings right off the airplane.

By and by I ran out of things to do on the ground, so I took off. The cornfield stubble and fenceposts whipped past on my right, and the rows of parked airplanes snapped by on my left. Eventually I noticed that the nose wheel was trying to come off the ground, and I lifted gently back on the yoke, and the airplane wobbled into a crabbing, lurching, wind-blown ascent.

I reached for the landing-gear lever, a little toy wheel on a metal stud, and lifted it out and up and listened while the clumsy wheels made their slow crunch-thunk-hammering transit into the belly. I glanced out the window to see that the wheels were gone and lifted away over Art Galt's yellow house and the pine trees at the next-door farm across Greenwood Road and saw the clouds coming at me fast. I set the power at twenty-five inches of manifold pressure and 2,500 RPM. Then I put my attention inside the cockpit and set the little artificial airplane in the round glass window for about 10 degrees nose-up. I spun the trim wheel to hold the plane in that attitude and glanced up to watch myself vanish into a cloud. The thin and swirling mist reminded me of the white fatty fabric of flesh seen through a microscope. I could see through it into a darkness beyond. The spaces closed around me. I found myself in a rushing whiteness that took my breath away, even though I could feel nothing, not even a bump, not even a change in sound as I plunged in. It was as if I entered the ghost of a living white tissue. I felt myself disappearing. I felt a chill as my body came apart, molecule by molecule. I can only look at the whiteness for so long, and then I have to return to the world of logic and order presented by the instrument panel or all will be lost in the delirious transubstantiation of the clouds. My brother Philip calls it Rapture of the Shallow.

The courage is not in the flying—never in the flying. It is in accepting ourselves as this small in a cloud so large and accepting the cloud as so small in the larger sky and the sky so small in the region of space we call home.

Now I hunched over the gauges, keeping the little airplane level, watching the directional gyro to keep on course, and scanning my altitude, airspeed, vertical speed, occasionally glancing at the engine gauges to see that everything was in the green. I called the air traffic controller to open my flight plan to Kenosha. "Chicago Center, Cessna five one nine two Victor just departed Galt Airport, out of one point five for three thousand."

A woman's voice reported back, "Cessna nine two Victor, radar contact, three west of Galt, climb and maintain three thousand, turn right zero six zero, vectors for the ILS."

"Nine two Victor," I said, "right to zero six zero and up to three." I rolled into the turn, watching the little airplane bank inside the glass like a child's Christmas toy that snows when it's turned upside down. As I turned, the directional gyro spun clockwise to north and then northeast. I perceived a brightening in my environment and looked up to see what was happening. I was simply moving closer to the top of the cloud layer, where more sunlight filtered through. I was like a fish swimming from the depths of the ocean toward the surface of the sea. But I knew from the controller's instructions that I would never reach sunlight; the tops were at 6,000 feet. There was no point in going higher for the short flight, only twenty-two nautical miles, or ten minutes in the Gutless Cutlass. In fact, Kenosha's ILS localizer went more or less over Galt, and when we practiced approaches in nice weather, we flew this beam right from takeoff.

The controller cleared me for the approach before I had gone much farther than the Wisconsin border; and by the time I had reached my assigned altitude, I was already setting up for the descent, like the shuttle from New York to Washington. I slowed the airplane by putting the nose up and reducing the power. When I reached the final approach fix, I put the lazy crunch-thunk-hammering landing gear down and began my descent down the glide slope—a second radio beam. The localizer provides lateral direction; the glide slope provides vertical guidance. The system is so accurate that I could descend to the exact spot on the runway where big white lines had been painted and then defaced by black tire marks. On the instrument panel, that process is managed by keeping two needles centered on an inscribed "O," which we call the Donut. One needle moves left and right, the other up and down—localizer, glide slope. The trick to a good approach is to have everything set up well ahead of schedule and then simply sit there, monitoring the needles, changing as little as possible. Pilots who like to fiddle with things make lousy approaches. Lazy pilots who like to watch the world go by tend to make nice smooth approaches. Most really good pilots move slowly all the time. You can see it in their walk. Their hands hang limp at their sides. Their heads turn slowly when they look left and right. In the clouds, a pilot can induce vertigo simply by tilting his head back to throw an overhead switch. Then the inner ear suddenly and insistently begins saying, "We're upside down. No, really. We're truly, for real upside down—hey! Wake up! Your instruments are all broken and you'd better turn this stupid plane right side up, or we're going to die!" And sometimes we listen to that evil voice. And we turn the plane, only to discover that now we're *really* upside down and we don't know how to fix it. The instruments don't work upside down.

Like Zen monks, we try to learn this essential lesson of life: Start slow and taper off. No sudden movements. Always believe the instruments. Of course, believing them is what kills us when the vacuum system fails, because then the instruments *are* lying. It's a cunning, baffling, and devious system.

I tried to concentrate on not doing anything as the white world swirled around me and the engine noise, now reduced to something that

seemed like quiet, churned along. I went through my checklists, routines designed to ensure that I was ready for the world below. I watched my altimeter
revolve like a clock ticking time in reverse. There is an Einsteinian quality to
the ride down a glide slope in the clouds. Nothing moves, and yet I was clipping along at better than a hundred miles an hour. I seemed suspended in time
and space. I had no faith that I was going to reach some sort of reality, a destination—certainly I didn't believe that there was an airport down there. I was
in the white tissue of a beast I could not describe, seeing molecules of cartilage
flinging past the window, cruising through its sinews, screwing through the
web of its flesh in my smoking machine.

I felt a jolt. Sometimes there are just jolts, but we're never enlightened
as to their meaning or origin. We say they're currents in the air, but they could
just as easily be tendons we cut as we screw our way through muscle and bone.

There was one last consideration to my instrument approach before I
could land. There is a minimum altitude beyond which we can't descend unless
the runway is in sight. Yes, after all the wizardry and gadgets, the radar and
technology, it all comes down to the eyeball, and that's true whether I'm in my
friendly little Cessna or a Boeing 747 lumbering home from London. If I
reached that altitude and I could not see the runway or its lights, then I would
have to miss the approach. I'd have to pour on the power and climb back up
and go somewhere else.

Most ILS approaches have a minimum altitude 200 feet above the runway, which is extremely low, since a typical descent rate is 500 feet per minute
or 100 feet every twelve seconds. At 200 feet, we have less than half a minute to
make a complete change in plans, reverse our direction from down to up, and
get out of there safely. If we are off the localizer to one side or the other, we are
closer to obstructions than we ought to be. If we have botched the approach
and we are not near the runway, we have no idea what might be beneath us. As
a consequence, part of preparing to arrive involves preparing to leave.

As I flew down the ILS, I wondered about the weather. I had to get
back to Galt. (I realize that makes no sense, going to all that trouble just to turn
around and go back, but that's the essence of general aviation.) Galt Airport has
an instrument approach, but it is a cruder style of guidance called a VOR
approach, which provides lateral information only. If the clouds were low
enough, I might not be able to return at all. And if Kenosha was below its min-
imums, then I'd have to go somewhere else where the weather was better. The
perfect flight is like the perfect crime: It doesn't exist except in novels.

The altimeter unwound down and down, showing 500 feet, now 400
feet, now 300 feet, and I felt my heart sinking as I tried to decide what I would
do. It's easy to see how the elements of a bad decision can assemble themselves

out of our idle musings: I don't have my toothbrush, I need to get home, I'm going to break out of the clouds any minute now, I'll just go a little lower . . . Out of the corner of my eye I saw a flashing light. I looked up from the instruments. The white fog burned my eyes as my strobe light flashed from the tip of my wing. I looked back at the gauges. That flash again. Was it my lights or something on the ground? I looked down at what I hoped was earth, and all at once the mists whirled into a vortex and split like fabric.

It looked like an electric rabbit running away. It vanished into ragged cloud, then reappeared, then ran away again. It happened over and over. The flesh of the beast unraveled around me. I was caught in a tissue of gray for a moment as the lights screamed out their directions to me, pointing, pointing, pointing the way, and then I sprung free of the vapor and came dripping out of the clouds over the sequenced flashing lights of Runway 6 at Kenosha Airport. It was as if the ether itself had sweated out a black cinder, and I rode it down toward the steaming earth.

I was in a kind of dun netherworld of weak light and misting rain that streaked my windshield and teared back toward the sides of the aircraft, as if my pretty blue Cutlass were crying with joy for our good fortune to have arrived at Kenosha, Wisconsin, where we had no business being on a day such as today, where we would do nothing more than drink a cup of coffee and take off again. Oh, joy! What a stupid idea!

I controlled the nearly irresistible urge to push the control wheel over and dive for the runway. The ILS went almost to the ground, and we were meant to follow it all the way down, the better to avoid pointy things sticking up at us. I left the airplane alone, and it did quite nicely without my help, flying all the way past the houses and farm fields and the rainy winter world, where it was obviously warmer by the lake, for this was definitely water on my windshield and not ice. We squatted over the white fixed-distance markers and I pulled back on the yoke to stop our descent rate. The airplane skimmed along above the concrete for a while and then gently touched down.

I'd had a grand plan to make many approaches and practice holding patterns, and it had been all very glorious and elaborate in my mind when I'd conceived it, but I abandoned all that and launched again without delay.

This is the way it goes sometimes: I had flown no more than five minutes out from Kenosha toward home when the clouds parted and a ray of light shone down on me as if I'd become part of a Rousseau painting I once saw in Paris: a scene of 1907, a Wright Flyer just over the river where fishermen were fishing, *Les Pêcheurs à la Ligne*. That dragonfly had barely hatched, and Henri Rousseau caught it on a cotton canvas not unlike the one that covered its wings. He had caught, too, in the tipping, precarious angle of the ship, the reckless, dire, and urgent impulse of the whole thing, which was opaque to so many people, just as a craving for heroin would be opaque to most.

I could see the winterscape below. I could even see in the distant vapors the traces of Wonder Lake on the shores of which Galt Airport lay. There was a hole in the clouds the size of McHenry County, and I flew within the grace of it, looking out the window at the ice-fishing shanties on Wonder Lake.

Closer to the airport, I passed over Art Galt's land, dotted with lakes. There were measureless sources of water here, pushing up from within the soggy earth beneath the ice. About 15,000 years ago, when the glaciers melted, they left a carpeting of gravel and rocks several hundred feet thick all across that area. And in that retreat, low hills of stone were dropped in steaming heaps like so much dung from a dying animal. Now it is possible to climb them and view the whole area from the treetops, just as we do when we lift off in an airplane, that first delicious glimpse when the mantle of the earth drops away and we see what seems like the whole of creation and we are filled with that dangerous illusion of elation and wonder at our own sovereignty.

As I banked into the pattern, I could see people fishing on the frozen eastern half of Lake Vera, a pond Art Galt created with a bulldozer from a natural spring pushing up out of the ground. He named it after his wife. As I banked over the snowy scene, I saw a child fall on the ice, and once again I was struck by my intoxicating perspective. On the ground we rarely see more than a block or so in any direction, and that which our vision encompasses is small, close, discreet, and personal. As soon as we lift away and clear the tops of the trees, our view opens to encompass whole counties full of people and houses, rivers and fires, and all their heroic calamity in a single glance. Here is a riding ring where people do dressage in the summer; and there is a house burning down; and there a deer runs across the open road, barely missing a truck. No wonder Satan sought to tempt Christ by taking him to a high place and showing him the view.

In the office, in response to questions about how it went, I shrugged nonchalantly and said, "Oh, it was all right. Nothing to it. No big deal." But I lied, of course, for effect. It was the deal of the century. I had gone when I didn't have to go, and I had no idea why. It seems that something more powerful than I was had hold of me, and it left me with the gnawing, uneasy sense that although I had come down out of the clouds, the danger had not passed.

Stranger to the Ground

BY RICHARD BACH

(Excerpted from the book *Stranger to the Ground*)

Despite the numerous Richard Bach titles that have followed the publication
of his now-classic *Stranger to the Ground* in 1963, I often find myself returning
to this engaging book with the feeling that it is his best work. Yes, I enjoyed
and appreciated aviation titles like *Nothing by Chance* and *A Gift of Wings* and
fables like *Jonathan Livingston Seagull*. But *Stranger to the Ground* puts me so into
the cockpit that I can practically feel the g-forces, hear the click of switches
being thrown, see the slashes of lightning on the horizon.

This excerpt is merely a taste of what you're served in the entire book,
but stands alone very well as a vivid aviation adventure piece. Allow me to set
the scene: The time is the early 1960s during the Berlin crisis, and tensions in
Europe are high between the United States and its allies and the Soviet Union
and its satellite communist states, in particular East Germany. Like many other
Americans, Bach has been called to active duty with the National Guard and is
assigned to fly a jet fighter over Europe. His mission on this particular night is
to fly his Republic F-84F Thunderstreak from the base at Wethersfield in England to Chaumont Air Base, France. For a fighter pilot, the assignment is a prosaic one—to deliver some important dispatches. There will be no Cold War
shooting action, but there are enemies and dangers in the skies ahead.

"Rhein Control, Air Force Jet Two Niner Four Zero Five, Wies-
baden." The City That Was Not Bombed.

137

Silence. Here we go again. "Rhein Control, Rhein Control; Air Force Jet . . ." I try once. Twice. Three times. There is no answer. I am alone with my instruments, and suddenly aware of my aloneness.

Click around with the radio channel selector under my right glove; perhaps I can talk to Barber Radar. "Barber Radar, Air Force Jet Two Niner Four Zero Five, over." Once. Twice. Three times. Nothing.

A flash in the clouds ahead. The air is still smooth, paving the way. Hold the heading. Hold the altitude.

A decision in my mind. If I were flying this crosscountry just to get myself home tonight, I would turn back now. I still have enough fuel to return to the clear air over Wethersfield. With my transmitter out, I cannot ask for a radar vector through the storms ahead. If it was not for the sack above the machine guns, I would turn back. But it is there, and at Chaumont there is a wing commander who is trusting me to complete my mission. I will continue.

I can use the radiocompass needle to point out the storms, if worst comes to worst I can dodge them by flying between the flashes. But still it is much more comfortable to be a spot of light on someone's radar screen, listening for sure direction about the white blurs that are the most severe cells of a thunderstorm. One more try, although I am certain now that my UHF radio is completely dead. Click click click to 317.5 megacycles. "Moselle Control, Moselle Control, Jet Zero Five." I have no hope. The feeling is justified, for there is no answer from the many-screened room that is Moselle Radar.

Turn back. Forget the wing commander. You will be killed in the storms.

Fear again, and it is exaggerating, as usual. I will not be killed in any storm. Someone else, perhaps, but not me. I have too much flying experience and I fly too strong an airplane to be killed by the weather.

Flash to the right, small flash to the left. A tiny tongue of turbulence licks at my airplane, making the wings rock slightly. No problem. Forty minutes from now I shall be walking across the ramp through the rain to Squadron Operations, Chaumont Air Base. The TACAN is working well, Phalsbourg is 80 miles ahead.

Friends have been killed. Five years ago, Jason Williams, roommate, when he flew into his strafing target.

I was briefing for an afternoon gunnery training mission, sitting on a chair turned backwards with my G-suit legs unzipped and dangling their own way to the wooden floor of the flight shack. I was there, and around the table were three other pilots who would soon be changing into airplanes. Across the room was another flight briefing for an air combat mission.

I was taking a sip of hot chocolate from a paper cup when the training squadron commander walked into the room, G-suit tossed carelessly over one shoulder.

"Anybody briefing for air-to-ground gunnery?"

I nodded over my cup and pointed to my table.

"I'm going to tell you to take it easy and don't get target fixation and don't fly into the ground." He held a narrow strip of paper in his hand. "Student flew into a target on Range Two this morning. Watch your minimum altitude. Take it easy today, OK?"

I nodded again. "Who was it?"

The squadron commander looked at the paper. "Second Lieutenant Jason Williams."

Like a ton of bricks. Second Lieutenant Jason Williams. Willy. My roommate. Willy of the broad smile and the open mind and the many women. Willy who graduated number four in a class of 60 cadets. Willy the only Negro fighter pilot I had ever known. It is funny. And I smiled and set down my cup.

I was amazed at myself. What is so funny about one of my best friends flying into a target on the desert? I should be sad. Dying is a horrible and terrible thing. I must be sad. I must wince, grit my teeth, say, "Oh, no!"

But I cannot keep from smiling. What is so funny? That is one way to hit the target? The '84 always was reluctant to change direction in a dive? The odds against the only Negro fighter pilot in all the USAF gunnery school at this moment flying into the ground? Willy's dead. Look sad. Look shocked. Look astounded. But I cannot keep from smiling because it is all so very funny.

The briefing is done and I walk outside and strap my airplane around me and push the throttle forward and go out to strafe the rocks and lizards on Range Number Three. Range Number Two is closed.

It happened again, a few months later. "Did you hear about Billy Yardley?" I had not heard from Bill since we graduated from cadets. "He flew into the side of a mountain on a weather approach to Aviano." A ringing in my ears. Billy Yardley is dead. And I smile. Again the wicked unreasoning uncontrollable smile. A smile of pride? 'I am a better pilot than Jason Williams and Billy Yardley because I am still alive'? Kenneth Sullivan crashed in a helicopter in Greenland. Sully. A fine man, a quiet man, and he died in a spinning cloud of snow and rotor blades. And I smile.

Somehow I am not mad or insane or warped, for I see it once in a while on the faces of others when they hear the ringing in their ears at the death of a friend. They smile, just a little. They think of a friend that knows now what we have wondered since we were old enough to wonder: what is behind the curtain? What comes after this world? Willy knows it, Bill Yardley knows it, Sully knows it. And I do not. My friends are keeping a secret from me. It is a secret that they know and that they will not tell. It is a game. I will know tonight or tomorrow or next month or next year, but I must not know now. A strange game. A funny game. And I smile.

I can find out in a minute. Any day on the range I can wait two seconds too long in the pullout from the strafing panel. I can deliberately fly at 400 knots into one of the very hard mountains of the French Alps. I can roll the airplane on her back and pull her nose straight down into the ground. The game can be over any time that I want it to be. But there is another game to play that is more interesting, and that is the game of flying airplanes and staying alive. I will one day lose that game and learn the secret of the other; why should I not be patient and play one game at a time? And that is what I do.

We fly our missions every day for weeks that become uneventful months. One day one of us does not come back. Three days ago, a Sunday, I left the pages of manuscript that is this book piled neatly on my desk and left for Squadron Operations to meet a flight briefing time of 1115. The mission before mine on the scheduling board was "Low-level," with aircraft numbers and pilots' names.

391—Slack

541—Ulshafer

Ulshafer came back. Slack didn't.

Before he was driven to Wing Headquarters, Ulshafer told us what he knew. The weather had gone from very good to very bad, quickly. There were hills ahead that stretched into the clouds. The two '84F's decided to break off the mission and return to the clear weather, away from the hills. Slack was in the lead. The weather closed in as they began to turn, and Ulshafer lost sight of his leader in the clouds.

"I've lost you, Don. Meet you on top of the weather."

"Roj."

Ulshafer climbed and Slack began to climb.

The wingman was alone above the clouds, and there was no answer to his radio calls. He came back alone. And he was driven, with the base commander, to Wing Headquarters.

The schedule board changed to:

51–9391—Slack AO 3041248

541—Ulshafer

A map was drawn, with a red square around the place where they had met the weather, southwest of Clemont-Ferrand. The ground elevation there changes from 1,000 feet to a jutting mountain peak at 6,188 feet. They had begun their climb just before the mountain.

We waited in Operations and we looked at our watches. Don Slack has another 10 minutes of fuel, we told ourselves. But we thought of the peak, that before we did not even know existed, and of its 6,188 feet of rock. Don Slack is dead. We call for the search-rescue helicopters, we fret that the ceiling is too low for us to fly out and look for his airplane on the mountainside, we think of all the ways that he could still be alive: down at another airport, with radio failure, bailed out into a village that has no telephone, alone with his parachute in some remote forest. "His fuel is out right now." It doesn't make any difference. We know that Don Slack is dead.

No official word; helicopters still on their way; but the operations sergeant is copying the pertinent information concerning the late Lieutenant Slack's flying time, and the parachute rack next to mine, with its stenciled name, *Slack,* is empty of helmet and parachute and mae west. There is on it only an empty nylon helmet bag, and I look at it for a long time.

I try to remember what I last said to him. I cannot remember. It was something trivial. I think of the times that we would jostle each other as we lifted our bulky flying equipment from the racks at the same time. It got so that one of us would have to flatten himself against a wall locker while the other would lift his gear from the rack.

Don had a family at home, he had just bought a new Renault, waiting now outside the door. But these do not impress me as much as the thought that his helmet and chute and mae west are missing from his rack, and that he is scheduled to fly again this afternoon. What arrogant confidence we have when we apply grease pencil to the scheduling board.

The friend whose parachute has hung so long next to mine has become the first recalled Air National Guard pilot to die in Europe.

A shame, a waste, a pity? The fault of the President? If we had not been recalled to active duty and to Europe, Don Slack would not be twisted against a French mountain peak that stands 6,188 feet high. Mrs. Slack could blame the President.

But if Don was not here with his airplane, and all the rest of the Guard with him, there might well have been many more dead Americans in Europe today. Don died in the defense of his country as surely as did the first of the Minutemen, in 1776. And we all, knowingly, play the game.

Tonight I am making a move in that game, moving my token five squares from Wethersfield to Chaumont. I still do not expect to fly into a thunderstorm, for they are isolated ahead, but there is always one section of my mind that is devoted to caution, that considers the events that could cost me the game. That part of my mind has a throttle in it as controllable as the hard black throttle under my left glove. I can pull the caution almost completely back to *off* during air combat and ground support missions. There, it is the mission over all. The horizon can twist and writhe and disappear, the hills of France can flick beneath my molded plexiglass canopy, can move around my airplane as though they were fixed on a spinning sphere about me. There is but one thing fixed in war and practice for war: the target. Caution plays little part. Caution is thrown to the 400-knot wind over my wings and the game is to stop the other airplane, and to burn the convoy.

When the throttle that controls caution is at its normal position, it is a computer weighing risk against result. I do not normally fly under bridges; the risk is not worth the result. Yet low-level navigation missions, at altitudes of 50 feet, do not offend my sense of caution, for the risk of scratching an airplane is worth the result of training, of learning and gaining experience from navigating at altitudes where I cannot see more than two miles ahead.

Every flight is weighed in the balance. If the risk involved outweighs the result to be gained, I am nervous and on edge. This is not an absolute thing that says one flight is Dangerous and another is Safe, it is completely a mental condition. When I am convinced that the balance is in favor of the result I am not afraid, no matter the mission. Carried to extremes, a perfectly normal flight involving takeoff, circling the air base, and landing is dangerous, if I am not authorized to fly one of the government's airplanes that day.

The airplane that I fly has no key or secret combination for starting; I merely ask the crew chief to plug in an auxiliary power unit and I climb into the cockpit and I start the engine. When the power unit is disconnected and I taxi to the runway, there is no one in the world who can stop me if I am determined to fly, and once I am aloft I am the total master of the path of my airplane. If I desire, I can fly at a 20-foot altitude up the Champs Elysées; there is no way that anyone can stop me. The rules, the regulations, the warnings of dire punishment if I am caught buzzing towns means nothing if I am determined to buzz towns. The only control that others can force upon me is after I have landed, after I am separated from my airplane.

But I have learned that it is more interesting to play the game when I follow the rules; to make an unauthorized flight would be to defy the rules and run a risk entirely out of proportion to the result of one more flight. Such a flight, though possible, is dangerous.

At the other extreme is the world of wartime combat. There is a bridge over the river. The enemy depends upon the bridge to carry supplies to his army that is killing my army. The enemy has fortified his bridge with anti-aircraft guns and antiaircraft missiles and steel cables and barrage balloons and fighter cover. But the bridge, because of its importance, must be destroyed. The result of destroying the bridge is worth the risk of destroying it. The mission is chalked on a green blackboard and the flight is briefed and the bombs and rockets are hung on our airplanes and I start the engine and I take off and I fully intend to destroy the bridge.

In my mind the mission is not a dangerous one; it is one that simply must be done. If I lose the game of staying alive over this bridge, that is just too bad; the bridge is more important than the game.

How slowly it is, though, that we learn of the nature of dying. We form our preconceptions, we make out little fancies of what it is to pass beyond the material, we imagine what it feels like to face death. Every once in a while we actually do face it.

It is a dark night, and I am flying right wing on my flight leader. I wish for a moon, but there is none. Beneath us by some six miles lie cities beginning to sink under a gauzy coverlet of mist. Ahead the mist turns to low fog, and the bright stars dim a fraction in a sheet of high haze. I fly intently on the wing of my leader, who is a pattern of three white lights and one of green. The lights are too bright in the dark night, and surround themselves with brilliant flares of halo that make them painful to watch. I press the microphone button on the throttle. "Go dim on your nav lights, will you, Red Leader?"

"Sure thing."

In a moment the lights are dim, mere smudges of glowing filament that seek more to blend his airplane with the stars than to set it apart from them. His airplane is one of the several whose *dim* is just too dim to fly by. I would rather close my eyes against the glare than fly on a shifting dim constellation moving among the brighter constellations of stars. "Set 'em back to bright, please. Sorry."

"Roj."

It is not really enjoyable to fly like this, for I must always relate that little constellation to the outline of an airplane that I know is there, and fly my own airplane in relation to the mental outline. One light shines on the steel length of a drop tank, and the presence of the drop tank makes it easier to visualize the airplane that I assume is near me in the darkness. If there is one type of flying more difficult than dark-night formation, it is dark-night formation in weather, and the haze thickens at our altitude. I would much rather be on the ground. I would much rather be sitting in a comfortable chair with a pleasant evening sifting by me. But the fact remains that I am sitting in a yellow-handled ejection seat and that before I can feel the comfort of any evening again I must first successfully complete this flight through the night and through whatever weather and difficulties lie ahead. I am not worried, for I have flown many flights in many airplanes, and have not yet damaged an airplane or my desire to fly them.

France Control calls, asking that we change to frequency 355.8. France Control has just introduced me to the face of death. I slide my airplane away from leader's just a little, and divert my attention to turning four separate knobs that will let me listen, on a new frequency, to what they have to say. It takes a moment in the red light to turn the knobs. I look up to see the bright lights of Lead beginning to dim in the haze. I will lose him. Forward on the throttle, catch up with him before he disappears in the mist. Hurry.

Very suddenly in the deceptive mist I am closing too quickly on his wing and his lights are very very bright. Look out, you'll run right into him! He is so helpless as he flies on instruments. He couldn't dodge now if he knew that I would hit him. I slam the throttle back to *idle,* jerk the nose of my airplane up, and roll so that I am upside down, watching the lights of his airplane through the top of my canopy.

Then, very quickly, he is gone. I see my flashlight where it has fallen to the plexiglass over my head, silhouetted by the diffused yellow glow in the low cloud that is a city preparing to sleep on the ground. What an unusual place for a flashlight. I begin the roll to recover to level flight, but I move the stick too quickly, at what has become far too low an airspeed. I am stunned. My airplane is spinning. It snaps around once and the glow is all about me. I look for references, for ground or stars; but there is only the faceless glow. The stick shakes convulsively in my hand and the airplane snaps around again. I do not know whether the airplane is in an erect spin or an inverted spin, I know only that one must never spin a swept-wing aircraft. Not even in broad light and clear day. Instruments. Attitude indicator shows that the spin has stopped, by itself or by my monstrous efforts on the stick and rudder. It shows that the airplane is wings-level inverted; the two little bars of the artificial horizon that always point to the ground are pointing now to the canopy overhead.

I must bail out. I must not stay in an uncontrolled airplane below 10,000 feet. The altimeter is an unwinding blur. I must raise the right armrest, squeeze the trigger, before it is too late.

There is a city beneath me. I promised myself that I would never leave an airplane over a city.

Give it one more chance to recover on instruments, I haven't given the airplane a chance to fly itself out.

The ground must be very close.

There is a strange low roaring in my ears.

Fly the attitude indicator.

Twist the wings level.

Speed brakes out.

I must be very close to the ground, and the ground is not the friend of airplanes that dive into it.

Pull out.

Roaring in my ears. Glow in the cloud around me.

St. Elmo's fire on the windscreen, blue and dancing. The last time I saw St. Elmo's fire was over Albuquerque, last year with Bo Beaven.

Pull out.

Well, I am waiting, death. The ground is very close, for the glow is bright and the roaring is loud. It will come quickly. Will I hear it or will everything just go black? I hold the stick back as hard as I dare—harder would stall the airplane, spin it again.

So this is what dying is like. You find yourself in a situation that has suddenly gone out of control, and you die. And there will be a pile of wreckage and someone will wonder why the pilot didn't eject from his airplane. One must never stay with an uncontrolled airplane below 10,000 feet.

Why do you wait, death? I know I am certain I am convinced that I will hit the ground in a few thousandths of a second. I am tense for the impact. I am not really ready to die, but now that is just too bad. I am shocked and surprised and interested in meeting death. The waiting for the crash is unbearable.

And then I am suddenly alive again.

The airplane is climbing.

I am alive.

The altimeter sweeps through 6,000 feet in a swift rush of a climb. Speed brakes *in*. Full forward with the throttle. I am climbing. Wings level, airspeed a safe 350 knots, the glow is fading below. The accelerometer shows that I pulled seven and a half G's in my recovery from the dive. I didn't feel one of them, even though my G-suit was not plugged into its source of pressured air.

"Red lead, this is Two here; had a little difficulty, climbing back through 10,000 feet . . ."

"TEN THOUSAND FEET?"

"Roger, I'll be up with you in a minute, we can rejoin over Toul TACAN."

Odd. And I was so sure that I would be dead.

The flashes in the dark clouds north of Phalsbourg are more frequent and flicker now from behind my airplane as well as in front of it. They are good indicators of thunderstorm cells, and they do not exactly fit my definition of "scattered." Directly ahead, on course, are three quick bright flashes in a row. Correct 30 degrees left. Alone. Time for twisted thoughts in the back of the mind. "You have to be crazy or just plain stupid to fly into a thunderstorm in an eighty-four F." The words are my words, agreed and illustrated by other pilots who had circumstance force them to fly this airplane through an active storm cell.

The airplane, they say, goes almost completely out of control, and despite the soothing words of the flight handbook, the pilot is relying only on his airplane's inertia to hurl it through and into smooth air beyond the storm.

But still I have no intention of penetrating one of the flickering monsters ahead. And I see that my words were wrong. I face the storms on my course now through a chain of logic that any pilot would have followed. The report called them "scattered," not numerous or continuous. I flew on. There are at least four separate radar-equipped facilities below me capable of calling vectors through the worst cells. I fly on. A single-engine pilot does not predicate his action on what-shall-I-do-if-the-radio-goes-out. The risk of the mission is worth the result of delivering the heavy canvas sack in the gun bay.

Now, neither crazy nor stupid, I am at the last link of the chain: I dodge the storms by the swerving radiocompass needle and the flashes of lightning that I see from the cockpit. The TACAN is not in the least disturbed by my uneasy state of mind. The only thing that matters in the world of its transistorized brain is that we are 061 miles from Phalsbourg, slightly to the left of course. The radiocompass has gone wild, pointing left and right and ahead and behind. Its panic is disconcerting among the level-headed coolness of the other instruments, and my right glove moves its function switch to *off.* Gratefully accepting the sedative, the needle slows, and stops.

Flash to the left, alter course 10 degrees right. Flash behind the right wing, forget about it. Flash-FLASH directly brilliantly ahead and the instrument panel goes featureless and white. There is no dodging this one. Scattered.

The storm, in quick sudden hard cold fury, grips my airplane in its jaws and shakes it as a furious terrier shakes a rat. Right glove is tight on the stick. Instrument panel, shock-mounted, slams into blur. The tin horizon whips from an instant 30-degree left bank to an instant 60-degree right bank. That is not possible. A storm is only air.

Left glove, throttle full forward. My airplane, in slow motion, yaws dully to the left. Right rudder, hard. Like a crash landing on a deep-rutted rock trail. Yaw to the right. My airplane has been drugged, she will not respond. Vicious left rudder.

The power, where is the power? Left glove back, forward again, as far as it will go, as hard as it will go. A shimmering blurred line where the tachometer needle should be. Less than 90 percent rpm at full throttle.

I hear the airplane shaking. I cannot hear the engine. Stick and rudders are useless moving pieces of metal. I cannot control my airplane. But throttle, I need the throttle. What is wrong?

Ice. The intake guide vanes are icing, and the engine is not getting air. I see intake clogged in grey ice. Flash and FLASH the bolt is a brilliant snake of incandescent noon-white sun in the dark. I cannot see. Everything has gone red and I cannot even see the blurred panel. I feel the stick I feel the throttle I cannot see. I have suddenly a ship in the sky, and the storm is breaking it. So quickly. This cannot last. Thunderstorms cannot hurt fighters. I am on my way to Chaumont. Important mission.

Slowly, through the bone-jarring shake of the storm, I can see again. The windscreen is caked with grey ice and bright blue fire. I have never seen the fire so brightly blue. My wings are white. I am heavy with ice and I am falling and the worst part of a thunderstorm is at the lowest altitudes. I cannot take much more of this pounding. White wings, covered in shroud. Right glove grips the stick, for that is what has kept my airplane in the sky for six years. But tonight the airplane is very slow and does not respond, as if she were suddenly very tired and did not care to live. As if her engine had been shut down.

The storm is a wild horse of the desert that has suddenly discovered a monster on its back. It is in a frenzy to rid itself of me, and it strikes with shocks so fast they cannot be seen. I learn a new fact. The ejection seat is not always an escape. Bailout into the storm will be just as fatal as the meeting of earth and airplane, for in the churning air my parachute would be a tangled nylon rag. My airplane and I have been together for a long time, we will stay together now. The decision bolts the ejection seat to the cockpit floor, the *Thunderstreak* and I smash down through the jagged sky as a single dying soul. My arm is heavy on the stick, and tired. It will be good to rest. There is a roaring in my ears, and I feel the hard ground widening about me, falling up to me.

So this is the way it will end. With a violent shuddering of airplane and an unreadable instrument panel; with a smothered engine and heavy white wings. Again the feeling: I am not really ready to end the game. I have told myself that this day would come to meet me, as inevitably as the ground which rushes to meet me now, and yet I think, quickly, of a future lost. It cannot be helped. I am falling through a hard splintering storm with a control stick that is not a control stick. I am a chip in a hurricane a raindrop in a typhoon about to become one with the sea a mass of pieces-to-be a concern of air traffic controllers and air police and gendarmerie and coroners and accident investigators and statisticians and newspaper reporters and a board of officers and a theater commander and a wing commander and a squadron commander and a little circle of friends. I am a knight smashed from his square and thrown to the side of the chessboard.

Tomorrow morning there will be no storm and the sun will be shining on the quiet bits of metal that used to be Air Force Jet Two Niner Four Zero Five.

But at this instant there is a great heavy steel-bladed storm that is battering and crushing me down, out of the sky, and the thing that follows this instant is another just like it.

Altimeter is a blur, airspeed is a blur, vertical speed is a blur, attitude indicator is a quick-rocking blurred luminous line that does not respond to my orders. Any second now, as before, I am tense and waiting. There will be an impact, and blackness and quiet. Far in the back of my mind, behind the calm fear, is curiosity and a patient waiting. And a pride. I am a pilot. I would be a pilot again.

The terrier flings the rat free.

The air is instantly smooth, and soft as layered smoke. Altimeter three thousand feet airspeed one-ninety knots vertical speed four thousand feet per minute down attitude indicator steep right bank heading indicator one seven zero degrees tachometer eighty-three percent rpm at full throttle. Level the white wings. Air is warm. Thudthudthud from the engine as ice tears from guide vanes and splinters into compressor blades. Wide slabs of ice rip from the wings. Half the windscreen is suddenly clear. Faint blue fire on the glass. Power is taking hold: 90 percent on the tachometer . . . thud . . . 91 percent . . . thudthud . . . 96 percent. Airspeed coming up through 240 knots, left turn, climb. Five hundred feet per minute, 700 feet per minute altimeter showing 3,000 feet and climbing I am 50 degrees off course and I don't care attitude indicator showing steady left climbing turn I'm alive the oil pressure is good utility and power hydraulic pressure are good I don't believe it voltmeter and loadmeter showing normal control stick is smooth and steady how strange it is to be alive windscreen is clear thud 99 percent rpm tailpipe temperature is in the green. Flash-FLASH look out to the left look out! Hard turn right I'll never make it through another storm tonight forget the flight plan go north of Phalsbourg 15,000 feet 320 knots flash to the left and behind, faint.

And strangely, the words of an old pilot's song: ". . . for I, am, too young, to die . . ." It is a good feeling, this being alive. Something I haven't appreciated. I have learned again.

Rpm is up to 100 percent. I am climbing, and 20,000 feet is below flash 21,000 feet is below. Blue fire washes across the windscreen as if it did not know that a windscreen is just a collection of broken bits of glass.

What a ridiculous thought. A windscreen is a windscreen, a solid piece of six-ply plate glass, for keeping out the wind and the rain and the ice and a place to look through and a place to shine the gunsight. I will be looking through windscreens for a long time to come.

Why didn't I bail out? Because the seat was bolted to the cockpit floor. No. Because I decided not to bail out into the storm. I should have bailed out. I definitely should have left the airplane. Better to take my chances with a rough descent in a torn chute than certain death in a crash. I should have dropped the external tanks, at least. Would have made the airplane lighter and easier to control. Now, at 32,000 feet, I think of dropping the tanks. Quick thinking.

Flash.

I flew out of the storm, and that is what I wanted to do. I am glad now that I did not drop the tanks; there would have been reports to write and reasons to give. When I walk away from my airplane tonight I will have only one comment to make on the Form One: UHF transmitter and receiver failed during flight. I will be the only person to know that the United States Air Force in Europe came within a few seconds of losing an airplane.

Flashflash. Ahead.

I have had enough storm-flying for one night. Throttle to 100 percent and climb. I will fly over the weather for the rest of the way home; there will be one cog slipping tonight in the European Air Traffic Control System, above the weather near Phalsbourg. The cog has earned it.

On Extended Wings

BY DIANE ACKERMAN

(Excerpted from the book *On Extended Wings*)

For every experienced pilot, whether their licenses are long expired or they are still flying as pilot in command of the latest and greatest aircraft in the skies, the weeks and months—even years, sometimes—that they are training for their original pilot's license are hallmark events in their lives. Meeting the challenges and the sheer exhilaration of learning to fly in light airplanes is an experience destined to stoke up the memory bank with a lifetime of worthy images—"instant replays" as they say in television football. Eventually, if all goes well and one is either skilled or tenacious enough, or a combination of all three, a test instructor will sign off a student, often with this comment, repeated on airport ramps all over the United States and no doubt around the world: "Congratulations. You've earned your pilot's license. Now you can *really* learn to fly."

And they mean every word. For learning to fly will become a lifetime experience.

The journey to obtain a pilot's license is one that can inspire the finest prose from many good writers. But when that journey is described by a writer of the experience and talent of Diane Ackerman, the reader can expect to soar to new heights of enjoyment.

Diane Ackerman's flying lessons take place at a small airport with a wreck-hungry patch of forest called "Sapsucker Woods" directly at the end of the runway. If you've ever flown a plane, or only dreamed about it, you will find that riding with Diane Ackerman in the cockpit will bring you images and emotions as keen as any you have ever garnered from printed pages.

D riving to the airport, I watch a plane skid through a 90-degree turn in the distance, loop twice, and grease across the sky. Then it roars into a hairpin climb, loops, and pulls out into a row of tight level rolls, screeches through another 90-degree turn, and does a fast diving run over the terminal, climbing out at the last moment into a series of tight spirals that continues the full length of the runway. Poleskie is out of his mind; his eyes must be marbles rolling around the cabin, his face must be flattened dough from the *g*'s he's pulling, and why doesn't he black out?

Doubling back at speed, the plane pivots around one wing like a gyroscope, crosses and recrosses the sky in front of me, with turns sudden enough to rip its wings off. We think like that, losing wings as an insect might, but the wings would probably keep flying for a while; it's the rest of the plane that would drop. Anyway, this is definitely it: Poleskie has gone bonkers at last.

Near the U.S. Air terminal, my vision jumps, and I suddenly realize how duped I've been. It isn't a real plane at a distance, but a model remote-controlled plane closer in. The Flying Club (someone has removed the "F" from their hangar sign, so that it now reads "LYING CLUB, Gov. Approved") is having a "Flyin' breakfast," as they frequently do, to raise money, boost morale, and draw new members. So the skies will be full of high-winged Cessnas giving visitors brief rides, and homebuilt or restored planes from the area, and perhaps even an ultralight or two. The air will be mobbed.

"It's too busy here today," Brad says, as he arrives in shorts, sipping at a can of root beer, rather than anything with caffeine. Small planes have no bathrooms (although, of course, men can improvise easier than women can). The cooler in the line shack has only fruit juices. *Breathtakingly humid all week,* the weather report said, and so far it's right, as both of our flushed, sweat-pouring faces show.

"We can do ground school," he says, "or we can go to Cortland."

Cortland? Of course, planes go places.

"Cortland!" I answer, and that sends us back indoors for a sectional map of New York State, and my first lesson in navigation.

A sectional map is a little like a two-dimensional version of the paste-and-water maps children make for school. The Adirondacks rise high and coffee-colored from the Sherwood-green forests and pale blue lakes. But there are also dark blue and magenta zones, delicately stippled, to show the limits of controlled airspace, unbroken magenta circles for terminal radar areas, unbroken blue lines for terminal control areas, magnetic compass roses in blue over some cities (such as Ithaca), and a stunning array of information about anything a pilot might conceivably use for navigation. The obvious: airports, in either blue (tower available) or magenta (no tower), showing which direction the runways run, how long they are, whether they're lighted or not, the elevation of the airfield, and what radio frequency is in use for both tower talk and, after hours, Unicom communication with other flyers in the area. Less obvious is the notion of site, not just the landforms, streams, lakes, and rivers, but forest towers, gravel pits, race tracks, glider areas, drag strips, saw mills (what does a saw mill look like from the air?), country clubs, ski areas, and, my favorite, twin drive-in theaters east of Utica, which look like two breasts, nipples pointing southeast and southwest. Power lines are also marked, and changes in elevation every seven miles or so. There is too much for any one eye gulp to capture, but such a palimpsest isn't meant for quick reference, unless, certain of a route or abysmally lost, one is hunting for special landmarks—an orchard or highway, perhaps, or something as large as the Hudson River. Otherwise, it's meant to be pored over, by someone less amazed than I to find towns called Sky Hook, Airy Acres, Jewett, Graphite, and Severance. I don't know what sort of people live in Zoar or Beckett or Canaan, but the itch to zoom over and nab a local for study, as aliens do in so many sci-fi stories, is mighty strong. Less than comforting is how many mistakes I find. Minor mistakes in radio frequency, spelling, services; three in less than five minutes' study is quite a lot, and you soon learn to use the map as what it is, a relic commemorating life as it was in Zoar only when the map was printed.

"As the crow flies" is no longer a tongue-in-cheek measure of the impossible we approximate by road. Our plane flies the birdways, and Cortland is only ten flight minutes away, up Route 13, which lies below us like a freshly ironed ribbon. The airport runway is 2,000 feet shorter than Ithaca's, so we enter the traffic pattern around it with a steep, slow descent in mind. With flaps on, the plane's nose tilts deep at the ground, and my hand is uneven and cold, an icepack on the stick, worrying it back and forth to an approximation of 70 knots, which I can't seem to find. 80 knots; 60 knots; too fast; dangerously slow. With a runway this short, you can't afford to float along it, but must land decisively right on the runway's threshold, or go round and try it again. There are no second chances if you land with half the brief runway behind you. My airspeed staggers all over the place, and I come in too fast, but, worst of all, at a ridiculous angle almost perpendicular to the center line. Touchdown, and we skid sideways on the wheels. I can hear the rubber screaming, as Brad curses in detail for the first time since I've known him. Quickly, he centers and straightens the nose. *Priorities, Diane! Land hard, land fast, drive it into the ground, but for God's sake don't land off the runway! Do you know how much damage we just did to the wheels? Do you know how fast we're landing? WHAT DO YOU THINK*

HAPPENS IF YOU SKID OFF THE RUNWAY AT SEVENTY MILES AN HOUR? If you don't do anything else right, you've got to keep the nose parallel to the center line!

Climb out to 1,800 feet, a left turn back into the traffic pattern, my hand gripping the stick never to let go. When I hunt for 70 knots, leaning stiffly forward to scout the skies and runway, Brad puts a hand gently against my shoulder, says quietly, almost tenderly, *Sit back, relax. It's fun to fly, remember?* Breath leaves my mouth before I realize that I've sighed. I slump back into the seat, and my hand magically relaxes, finds the right airspeed. It's so easy, so easy. The runway rises below me, and with a paradoxical all-out effort I am concentrating only on being relaxed, not thinking about it, but releasing my body to the deepest calm I can imagine, and there is no problem with the stick. It doesn't bumble in my hand, it doesn't yank at my wrist, it moves by finger-light touch so delicate it doesn't seem to move at all. *Perfect approach,* Brad says. *Keep it coming down. Beautiful. Beautiful. Are you too high or too low?* Suddenly I see what I didn't see before, that I'm a little too high, and reach down for the flaps already on at two notches, still concentrating on the calm that will keep my hands steady. Then something goes wrong. The plane sinks a fast 50 feet, an elevator car dropping. Brad grabs control to steal us out of danger. I had done the exact opposite of what I meant to: took off the flaps I already had on instead of adding another notch. I had caged my mind on one thought—the airspeed, relaxing—and left no nomadic problem-solving part of it free to travel across the wide tundra of accident and improvisation. There is no need for Brad to scold me for such a mistake, so obvious and so stupid; I berate myself. He's remarkably quiet. *You're tired,* he says matter-of-factly, *let's go back to Ithaca.* He pushes me away from the instrument panel that I'm once again leaning stiffly against. *Sit back.* And deep into the seat. *Relax.* Everything falls into place, the controls, the direction. It was fun flying a new airport. Even if I did stall the plane on the taxiway and waggle across the center line every landing. I'd been able to relax a few moments at the wheel, able to savor the touchable sky again.

In the distant shimmer, a long white rectangle looks like white enamel, a sink or a ledge. The airport? My eyes fix it, try to read its contours. But, no, it is a building at the edge of the airport, whose runway jumps into focus dead in front of me. I *didn't* see it. The eyes are so stubborn about traveling around the landscape. They want to cling to something safe, not prowl. Late in the afternoon, the sky is still full of visiting aircraft. But I don't hear the voice of a woman in a Cessna, who flies every evening at twilight into our airport for touch-and-goes. I've come to recognize her careful, fastidious voice on the radio as she announces that she's a Cessna southbound from Cortland,

and then states her position at each leg of the traffic rectangle, in a voice so fabricatedly calm, deliberate, unconfusing that I can almost see her neat clothing and hairdo, guess what she does for a living (high-skill, but essentially routine profession; a lawyer, perhaps, schoolteacher, or GP). Her clarity is almost an affectation, a self-definition. We are not the only creatures declaring their whereabouts in the sky. Below us, Sapsucker Woods is full of self-announcement, by frog, squirrel, chipmunk, and bird, each declaiming its bird- or frogness with sound. Bats in Peru have learned to hunt the plumpest frogs by the fullness of their croak. A vast song means death in Peruvian marshes, but not here, where haze is a mainstay of each evening, and scouting the skies for a small plane flying edge-on is like looking at Saturn when its ring plane is an exact tally line midplanet: invisible wide ribbons of snowball and rock.

A small low-winged training plane enters the airport traffic to my right, but I cannot tell whether it's coming toward us or going away from us. Even in daylight. *Where are you going?* Brad asks in disbelief as I turn the wrong direction for the traffic pattern, an invisible shoe box over the airport. *Which runway did you leave from?* he says acidly. I laugh to myself, knowing I've done this before in a car—turned off a highway into a gas station, and left in the wrong direction. A steep turn brings me back into the right pattern, and I land surprisingly well. Brad is appreciative, generous even, until the tower requests that I use the first taxiway, one rushing close to me, and, slamming brakes on, I skid around a corner. The wheels have had a rough afternoon, but I haven't. Brad has managed to straitjacket his temper today, even worked at relaxing me; it can only be a new strategy, one I've been pleading for. I feel my confidence climbing back again, and the most delicious excitement that makes me pace and twitch from rekindled thrill as we plan Tuesday's cross-country flight clear to Teterboro Airport.

By nightfall, the sky is a thick, dark curd; thunderstorms are predicted for early morning, with poor visibility all the way to Teterboro Airport, where I'm hoping to fly on my first cross-country. Six times in the night I wake, worrying about the weather, and then a seventh, when I give up being anxious, and lie awake thinking about weather in general.

Clouds are such strange apparitions to be staggering hugely across the sky, making weather mischief, foretelling calms, taken equally for granted by Innuit and Quechua. Look up into the blue awning, and they are almost always there, above the rain forests, above the mesas, above the store in Secaucus, New Jersey, called Hubcap World, above the downed RAF tail gunner in World War

II (who fell from his plane into deep snow and lived, to be discovered by a mystified German patrol), above the skies in which I fly, and sometimes below them. What an odd predicament for matter to get into. Jagged ice looking soft, tiny motes looming large, all of it sea-changing from fireball to thunderhead to diced ice-crystal smears iridescent as live fish. Variety is the pledge that matter makes to itself to try all the spindrift possibilities of a form. Take the idea of *ice crystal,* rotate it a thousand ways, and float it in the wide blue banner overhead.

Last February, milk bottles were exploding on doorsteps in England, the North Atlantic began to freeze, and, in Pittsburgh, subzero winds sounded like incoming jets, as they filled the streets with snow-djinns and mauled uncovered flesh. *Dangerously cold,* the weather people warned, *Don't go outside.* One evening I watched six cars, lobotomized by cold during the two-hour hockey match, hulk lifeless in the parking lot across the street from my apartment, then vanish from the landscape, whited over. Only the streets were visible: black macadam dashed with white lane dividers and zebra crossings, patterns we've all seen before, some in dreams: the timekeeping on antler bones of Cro-Magnon hunters, people into whose facemasks I've stared in museums, startled to think that we've not always been here, certainly not during the Riviera of the world when time was sweat and food the landscape. Only in the fiercest ice age did we thrive, inventing cold-defying arts like sewing and fire tending. We haven't always been here, but we've always been part of wintry unrest, always honed to iciest craft by ice. I keep trying to capture it, the quarry in my cells, the time before time was chronic, when fire was something living that could only be captured, art was a potion, and mulled by the harsh stratagems of winter, we drew ibex, fish, and flower on cave walls, to reconjure the spring, each hair luminously recalled, every fin exact, every gesture perfect.

At 7:30 A.M. Brad phones to say that thunderstorms all the way south have turned the sky into a pinball game. Even if we could fly, we wouldn't see any landmarks below; it would all be on instruments. Instead, I take U.S. Air into Kennedy, crabby and impatient, a backseat driver. A day later, he and his girl friend, Melissa, who have flown the training plane down for a jaunt, meet me in Manhattan, so I can fly back from Teterboro, a short cab ride across the Hudson River and one of the best kept secrets in New York. Not until you actually turn down the airport road is there a sign to tell you where the busy airport lies, packed with private jets and props, and charter planes available at close to standard carrier price, but at a less-congested, less-delay-ridden, easier-to-reach airport. In the front seat, next to the cab driver, Brad directs which routes to take through rush-hour traffic. He is all flying instructor.

"Turn right just after the tunnel, and follow that until you come to the junction with one and nine," he says in a tone that was lately saying *Right, take it up to three thousand feet and level off at a hundred and ten.* "Right, right, keep it coming RIGHT . . .," he says to the driver, an old unflappable man wearing a peaked cap. "Keep it going right, I said, RIGHT." There are no controls for him to seize. In the backseat, Melissa and I are laughing as Brad instructs, revises, chides, and urges, clear out to Teterboro. Cowed, the driver does everything he's commanded to.

Nothing like the congested aerial Calcutta of Kennedy or La Guardia, Teterboro traffic is still thickly abuzz. A conga line of planes inches along the taxiway until we are only number six in line, creeping up close behind a twin-engine prop, to use some of its wash to cool down our overheating engine. Brad does all the radio, because the electronic jabber is relentless. The air traffic controller works constantly; his voice is the white glove of a traffic cop on a busy New York street corner, windmilling, sweeping, bringing things to a halt, then ushering a car through, halting again, and directing cross traffic.

"Uh, Cessna, wag your wings," the voice asks. We look to the sky down-wind, where two planes are nose to rump at 1,000 feet. One of them wags its wings. "Right. Cessna, turn base, you're number two behind the Aztec . . . eight-seven Zulu, you're cleared to land . . . Number one, on six, what's your sign?"

Brad punches in the button on the mike. "Seven-one-four," he says.

"Warrior seven-one-four, taxi into position and hold," the tower continues, "Tomahawk five-niner Sierra, taxi to six . . ."

We take to the runway, and wait, feet on the brakes, a long sprawl of concrete in front of us, whose middle a light, guppy-shaped airplane skitters across, from one taxiway to another. The Tomahawk. Fifteen minutes later, we are still waiting, thighs twitching from the strain of holding the brakes, as planes land and taxi across our runway. Oil temperature has risen steeply. Soon we'll have to turn the engine off, to cool down. The radio chat is relentless, full of inflection and reply, a numerical catechism. But the numbers change quickly, and even the air traffic controller forgets which plane is number one to take off. What stays constant is the rhythm, the inquiry and reply, the short clauses, the reverse synecdoche of words that stand for letters: Alpha, Bravo, Charlie, Delta, Echo, Foxtrot, Golf, Hotel, India, Juliett, Kilo, Mike, November, Oscar, Papa, Quebec, Romeo, Sierra, Tango, Uniform, Victory, Whiskey, X-Ray, Yankee, Zulu. It's a strange miscellany of words for piloting by. Alpha suggests the waves, electronic markers of mental calm. Echo, a form of location by sub-marine, porpoise, or bat. Juliett is spelled wrong, though Romeo is right. Delta, India, Uniform, Victory, Yankee, and Zulu all smack of British and American imperialism. Foxtrot and Tango say more about when the list came into fashion than the others do. Golf is the puzzler. Why not Garnet, Goldfish, Garbanzo (too ethnic), Gargoyle, Galaxy, Galley, or Garbo? Golf seems such a yawn amid foxtrots and zulus. And isn't Sierra a confusing word when one is over the Sierra Mountains?

"Teterboro Tower, Focke-Wulf one-six-seven-five Quebec, five miles southeast, with X-Ray." Brad and I look at each other. There is a rebuilt German World War Two trainer in the sky somewhere over Queens, and "with X-Ray" means it has listened to a special channel of airport information, and thus knows the runway in use, weather conditions, and other ephemera the tower won't have to repeat. In the distant heat mirage, another three planes taxi across the runway on command from the tower. Two planes land, minutes apart. The cockpit temperature rises into the high eighties.

"Seven-one-four cleared for takeoff," we hear, casual as a wave through.

"Let's go!" Brad says, and I slide the throttle full open for takeoff, keeping low to the ground to pick up cooling-off speed before we climb out.

Rising high over the runway, we are higher still over nothing but housing developments. I've only ever seen them from above in Monopoly sets, or from the sound-stagelike cabin of a large airplane, and my pulse starts to run. What if we crash now; imagine the deaths. There is no place to land safely in an emergency, no field or empty road or forest. The best we could do is look for a wide truck going around 45, our stalling speed, and try to land right on top of it. A radio tower off to our left and close at eye level. Steer clear of it! For what seems the length of a movie, small houses set in even perpendicular rows pour underneath us, the same house side by side, street by street, mile by mile.

"Keep your eyes moving around the sky, for other planes," Brad cautions. "It's a busy airport. Keep looking. Check your speed, check your altitude, watch out for that radio tower, swing good and wide of it. Climb out at ninety knots, I said, to cool down. Ninety! Not eighty. You're acting like a trained monkey! Ninety!"

Then the city disappears into green felt, forest, and farmland, carpet and wood floor. We pull out the map, and my eyes stagger over it in small doses, then back to airspeed and altitude and heading, but already I've lost track of the site I found on the map, finding it again too slowly to get back to the instruments. Like separate stalks of broccoli, trees are bushy-topped, packed tight around the calm rigor of plowed fields, or sprawling inexhaustibly lush to eye's limit, more like a carpet of moss than a forest. Martin Heidegger saw in trees the slender daily link between earth and sky, carrying our eyes always from the mundane to the ethereal and back again. Green cathedrals. The limits of growth. That all changes when you view trees from above in summer, barkless, branchless, unaspiring, indecipherably maple, ash, or hickory, but only a deep-piled green or brown-green fur, as if the earth had a hide thick as a sheep's or wild pony's. The undulation amazes you, hills swelling close, touchably present, then plunging into what can only be a valley. A *valley,* you say to yourself, because it is the first time in your life you have ever fully understood it. If you were asked to define one, this is what you would remember, not a gouge in the land, but a falling away, a gentle pocket between two sloping hills or stiff, pluckable mountains.

"What's our next landmark?" Brad asks.

I consult the map. "A river."

"Where do rivers occur?"

"In valleys."

"Right. Where are you likely to find a valley?"

"How about over that mountain?" I point to a long, reclining figure in the distance.

"Sounds good. Let's look." The reclining figure becomes a rumple of clothes on a bed, flowing toward us, passing under us, and then a valley appears, through it a river snaking from one edge of vision to the other, disappearing behind hills, reemerging onto flatlands as bits of wire. How easy it would be to overlook. Northwest lies a large lake, one of many on the chart, all different sizes, but similar in shape, and I try to decide which one it is.

"If we passed over the transmission lines there, and the island in the river, when we were supposed to be north of it, and there's no city visible to the right, where this one on the map is supposed to be, except that we must be too far west, then is that lake this one to the west, or that one to the northwest, in which case I'm headed wrong to keep on the course we plotted." I fumble the map farther open until part of it blocks my view of the window. Brad pulls some of its accordioned folds back down again.

"Is that a question?" he says. "Look for landmarks. That watch tower, that bend in the river. What is the biggest landmark you've got between here and here?" He frames the latitudes with his hand. Then I see them, lozenges the color of Brownie uniforms, a mountain range. He points toward the sun, reddening up just now over an uneven horizon. Surely we'll have to climb, to breeze over the peaks. Wolfgang Langewische, the grand old man of flying technique, offers this simple rule of eye: if something appears to be moving closer to you, you'll miss it; if it's moving farther away from you, you'll hit it. They are moving closer, those pastilles on the horizon.

"How far away is it?" Brad asks.

"Thirty-five miles?"

He nods. "About that."

Moving fast in the opposite direction, thick puffy clouds pour overhead like time-lapse photography. We are somewhere between the moon and New York City, as the song goes, and if we are on course, Binghamton is just ahead, and a U-shaped section of the Susquehanna River, through which we should pass like an arrow through a valentine heart. Anyway, that's what the map shows. But the thing about a mountain is that it's a visual impasse, too. A range reveals only one valley at a time, unless you are high enough for a wide perspective. Low down, under 5,000 feet, there is a startling feel for how round the world is; one can never see past its gentle bulging. In the distance, a river spirals through the landscape, and I follow its meander through dense settlements to open field and back to settlements again. Floating over its middle, a saddle-shaped basin, I glance at the map, then down below, then at the map again.

That's it. The piece of world the chart maps. I am jumping all around, thunderstruck, trying to touch landscape with my eyes. Brad casts an indulgent glance back at Melissa. He imagines I'm thrilled to have identified the right landmark, to be on course, imagines I'm easily wowed again by a powder puff–like beauty, the blue symmetry of the river winding around the city.

But I am thinking about hydrogen, the simplest atom, consisting of nothing but a nucleus and one electron, or one positive charge and one negative, if you like, the positive 1,800 times as massive as the negative, the electron a mere insinuation of matter, its whereabouts iffy, somewhere within a radius of about 1.15 angstroms from the nucleus, a range scientists like to depict as shadow. Who can say where the electron really travels? You watch it the way you watch a cat run behind a slat fence. I am thinking about how hydrogen gave rise to everything in the Universe: limestone caves sweaty and cool in the summer, gladioli, ocelots, adenoids, jealousy, bombs, pulsars, star-nosed moles, pouting, the fetlock stars by which racehorses are frequently identified, video games, golden-shouldered parakeets, desire. Thinking about the cosmos before the Big Bang, when the Universe was all in one place and solid, a hard local object in an endless ether, which exploded into a prowling, radiant fog of hydrogen and helium, which collapsed into stars, some hot enough in their innards to forge the elements, and into planets, some cool enough to harbor life, in part because one of the elements produced was carbon, a molecular wild card. What a long, dicey, unlikely journey from that tough silky ball of hydro-gen to this sac of chemicals that can contemplate itself, holding now a two-dimensional map in one hand, and translating it into the three-dimensional view far below, as we fly over the planet in a sheath of metal going 125 nauti-cal miles per hour. No, I am not thinking of beauty, but of chance, though they have much in common, and how long ago, in some early chaos of the Uni-verse, my atoms became possible, out of nothing more elaborate than hydrogen and motion. As I look from the flat map to the 3-D world below, there is too much to be said to say anything. If I change course to 104.5 degrees, the angle of bond of oxygen and hydrogen in a water molecule, we'll end up far east, in the Catskill Mountains somewhere, at altitudes higher than I've flown so far, with no airports to put down in when our gas gives out. Anyway, Melissa and Brad are tired, Ithaca is only a few minutes away, and there's no point in chang-ing course so late, with the sun well down, and the light already fading.

From all accounts, this is the monster, the chief pilot with a jackboot for a voice and a temper that will fry you in your seat. Dark-haired, grinning, he leans against the wing, his arms folded, his body one long casual angle. Though it's only noon, his beard is already starting to grow in, and his cut-off blue jeans are frayed into thread haloes just above each knee. His eyes are hard brown lenses of aviator sunglasses—shaped like an insect's, you are thinking—and there is something at once inquisitional and nonchalant about his expres-sion. Everyone has warned you about him: ex-student (*Bob screams like crazy*), new wife (*Bob's a sweet, quiet guy, but when he climbs into a plane he becomes a gestapo*), instructor (*If you think I have a temper, you ought to fly with our chief pilot; now* THAT'S *a temper*), secretary (*I'll just schedule you with one of the "less aggres-sive" instructors*). But with Brad en route to Atlanta for the weekend, if you wish to fly, it will have to be with the man grinning in front of you, as he speaks in a tone willing, ready, but also profoundly bored. He has been here before, arms folded, the chord of the wing pressing across the back of his waist, a student pilot pacing nervously in front of him, has acted out this petty drama so many times, it no longer challenges or ignites him. A plane is landing, and then an-other, a high-winged Cessna, a U.S. Air jet. Bob is the first instructor who doesn't turn to watch, doesn't use it as instruction. The clouds are rumpled and smeared overhead in a strange combination against gunmetal blue; it's how a child would make Christmas angels out of cotton, cardboard, and paste. But he is not noticing the clouds or the hot, dusty breeze, or any other significa of the day, or anything special about what we are doing standing beside the accumulated power of 150 horses three miles northwest of a long lake pointing roughly toward the Bahamas. Brad is always tense and meticulous about ground school and preflight, all attention, all verbal run-up and mind-set. But Bob checks only to see that there's gas and oil, then brushes the air with one hand.

"Let's fly," he says.

"You've got quite a rep," I say, swinging into the left seat.

He laughs. "I'm only aggressive when someone doesn't work," he says, sliding his seat into position. "You've got to experiment. Try something. If it doesn't work, try something else. Try everything else. But keep experimenting, that's the only way you'll learn." He buckles the shoulder harness and relaxes into his seat, crosses his arms and waits, glances outside at the airport parking lot, into the hangar. I am keeping him from his newspaper or trimming his nails. By now the plane is turned on, the engine idling smoothly, and nothing remains but calling the tower. I can see the shadowy angle of his open eyes behind the glasses, as he watches a Bonanza taxi into a nearby bay.

"Shall I call the tower?" I ask.

"Do whatever you want to, I'm just along for the ride," he says, and I tell the tower who we are, and that we'd like to do touch and goes, then we taxi toward runway 14. The tower is all gab today, asking me to wait before I get to taxiway Charlie, so the U.S. Air jet can leave first, then clearing me to taxi up to runway 14 but adding something I can't quite decipher.

"What did he say?" I ask Bob.

Bob shrugs. "Ask him to repeat himself. That's what he's there for."

By now the jet has rushed down most of the runway, leaving a fine black residue in the air behind it. I hold the ridge of the microphone against my upper lip, press in the button with my thumb.

"Ithaca tower, twenty-two Juliett, say again please."

After a moment, the cockpit fills with voice: "Twenty-two Juliett, taxi to one-four, caution wake turbulence."

"Twenty-two Juliett," I reply, to signal my understanding, then ask Bob to make sure: "They're telling me to wait three minutes until the wake clears?"

"When does the wake begin?"

I think of boats flying through the water. "At takeoff."

"Right. You should have been watching for the moment of rotation, and then planning to take off before you reached that spot on the runway."

The tower calls again, nagging. Why aren't I busy taxiing, and do I plan a run-up, to test the engine, before takeoff? I do, press the heart-shaped mike against my mouth, and say so. After run-up, request clearance for take-off, but the tower tells me to wait exactly where I am because a Cessna is on final. A glance left: 300 feet from the runway threshold, a high-winged plane is gliding in slowly. I'd forgotten to look. When it lands, the tower tells me to *taxi into position and hold,* then, once I'm square on the numbers, clears me for takeoff. It seems like it's been a long conversation, all at the level of pouring pebbles.

"Ready?" I ask Bob.

He nods. "Wheel all the way back."

In seconds we are airborne, though only just, inches above the runway and climbing into a low stall, then falling to the runway again and taking off, touching down and taking off. Bob is laughing.

"Think of Yosemite Sam—pointing his six-guns down, shooting at the ground, and saying, 'Dance, pardn'r.' I want you to dance the controls, dance your feet, dance your hand on the wheel. Keep it all in motion until you see what works, keep it fluid."

Opposite the spot on the runway where I wish to land, I put on the fuel pump, and cut off my power entirely, then slow up to the proper airspeed for gliding in.

"Are you too high, too low, or just right?" he asks.

"Don't know yet. I'll wait a second to see."

"Okay. You get a little nervous, huh?"

"Why?" I ask, though the answer would be obvious if I could only see myself leaning stiffly forward in the seat, my mouth open and puffing gently, as if I could blow the plane down to earth like a soap bubble.

"Your hands are shaking like a machine gunner's." He sounds so calm.

I am too high. I put on one, then, two, then three notches of flaps. I am too low, reach for the throttle. His hand gets to mine first, pulling it away. As I remove the flaps I've only just put on, I am still too low, so I reach for the throttle again, and this time he doesn't stop me, as I add a spurt of power, enough to clear the row of red lights fringing the runway. *Why has he not been screaming?* I wonder, just as he begins to talk nonstop.

"Where are your feet? Rudders, rudders! Keep it coming down. Dance your feet. Dance your hand. Keep working on it. Don't give up. Keep working it down. Move your feet!"

Leveled off too high, the plane drops to the runway, then lifts off again; as I ram the nose down, it bounces up again, and I pull the nose up; it falls back down and reels up again.

"You're using a sledge hammer," Bob says. "Use a ballpeen instead." Finally the plane sticks to the runway, but only after a long roller coaster that's undoubtedly left the tower men cackling. "Listen," Bob says quietly, "you can't *force* a plane to land. It can only land itself. *You* can't land it. You have to permit it to dissipate its energy. That's the only way it can land."

I am thinking about poems, and how the best of them land, dissipating their energy completely only by the last line, how you cannot force them into a new attitude in the closing seconds, or try to sum up all the preceding commotion, or leave them hanging, level, at 6 feet off the ground, or plunge them toward a point they should have been drifting toward all along. And I am thinking of the balsa airplane I fly in the backyard, how it swoops and stalls and works its way down to a smooth landing, all by itself, through one excursion and crescendo after another.

"Take off." He sits back and folds his arms again.

We are sprinting down the runway, as I glide the wheel lightly in and out until it grabs hold, and we are up. My feet are dancing, too. How can this combined apoplectic seizure lift 2,000 pounds of metal into the air? It seems like such a twittery act, Obeah, not physics. *Voodoo,* an early plane was called; *Vampire,* another one; only later, in aviation's no-nonsense maturity, were planes named things like *Cruisemaster, Airtourer, Agtruck,* or *Traveller,* and the full run of Indian names, *Seneca, Apache, Cherokee,* etc., but nothing like the original names that tagged the hocus-pocus of flying, the inexplicable marvel behind even the cleanest-boned and most lucid explanation, the mystification that comes only after complete understanding, when you realize that there are questions left that have no inflection at the end, no words like "Why" or "How" or "What" to begin them, but a nuggety amazement at the *thatness* of life, that its forces and processes and forms should be what they are, even if you understand how they are. The rouge of mystery under the whiteface of explanation.

More takeoffs and landings, with Bob speaking only at the outer edges of a mistake, never en route to it, or preventatively. In between, he is cool, nonchalant as a paratrooper sitting on a stool at a drugstore counter. My feet are lazy when I land; I haven't figured out how to tap-dance and juggle at the same time. I am axing the runway to death, instead of letting the plane slice through the air at its own tempo. But he is not haranguing me. At least I am trying, ad lib, to land more smoothly than a dump truck pouring a crash of hot metal onto the runway. *Keep it dancing, keep it moving,* he urges as I come in to land again. And I'm reminded how different he and Brad are as instructors and in attitude, even though they're both obsessed with combating the same horror: passivity. Everything else is acceptable—error, bad judgment, ignorance, lack of skill—everything but passivity. Otherwise, life stops; life is motion. Otherwise chance gives way to fate. Lack of action is for zombies, for life's passengers, not for those the Universe doesn't scare into stupefaction. Their implicit attitude is clear: born astride the grave, as Beckett says, we move from one helpless state to another. But between them runs all the pant and lather of a life, spent mainly responsively and responsibly, or spent mainly in a long dulling wait—waiting to be told what to do, waiting to be buffeted by one breeze after another, waiting limply between two wakes.

The art of flying is overcoming the lure of passivity. It's that simple, but also that chilling. Some pilots, like Brad, see it as a form of agitation: get out and assassinate, form parties, take action. And some, like Bob, are more ontological: just keep attentive, keep moving. But the enemy is only ever the same, with its potion of inertia and mask of quiet. Quiet is a form of trauma; there is no real quiet on earth. But given the hurdy-gurdy raucous spell of life, inexplicable, explosive, jetting away at every angle, the idea of quiet is so riveting; it's tempting just to lie still and wait, mentally camouflaged, for the next command.

PART THREE

The Wings of War

I know that I shall meet my fate
Somewhere among the clouds above;
Those that I fight I do not hate,
Those that I guard I do not love;

Nor law, nor duty bade me fight,
Nor public men, nor cheering crowds,
A lonely impulse of delight
Drove to this tumult in the clouds . . .

—William Butler Yeats

"An Irish Airman Foresees His Death" (excerpt)

Winter's Morning

BY LEN DEIGHTON

(Excerpted from the book *Eleven Declarations of War*)

The first battles to be fought in the skies took place over the trenches that lined the French countryside in World War One. Airplanes with names like Bristol, Fokker, and Clerget flew into history, along with pilots like Manfred von Richthofen (The Red Baron) and Captain Eddie Rickenbacker. Despite Hollywood's portrayal of this kind of fighting as gentlemanly duels, with honor and glory for all, the real thing was something quite different. Most kills—or "victories," as they were euphemistically called—were achieved by a sudden diving ambush made before the other pilot ever spotted your plane. Squaring off in a proper dogfight, with Hollywood-type salutes, was not the stuff from which aces were made. In reality, you spotted the other guy before he spotted you, dived on him, and gunned him down, probably in flames, without a parachute.

The English writer Len Deighton has not only written many widely acclaimed spy thrillers—*Funeral in Berlin, The Ipcress File*—he is also a war historian of the first rank with nonfiction titles such as *The Battle of Britain* and *Airshipwreck*. Although it is fiction, this short story from the collection *Eleven Declarations of War* presents the action in the cockpits of World War One as it no doubt really happened in 1914, long before Hollywood's cameras came along to soften the reality of death in the skies.

Major Richard Winter was a tall man with hard black eyes, a large nose and close-cropped hair. He hated getting out of bed, especially when assigned to dawn patrols on a cold morning. As he always said—and by now the whole Officers' Mess could chant it in unison—"If there must be dawn patrols in winter, let there be no Winter in the dawn patrols."

Winter believed that if they stopped flying them, the enemy would also stop. In 1914, the front-line soldiers of both armies had decided to live and let live for a few weeks. So now, during the coldest weather, some squadrons had allowed the dawn patrol to become a token couple of scouts hurrying over the frosty wire of no-man's-land after breakfast. The warm spirit of humanity that Christmas 1914 conjured had given way to the cold reality of self-preservation. Those wiser squadrons kept the major offensive patrol until last light, when the sun was mellow and the air less turbulent. At St. Antoine Farm airfield, however, dawn patrol was still a grueling obligation that none could escape.

"Oatmeal, toast, eggs and sausage, sir." Like everyone else in the Mess tent—except Winter—the waiter spoke in a soft whisper that befitted the small hours. Winter preferred his normal booming voice. "Just coffee," he said. "But hot, really hot."

"Very good, Major Winter, sir."

The wind blew with enough force to make the canvas flap and roar, as though at any moment the whole tent would blow away. From outside they heard the sound of tent pegs being hammered more firmly into the hard chalky soil.

A young Lieutenant sitting opposite offered his cigarette case, but Winter waved it aside in favour of a dented tin from which he took cheap dark tobacco and a paper to fashion a misshapen cigarette. The young officer did not light one of his own in the hope that he would be invited to share in this ritual. But Winter lit up, blew the noxious smoke across the table, coughed twice and pushed the tin back into his pocket.

Each time someone entered through the flap there was a clatter of canvas and ropes and a gust of cold air, but Winter looked in vain for a triangle of grey sky. The only light came from six acetylene lamps that were placed along the breakfast table. The pump of one of them was faulty; its light was dull and it left a smell of mould on the air. The other lamps hissed loudly and their eerie greenish light shone upon the Mess silver, folded linen and empty plates. The table had been set the previous night for the regular squadron breakfast at 8 a.m., and the Mess servants were anxious that these three early-duty pilots shouldn't disarrange it too much.

Everyone stiffened as they heard the clang of the engine cylinder and con. rod that hung outside for use as a gas warning. Winter laughed when Ginger, the tallest pilot in the squadron, emerged from the darkness rubbing his head and scowling in pain. Ginger walked over to the ancient piano and pulled back the edge of the tarpaulin that protected it from damp. He played a silly melody with one finger.

"Hot coffee, sir." The waiter emphasized the word "hot," and the liquid spluttered as it poured over the metal spout. Winter clamped his cold hands round the pot like a drowning man clinging to flotsam. He twisted his head to see Ginger's watch. Six twenty-five. What a time to be having breakfast: it was still night.

Winter yawned and wrapped his ankle-length fur coat round his legs. New pilots thought that his fur overcoat had earned him the nickname of "the Bear," but that had come months before the coat.

The others kept a few seats between themselves and Winter. They spoke only when he addressed them, and then answered only in brief formalities.

"You flying with me, Lieutenant?"

The young ex-cavalry officer looked around the table. Ginger was munching his bread and jam, and gave no sign of having heard.

"Yes, sir," said the young man.

"How many hours?"

Always the same question. Everyone here was graded solely by flying time, though few cared whether the hours had been spent stunting, fighting or just hiding in the clouds. "Twenty-eight and a half, solo, sir."

"Twenty-eight *and a half*," nodded Winter. "Twenty-eight *and a half!* Solo! Did you hear that, Lieutenant?" The question was addressed to Ginger, who was paying unusually close attention to the sugar bowl. Winter turned back to the new young pilot. "You'd better watch yourself."

Winter divided new pilots into assets and liabilities at either side of seventy hours. Assets sometimes became true friends and close comrades. Assets might even be told your misgivings. The demise of assets could spread grief through the whole Mess. This boy would be dead within a month, Winter decided. He looked at him: handsome, in the pallid, aristocratic manner of such youngsters. His tender skin was chapped by the rain and there were cold sores on his lip. His blond hair was too long for Winter's taste, and his eyebrows girlish. This boy's kit had never known a quartermaster's shelf. It had come from an expensive tailor: a cavalry tunic fashionably nipped in at the waist, tight trousers and boots as supple as velvet. The ensemble was supplemented by accessories from the big department stores. His cigarette case was the sort that, it was advertised, could stop a bullet.

The young man returned Richard Winter's close examination with interest. So this rude fellow, so proud of his chauffeur's fur coat, was the famous Bear Winter who had twenty-nine enemy aircraft to his credit. He was a blotchy-faced devil, with bloodshot eyes and a fierce twitching eyebrow that he sometimes rubbed self-consciously, as if he knew that it undid his carefully contrived aplomb. The youngster wondered whether he would end up looking like this: dirty shirt, long finger-nails, unshaven jaw and a cauliflower-knobbly head, shaved razor-close to avoid lice. Except for his quick eyes and occasional wry smile Winter looked like the archetypal Prussian *Schweinhund.*

Major Richard Winter had been flying in action for nearly two years without a leave. He was a natural pilot who'd flown every type of plane the makers could provide, and some enemy planes too. He could dismantle and assemble an engine as well as any squadron fitter, and as a precaution against jams he personally supervised the loading of every bullet he would use. Why must he be so rude to young pilots who hero-worshipped him, and would follow him to hell itself? And yet that too was part of the legend.

The young officer swallowed. "May I ask, sir, where you bought your magnificent fur coat?"

Winter gulped the rest of his coffee and got to his feet as he heard the first of the scout's engines start. "Came off a mug I shot down in September," he bellowed. "It's from a fashionable shop, I'm told. Never travelled much myself, except here to France." Winter poked his fingers through four holes in the front. Did the boy go a shade paler, or had he imagined it in the glare of the gas lights? "Don't let some smart bastard get your overcoat, sonny."

"No, sir," said the boy. Behind him Ginger grinned. The Bear was behaving true to form. Ginger dug his knife into a tin of butter he'd scrounged from the kitchen and then offered it to the cavalry officer. The boy sniffed the tin doubtfully. It smelled rancid but he scraped a little on to his bread and swamped it with jam to hide the taste.

"This your first patrol?" asked Ginger.

"No, sir. Yesterday one of the chaps took me as far as Cambrai to see the lie of the land. Before that I did a few hours around the aerodrome here. These scouts are new to me."

"Did you see anything at Cambrai yesterday?"

"Anti-aircraft gunfire."

Winter interrupted. "Let's see if we can't do better than that for you today, sonny." He leaned close to the boy and asked in his most winning voice, "Think you could down a couple before lunch?"

The boy didn't answer. Winter winked at Ginger and buttoned his fur coat. The other motors had started, so Winter shouted, "That's it, sonny. Don't try to be a hero. Don't try to be an ace in the first week you're out here. Just keep under my stinking armpit. Just keep close. Close, you understand? Bloody damn close." Winter flicked his cigarette end on to the canvas floor of the tent and put his heel on it. He coughed and growled, "Hurry up," although he could see that the others were waiting for him.

From the far side of the wind-swept tarmac, Major Winter's Sergeant fitter saw a flash of greenish light as the Mess tent flap opened and the duty plots emerged. Winter came towards him out of the darkness, walking slowly because of his thick woollen underwear and thigh-length fleece boots. His hands were tucked into his sleeves for warmth, and his head was sunk into the high collar that stood up around his ears like a cowl. Exactly like a monk, thought the Sergeant, not for the first time. Perhaps Winter cultivated this resemblance. He'd outlived all the pilots who had been here when he arrived, to become as high in rank as scout pilots ever became. Yet his moody introspective manner and his off-hand attitude to high and low had prevented him from becoming the commanding officer. So Winter remained a taciturn misanthrope, without any close companions, except for Ginger who had the same skills of survival and responded equally coldly to overtures of friendship from younger pilots.

The Sergeant fitter—Pops—had been here even longer than the Bear. He'd always looked after his aeroplane, right from his first patrol when Winter was the same sort of noisy friendly fool as the kid doing his first patrol this morning. Aeroplanes, he should have said: the Bear had written off seven of them. Pops spat as the fumes from the engine collected in his lungs. It was a bad business, watching these kids vanish one by one. Last year it had been considered lucky to touch Pops's bald head before take-off. For twelve months the fitter had refused leave, knowing that the pilots were truly anxious about their joke. But Pops's bald head had proved as fallible as all the other talismans. One after another the faces had been replaced by similar faces until they were all the same pink-faced smiling boy.

Pops spat again, then cut the motor and climbed out of the cockpit. The other planes were also silent. From the main road came the noise of an army convoy hurrying to get to its destination before daylight made it vulnerable to attack. Any moment now artillery observers would be climbing into the balloons that enabled them to see far across no-man's-land.

"Good morning, Major."

"Morning, Pops."

"The old firm, eh, sir?"

"Yes, you, me and Ginger," said Winter, laughing in a way that he'd not done in the Mess tent. "Sometimes I think we are fighting this war all on our own, Pops."

"We are," chuckled Pops. This was the way the Bear used to laugh. "The rest of them are just part-timers, sir."

"I'm afraid they are, Pops," said Winter. He climbed stiffly into the cramped cockpit and pulled the fur coat round him. There was hardly enough room to move his elbows and the tiny seat creaked under his weight. The instruments were simple: compass, altimeter, speedometer and rev-counter. The workmanship was crude and the finish was hasty, like a toy car put together by a bungling father. "Switches off," said Pops. Winter looked at the brass switches and then pressed them as if not sure of his vision. "Switches off," he said.

"Fuel on," said Pops.

"Fuel on."

"Suck in."

"Suck in."

Pops cuddled the polished wooden prop blade to his ear. It was cold against his face. He walked it round to prime the cylinders. That was the thing Pops liked about Winter: when he said off, you knew it was off. Pops waited while Winter pulled on his close-fitting flying helmet; its fur trimmed a tonsure of leather that had faded to the colour of flesh.

"Contact."

"Contact." Pops stretched high into the dark night and brought the blade down with a graceful sweep of his hands. Like brass and percussion responding to a conductor, the engine began its performance with a blinding sheet of yellow flame and a drum roll. Winter throttled back, slowing the drum and changing the shape and colour of the flame to a gaseous feather of blue that danced around the exhaust pipes and made his face swell and contract as the shadows exploded and died. Winter held blue flickering hand above his head. He felt the wheels lurch forward as the chocks were removed and he dabbed at the rudder bar so that he could see around the aircraft's nose. There was no brake or pitch adjustment and Winter let her gather speed while keeping the tail skid tight down upon the ground.

They took off in a vic three, bumping across frozen ruts in the balding field with only the glare of the exhausts to light their going. It was easy for Winter; as formation leader he relied on the others to watch his engine and formate on him accordingly. At full screaming throttle they climbed over the trees at the south end of the airfield. A gusty crosswind hit them. Winter banked a wing-tip dangerously close to the tree tops rather than slew into the boy's line of flight. Ginger did the same to avoid his Major. The boy, unused to these heavy operational machine with high-compression engines, found his aircraft almost wrenched from his grasp. He yawed across the trees, a hundred yards from the others, before he put her nose up to regain his position in formation. Close, he must keep close. Winter spared him only a brief glance over the shoulder between searching the sombre sky for the minuscule dots of other aeroplanes. For by now the black lid of night had tilted and an orange wedge prised open the eastern horizon. Winter led the way to the front lines, the others tight against his tailplane.

The first light of the sun revealed a land covered by a grey eiderdown of mist, except where a loose thread of river matched the silver of the sky. Over the front line they turned south. Winter glanced eastwards, where the undersides of some low clouds were leaking dribbles of gold paint on to the earth. As the world awakened stoves were lit and villages were marked by dirty smoke that trailed southwards.

Major Winter noted the north wind and glanced back to see Ginger's aeroplane catch the first light of the sun as it bent far enough over the horizon to reach them at fifteen thousand feet above the earth. The propeller blades made a perfect circle of yellow gauze, through which reflections from the polished-metal cowling winked and wavered as the aeroplanes rose and sank gently on the clear morning air.

Here, on the Arras section of the front, the German and French lines could be clearly seen as careless scrawls in the livid chalk. Near the River Scarpe at Feuchy, Winter saw a constant flicker of artillery shells exploding: "the morning hate." Pinheads of pink, only just visible through the mist. Counter-battery fire he guessed, from its concentration some way behind the lines.

He pulled his fur collar as high round his face as it could go, then raised his goggles. The icy wind made his eyes water, but not before he had scanned the entire horizon and banked enough to see below him. He pulled the goggles down again. It was more comfortable, but they acted like blinkers. Already ice had formed in the crevices of his eyes and he felt its pin-pricks like daggers. His nose was numb and he let go of the stick to massage it.

The cavalry officer—Willy, they called him—was staring anxiously at the other two aeroplanes. He probably thought that the banking search was a wing-rocking signal that the enemy was sighted. They read too many cheap magazines, these kids; but then so had Winter before his first posting out here: *Ace of the Black Cross, Flying Dare-Devils, True War Stories.*

Well, now Winter knew true war stories. When old men decided to barter young men for pride and profit, the transaction was called war. It was another Richard Winter who had come to war. An eighteen-year-old child with a scrapbook of cuttings about Blériot and the Wright brothers, a roomful of models which his mother wasn't permitted to dust and thirteen hours of dangerous experiments on contraptions that were bigger, but no more airworthy, than his dusty models. That Richard Winter was long-since dead. Gone was the gangling boy whose only regret about the war was leaving his mongrel dog. Winter smiled as he remembered remonstrating with some pilots who were using fluffy yellow chicks for target practice on the pistol range. That was before he'd seen men burned alive, or, worse, men half-burned alive.

He waved to frightened little Willy who was desperately trying to fly skilfully enough to hold formation on his bad-tempered flight commander. Poor little swine. Two dots almost ahead of them to the south-east. Far below. Ginger had seen them already but the boy wouldn't notice them until they were almost bumping into him. All the new kids were like that. It's not a matter of eyesight, it's a matter of knowledge. Just as a tracker on a safari knows that a wide golden blob in the shadow of a tree at midday is going to be a pride of lions resting after a meal, so in the morning an upright golden blob in the middle of a plain is a cheetah waiting to make a kill. So at five thousand feet, that near the lines, with shellfire visible, they were going to be enemy two-seaters on artillery observation duty. First he must be sure that there wasn't a flight of scouts in ambush above them. He looked at the cumulus and decided that it was too far from the two-seaters to be dangerous. Brownish-black smoke patches appeared around the planes as the anti-aircraft guns went into action.

Winter raised his goggles. Already they had begun to mist up because of the perspiration generated by his excitement. He waggled his wings and began to lose height. He headed east to come round behind them from out of the sun. Ginger loosed off a short burst of fire to be sure his guns were not frozen. Winter and the boy did the same. The altitude had rendered him too deaf to hear it as more than a ticking, as of an anxious pulse.

Winter took another careful look around. Flashes of artillery shells were bursting on the ground just ahead of the enemy planes' track. The ground was still awash with blue gloom, although here and there hillocks and trees were crisply golden in the harsh oblique light of morning. The hedges and buildings threw absurdly long shadows, and a church steeple was bright yellow. Winter now saw that there were four more two-seaters about a mile away. They were beginning to turn.

Winter put down his nose and glanced in his mirror to be sure the others were close behind. The airspeed indicator showed well over a hundred miles an hour and was still rising. The air stream sang across the taut wires with a contented musical note. He held the two aeroplanes steady on his nose, giving the stick and rudder only the lightest of touches as the speed increased their sensitivity.

Five hundred yards: these two still hadn't seen their attackers. The silly bastards were hanging over the side anxious not to get their map references wrong. Four hundred.

The boy saw them much later than Ginger and Winter. He stared in wonder at these foreign aeronauts. At a time when only a handful of madmen had ever tried this truly magical science, and when every flight was a pioneering experiment to discover more about this new world, he hated the idea of killing fellow enthusiasts. He would much rather have exchanged anecdotes and information with them.

Ginger and Winter had no such thoughts. Their minds were delivered to their subconscious. They were checking instruments, cocking guns and judging ever-changing altitudes, range and deflection.

If that stupid kid fires too early . . . damn him, damn him! Oh, well. Ginger and Winter opened fire too. Damn, a real ace gets in close, close, close. They'd both learned that, if nothing else. Stupid boy! The artobs leader pulled back on the stick and turned so steeply as almost to collide with the two-seater to his left. He knew what he was doing; he was determined to make himself a maximum-deflection shot. Winter kept his guns going all the way through the turn. The tracer bullets seemed unnaturally bright because his eyes had become accustomed to the morning's gloom. Like glowworms they were eating the enemy's tailplane. This is what decided a dogfight: vertical turns, tighter and tighter still. Control stick held into the belly, with toes and eyes alert so that the aeroplane doesn't slide an inch out of a turn that glued him to the horizon. It was sheer flying skill. The sun—a watery blob of gold—seemed to drop through his mainplane and on to his engine. Winter could feel the rate of turn by the hardness of his seat. He pulled even harder on the stick to make the tracers crawl along the fuselage. The smell from his guns was acrid and the thin smoke and heat from the blurring breechblocks caused his target to wobble like a jelly. First the observer was hit, then the pilot, throwing up their hands like badly made marionettes. The two-seater stalled, falling suddenly like a dead leaf. Winter rolled. Two more aeroplanes slid across his sights. He pushed his stick forward to follow the damaged two-seater down. Hearing bullets close to his head, he saw the fabric of his upper plane prodded to tatters by invisible fingers which continued their destruction to the point of breaking a centre-section strut and throwing its splinters into his face. His reflexes took over and he went into a vertical turn tighter than any two-seater could manage. Aeroplanes were everywhere. Bright green and blue wings and black crosses passed across his sights, along with roundels and dark green fabric. One of them caught the light of the sun and its wings flashed with brilliant blue. All the time Winter kept an eye upon his rear-view mirror. A two-seater nosed down towards his tail, but Winter avoided him effortlessly. Ginger came under him, thumping his machine-guns with one of the hammers which they all kept in their cockpits. He was red-faced with exertion as he tried to clear the stoppage by force. At this height every movement was exhausting. Ginger wiped his face with the back of his gauntlet and his goggles came unclipped and blew away in the air stream.

Winter had glimpsed Ginger for only a fraction of a second but he'd seen enough to tell him the whole story. If it was a split round he'd never unjam it. Trees flashed under him. The combat had brought them lower and lower, as it always did.

The new boy was half a mile away and climbing. Winter knew it was his job to look after the kid but he'd not leave Ginger with a jammed gun. A plane rushed past before he had a chance to fire. Winter saw one of the two-seaters behind Ginger. My God, they were tough, these fellows. You'd think they'd be away, with their tails between their legs. Hold on, Ginger, here I come. Dive, climb, roll; a perfect Immelmann turn. The world upside down; above him the dark earth, below him the dawn sky like a rasher of streaky bacon. Hold that. He centred the stick, keeping the enemy's huge mainplane centred in his sight. Fire. The guns shook the whole airframe and made a foul stink. He kicked the rudder and slid down past the enemy's tail with no more than six feet to spare. A white-faced observer was frozen in fear. Up. Up. Up. Winter leaned out of his cockpit to see below him. The new boy is in trouble. One of the two-seaters is pasting him. The poor kid is trying for the cloud bank but that's half a mile away. Never throttle back in combat, you fool. White smoke? Radiator steam? No, worse: vaporizing petrol from a punctured tank of fractured lead. If it touches a hot pipe he'll go up like a torch. You should have kept close, sonny. What did I tell you. What do I always tell them. Winter flick-rolled and turned to cover Ginger's tail.

Woof: a flamer. The boy: will he jump or burn? The whole world was made up of jumpers or burners. There were no parachutes for pilots yet, so either way a man died. The machine was breaking up. Burning pieces of fuel-soaked wreckage fell away. It would be difficult to invent a more efficient bonfire. Take thin strips of timber, nail them into a framework, stretch fabric over it and paint it with highly inflammable dope. Into the middle of this build a metal tank for 30 gallons of high-grade fuel. Move air across it at 50 m.p.h. Winter couldn't decide whether the boy had jumped. A pity, the chaps in the Mess always wanted to know that, even though few could bear to ask.

The dogfight had scattered the aeroplanes in every direction, but Ginger was just below him and a two-seater was approaching from the south. Ginger waved. His gun was working. Winter side-slipped down behind a two-seater and gave it a burst of fire. The gunner was probably dead, for no return fire came and the gun rocked uselessly on its mounting. The pilot turned steeply on full throttle and kept going in an effort to come round in a vertical turn to Winter's rear. But Ginger was waiting for that. They'd been through this many times. Ginger fired as the two-seater was half-way through the turn, raking it from engine to tail. The whole aeroplane lurched drunkenly, and then the port mainplane snapped, its main spar eaten through by Ginger's bullets. As it fell, nose-down, the wings folded back along the fuselage like an umbrella being closed. The shapeless mess of broken struts and tangled steel wire fell vertically to earth, weighted by its heavy engine which was still roaring at full throttle. It was so low that it hit the ground within seconds.

Winter throttled back and came round in a gentle turn to see the wreckage: not a movement. It was just a heap of junk in a field. Ginger was circling it, too. From this height the sky was a vast bowl as smooth and shiny as Ming. They both looked round it but the other two-seaters had gone. There were no planes in sight. Winter increased his throttle and came alongside Ginger. He pushed his goggles up. Ginger was laughing. The artillery fire had stopped, or perhaps its explosions were lost in the mist. They turned for home, scampering across the trees and hedges like two schoolboys.

Winter and Ginger came over the airfield in echelon. Eight aeroplanes were lined up outside the canvas hangars that lacked only bunting to be a circus. A dozen officers fell over themselves scrambling out of the Mess tent. One of them waved. Winter's machine, painted bright green with wasp-like white bands, was easily recognized. Winter circled the field while Ginger landed. He'd literally lived in this French field for almost a year and knew each tree, ditch and bump. He'd seen it from every possible angle. He remembered praying for a sight of it with a dead motor and a bootful of blood. Also how he'd focused on blurred blades of its cold dewy grass, following a long night unconscious after a squadron booze-up. He'd vomited, excreted, crashed and fornicated on this field. He couldn't imagine being anywhere else.

For the first time in a month the sun shone, but it gave no warmth. As he switched off his engine the petrol fumes made the trees bend and dance on the heavy vapour. Pops hurried across to him but couldn't resist a quick inspection of the tail before saluting.

"Everything in order, Herr Major?"

Winter was still a little deaf but he guessed what the Sergeant was saying. He always said the same thing. "Yes, Sergeant. The strut is damaged but apart from that it probably just needs a few patches."

Winter unclipped his goggles, unwound his scarf and took off his leather helmet. The cordite deposits from his Spandaus had made a black band across his nose and cheeks.

"Another Englishman?" said Pops. He warmed his hands before the big Mercedes engine, which was groaning softly.

"Bristols: one forced down, one destroyed. We lost the new young officer, though." Winter was ashamed that he didn't know the boy's name, but there were so many of them. He knew he was right to remain unfriendly to all of them. Given half a chance new kids would treat him like some sort of divinity, and that made him feel like hell when they went west.

Winter wiped the protective grease from his face. He was calm. Briefly he watched his own unshaking hand with a nod of satisfaction. He knew himself to be a nerveless and relentless killer, and like any professional assassin he took pride in seeing a victim die. Only such men could become aces.

Piece of Cake

BY DEREK ROBINSON

(Excerpted from the novel *Piece of Cake*)

Later, when what was called "The Battle of Britain" was over, although the war with the Germans raged on, Prime Minister Winston Churchill would say: "Never in the field of human conflict was so much owed by so many to so few."

He was referring to the Royal Air Force and their heroic defense of the British Isles against Luftwaffe attacks in the summer of 1940. But even before the all-out German assault against England, the RAF, particularly Fighter Command, had been in action in the defense of France. The valiant efforts of the Hurricane pilots against the flyers of the Luftwaffe's Messerschmitts, Heinkels, and other aircraft could not save France, which fell in June. The British squadrons in France suffered grievous losses, and Air Chief Marshal H. C. T. Dowding agonized over every single one because he knew the ultimate battles, in the skies over England, were yet to be fought.

Perhaps the most detailed and vivid in-cockpit descriptions of what the air battles were like for the RAF fighter pilots appear not in a work of nonfiction, but in a novel—Derek Robinson's *Piece of Cake,* published in the United States by Knopf in 1984. *Piece of Cake* is fiction, but its details are authentic—the story of an RAF Hurricane Squadron in the time period from September, 1939, through their posting to France, then back to England at the fall of France in early June, 1940. This is a novel of flying and action that few can match, in my opinion. Now out of print in the United States as I write these words, Derek Robinson's book certainly deserves to be reissued, for it could become an enduring classic.

Derek Robinson would probably be the first to agree with legendary Saint-Exupéry and others who had described dog-fighting as a form of murder in the skies. Beginning in World War One, the real art of dog-fighting was a sudden sneak attack on an enemy, blowing him away before he ever realized you were there. Most "kills" or "victories" were achieved that way in the fights over the trenches of the Marne and the Somme, and were still being done that way as the German armies rolled through France in 1940.

Robinson's prose is unrelenting in capturing the intensity of aerial combat. You will be shaken by what these pilots endured.

The chief characters in the drama presented here—which took place during May of 1940, in some of the fiercest fighting over France—are the Hurricane Squadron's commander, Rex, and the pilots under his command, including an American with the nickname CH3. There's quite a lot of kidding around going on here, attempts at comic relief, I suppose, while pressing on against the odds. It's amazing they could find anything to joke about amidst what they were going through.

Halfway across Belgium, Hornet squadron met the waiting Messerschmitts. They were 109's, high in the sky, so high that the formation was just a speckle of dots.

As soon as he had passed the Battles Rex had reduced speed to stay just a few miles ahead of them. Now he was flying at eight thousand feet where the cloud was thin and he could see above him. The squadron was tucked-up nicely and performing well. That exhibitionist Yank kept zigzagging about at the rear, which meant he would probably run out of fuel on the way home, but otherwise things were going well.

The speckle of dots split in two, and one part fell away. "Hilltop aircraft," Rex called, "bandits at one o'clock, up we go." They climbed towards the enemy. The dots took shape as tiny crosses, the crosses grew tails; light gleamed on cockpits and prop-discs. "Pick your targets," Rex said calmly. *How the hell can I?* Fanny Barton thought. *You're slap in front of me, you great turd.* All the same he chose a 109 at the left rear and hoped it was being flown by a panicky cretin who would stall and pick his nose and get himself shot down before he could . . .

The 109's almost vanished. One moment they were diving, the next they were head-on, knife-edge wings nearly invisible. And then they were sheering off, climbing away. "No stomach," Rex grunted. Barton watched the enemy make height, and let his muscles unclench. He felt weak, and he took a deep breath of oxygen. *Now why did they do that?* he asked himself, and the answer came back: *Because they know where we're going and what we're doing and they want the Battles.*

Three more times they met bands of German fighters. They were at various heights, in various strengths: a dozen 110's, a handful of 109's, a mixed bunch of both, maybe twenty-strong. Each time the enemy wandered over, had a look, and lost interest. "Don't worry, we'll catch them on the way back," Rex said. Cox, at Red Three, glanced unhappily at his leader.

They droned on. There was nothing but a sea of dirty cloud beneath them. Occasionally it split open and revealed an underwater glimpse of a lot of even dirtier cloud. At the tail of the formation, Fitz Fitzgerald realised that he had stopped feeling afraid. Ever since they took off he had been frightened, and whenever the enemy came near he had begun trembling so much that he had to force himself to breathe; but now, suddenly, he seemed to have run out of fear. It didn't make him any happier. Anyone who'd been shot down and wasn't afraid it might happen again must be very stupid.

"Right, Hilltop aircraft, we're there," Rex announced.

Beautiful, Moran thought. *Now turn around and go straight home and don't talk to any strangers.*

They flew a wide circle. "We'll just pop down and have a look," Rex said. Moran glanced sideways at Fitzgerald and threw up a hand in disgust. "The Battle boys might need some help," Rex added. Somebody pressed his transmission switch and blew a raspberry.

The descent through cloud seemed endless. The lower they went the thicker it got; and then abruptly it rose like a theatre curtain and they were in clear air. Eight hundred feet below, a broad band of water cut across the landscape. "Albert Canal!" Rex said triumphantly. "Right: fingers out."

They followed the canal, throttles wide open, exhausts trailing smoke, and saw flak bursting a few miles ahead. The nearer they got to it the more there was, each burst spawning two more, doubling and redoubling until the sky was blotched with blackness, flecked with small white puffs, streaked with red and green and orange. "Holy shit," someone said quietly. "Shut up!" Rex barked. "Radio discipline!"

Baggy Bletchley had been right about one thing. The cloud was a godsend to the Battles. It had hidden them from the enemy fighters. It had also forced them desperately low on their bombing runs. The Hornet pilots could see three Battles at about five hundred feet: slim monoplanes flying straight and level through the barrage like blind men walking down the middle of a busy road.

One exploded. A thick line of flak appeared as if someone had shaken a loaded pen at the sky and the Battle just touched a blot and blew up: a flicker of incandescence that pulverised three men in the time it took to draw breath. Almost at once another plane was hit, and it angled steeply downwards as if seeking out the source of the hurt. The third bomber was on fire. It dropped its bombs and tried to climb away, but though the nose went up the plane did not. Flak raged around it, obsessed with annihilation. Still the Battle slogged on. The Hornet pilots saw its bombs burst in a long row, nowhere near the bridge, and then the plane sank and hit the ground, and the flames claimed it with a rush.

The Hurricanes swept through the dying flak like cavalry fording a stream. Rex led them up through the cloud and headed west. Every machine had been holed. "Close up," Rex said to them. "I can't afford stragglers."

Almost at once, CH3 called him: "Bandits behind, bandits behind, coming down now."

Rex wheeled the squadron. They were still in the spearhead formation and they needed a wide turning-circle. Patterson, at Yellow Three, was looking from Barton's wingtip to Cox's tailplane and back again when the corner of his eye glimpsed a double file of Messerschmitt 110's hurtling at him from the side. Hot flame bubbled out of the leading pair. The planes split left and right. Two more sprang forward, pumping fire, heaved apart, were replaced. Bullet-streams flickered and slashed across the Hurricanes. Patterson felt them rip through his fuselage and he shouted with fear and anger; by then the 110's were gone, the squadron had made its turn, and there in front was a pack of 109's, pouring down, head-on.

Rex immediately hauled back and climbed. Cox and Cattermole went with him and fired when he fired. Yellow Section, chasing too hard, swerved outward to avoid them. Blue Section instinctively followed Yellow. Within a second the squadron had scattered and the plunging 109's were hosing its exposed flanks with fire. For an instant the air was brightly stitched with tracer. Then the 109's sliced past the wallowing Hurricanes and howled off into nowhere. "Regroup!" Rex was shouting. "Hilltop, regroup! Hilltop, regroup!"

It was a struggle. They were all over the sky. Barton was undamaged but Patterson was not: control cables hit, probably: the plane wouldn't do what he told it, needed continuous full opposite rudder to stop it swinging left. Gordon couldn't see through his bullet-crazed windscreen and his cockpit was filling with black-green smoke, until he got the canopy open. Moran was all right apart from a perforated wing that vibrated during turns. Fitzgerald's propeller was making a noise like a rusty saw hitting a rusty nail: it had probably stopped a couple of bullets. CH3 had seen the enemy coming and climbed above them. Lloyd too was intact.

The man in real trouble was Moke Miller. He had no fingers on his left hand.

The burst from the 109 that chopped through his knuckles also hit him in both thighs, missing the bone but tearing great holes in the muscle. At the same time, a spent bullet ricocheted around the cockpit and smashed into his mouth. This hurt most of all. The agony of torn lips and tongue and broken teeth was too great and he blacked out. At once his mouth filled with blood and he began to choke. The choking brought him back to consciousness: he coughed and spat, and lifted his left hand to wipe off the mess. But the glove seemed to be hinged. It flapped open the wrong way. As he looked, blood ran out in a brilliant red stream. Pain raged through the hand, a flame that kept flaring bigger and hotter until once again his brain rejected it, his vision fogged, his ears went deaf, and the cockpit receded as if he were falling backwards.

When he could see again, the horizon was moving rapidly from left to right. There seemed to be no end to this rotation.

After a while he noticed that the Hurricane was flying a continuous bank. It was therefore making a circle. He looked down and saw his feet on the rudder-pedals and his right hand gripping the control column. That explained everything. He moved his feet and screamed at the pain in his legs. The Hurricane levelled out. He was sick, vomit and blood and bits of broken teeth slopping down his front. Needle-sharp sparks of light danced furiously before his eyes. His ears were full of a loud beehive buzz. The dancing lights faded to soft purple blooms and the buzz died and a voice was speaking to him.

". . . any damage? How's your radio? Blue Leader to Blue Three, over."

Miller looked out and saw Flip Moran flying alongside. He tried to speak but his mouth was too broken so he shook his head instead. Gobs of blood flew off and stained the Perspex.

"Are you hit, Blue Three?"

Miller showed him his left hand. That was a mistake. The hand burned like a furnace. The needle-sparks rushed back to the dance and the beehive-buzz surged.

"Stay with me, Blue Three. We'll see you home."

The squadron regrouped and flew west. Miller did very well: he switched on his oxygen, he tucked his left hand under his chin so as to cut down the blood flow, he coughed out most of the debris, he even got rid of his goggles, which were spattered with blood. For twenty miles he tagged along at the rear, seeing nothing but Moran's plane to his right. Then CH3 called: "Bandits behind, bandits behind. 109's coming down."

Rex wheeled the squadron again, but Miller flew on. Violent tactics like that were beyond him now. Moran hesitated, then turned and joined him. He gestured downwards: down where the cloud offered cover. Miller fed this information into his stumbling, fumbling brain. Eventually his good right arm responded. They went down.

This time Rex managed to face the enemy fighters before they could strike. The diving 109's still looked no bigger than skylarks. Fanny Barton marvelled at CH3's eyesight as he braced himself for another split-second whizz-bang head-on attack. It never came. The Messerschmitts curled away and went into a wide orbit—all except one, which pulled up to a climbing stall, toppled sideways, and went down again in a slow, flat spiral, trailing smoke.

"Decoy," Rex said. "Old trick."

Hornet squadron held height and held formation and watched the decoy, while the other 109's orbited and watched them. Eventually the decoy gave up and climbed back to rejoin his friends.

"Let's go," Rex said. They turned and flew westward. The 109's followed. Twice more, CH3 called a warning; twice more, Rex wheeled the squadron and the attackers sheered off. Boy Lloyd, trailing along behind, had time to watch and wonder. It seemed a stupid way for the Germans to behave; they were just making nuisances of themselves. It reminded him of a yarn he had read when he was a boy: all about British infantry in the desert forming a hollow square to beat off the fuzzy-wuzzies. Every time Rex wheeled the squadron the Huns got cold feet.

Lloyd was watching the enemy and wondering what sort of pansies they were when CH3 called: "Bandits behind, bandits behind!"

Simultaneously the 109's tipped into a dive. Lloyd stared up. He watched them for perhaps two seconds. When he looked down the Hurricanes had disappeared. He fingered his transmission switch, thinking Rex ought to know about those 109's. That made three seconds. He was still fingering it when tracer started flickering past. He blinked, and actually heard the first meaty *thud-thud-thud* of cannonfire. Three and a half seconds after CH3's warning, those cannonshells smashed through the tail unit, whizzed along the fuselage, punched a dozen holes in the seat-back, and blew most of Boy Lloyd's chest and stomach all over the instrument panel.

The shells still had enough energy to bash through the reserve petrol tank just beyond the panel and rip open the glycol header tank in front of that. Some shells went on and battered the engine. Fuel gushed backwards and ignited. It made a long, brilliant streamer, as thin and bright as a knight's pennant. Then one of the wings came off and the Hurricane was just a piece of falling junk.

The enemy aircraft that destroyed it was a 110, the last of three that had come out of the sun. The other two dived too far and too fast; when Rex wheeled they overshot their target. But the third had had time to see a lone Hurricane flying straight and level and looking the wrong way, and he gratefully slid into place dead astern. CH3, still weaving, saw it happen. He also saw the departing 110 fly across the diving 109's and force them to swerve away: the German pilot, he guessed, had been too busy watching his victim go down. It was Hornet squadron's first bit of luck all day.

Ten minutes later they caught up with Moran and Miller, limping along half-in, half-out of cloud. Almost immediately they ran into a dozen 109's, evidently on their way home. There was a short, savage skirmish and Flash Gordon's engine died in the middle of it. He drifted through a couple of dogfights, shouting angrily at people to get out of his way, when a bucketful of bullets pounded past his head and the stink of petrol surged everywhere, and he bailed out. He fell head-first. A Hurricane streaked beneath him, so close that the wash of air sent him tumbling. He tumbled for a very long way, sprawling and spinning like a bad acrobat, until he plunged into cloud and pulled the rip-cord.

The 109's broke off and cruised away. What was left of Hornet squadron caught up with Miller and Moran again. They went down through the cloud and Rex called Amifontaine. By great good fortune it was only ten miles away.

"Blue Three pancakes first," Rex ordered.

Moran escorted Miller down while the others circled and watched.

Miller knew he had a problem. His problem was that he could do only one thing at a time, and doing that one thing was a sluggish business. His legs had stopped hurting. He felt nothing from the waist down. His left hand was just a stump with a flap on it, sloppy with blood. Maybe it would move the throttle lever, maybe not. His right arm was okay, provided he gave it plenty of time. That was another problem. Landing a Hurricane meant doing several things at once, quite quickly.

Also he was very sleepy. His mouth had formed a thick shield of dried blood, behind which the pain merely flickered occasionally. He watched Moran and did what Moran did, if he could.

"Lose more speed, Moke," Moran said. They were far too fast. "Throttle back, Moke. Lose speed." No flaps down, no wheels. Far too fast.

Miller poked at the throttle with his bloody stub and failed to shift it. He looked at Moran and the other Hurricane distorted hugely, like a bad reflection, so he looked away. Moran was shouting at him again. The drome was down there, somewhere. He didn't know where. He took his right hand off the joystick, swung it, and based the throttle with his fist, hard.

Too hard. The engine faltered, the Hurricane nearly stalled. Moran bawled something and Miller gave the stick an almighty shove.

The Hurricane teetered for a moment. Moran flinched: she was going to fall on her tail. Instead she dropped her nose and swooped; swooped like a hawk; but there was only a hundred feet of air beneath her and she needed twice that. She hit the ground hard and broke her back. The wreckage slithered fifty yards, crushed and concertinaed so badly that at first the crash crew couldn't even find Miller.

In the ambulance his left glove finally dropped off. Not much blood was coming out of the remains of the hand. Not much blood was left.

Rex had left the basement to go to the station sickbay. Bletchley was waiting there. "How do you feel?" Bletchley asked.

"A bit stiff. The neck especially."

"Better let them have a look."

Rex took off his uniform. A medical orderly cut away his shirt: it was ripped and punctured at the back and stiff with dried blood. He lay face-down while an RAF doctor gently swabbed his back, working from the neck to the buttocks. "I make it twenty-three incisions, but I may have missed one or two," he said. "So that's at least twenty-three items of rubbish to be got out."

"Bits of flak," Rex said. "Hell of a bang outside the office."

The doctor grunted. "Look: I don't know how big these bits are, or exactly where they are. Some of them might not want to come out. You're going to need an anaesthetic. Rather a lot of anaesthetic."

Rex turned his head to look up at him. "If you do that I shan't be able to fly again today."

"You're in no condition to fly now."

"I'll be the judge of that. Sir: when's this big show of yours at Sedan?"

"If we don't do it soon," Bletchley said, "then the answer is never."

Rex took a good grip of the horizontal rail at the head of the bed. "Go ahead and winkle it all out," he told the doctor. "If it hurts too much I'll let you know."

Bombs crumped on and around the aerodrome, and anti-aircraft guns pounded away, while the doctor probed Rex's back, and dragged out ragged slivers and little torn chunks of metal. Rex thrust his face into the pillow. At the end, the pillow was drenched with sweat and the rail at the bedhead was bent like a set of handlebars.

Gangs of airmen were shovelling earth into the smoking bomb-craters that dotted Amifontaine. A couple of buildings were on fire, but they were only billets; they were left to burn. Low cloud and strong winds had hampered the German bombers, and they had missed all the Hurricanes except one: Miller's wreck had been blown to bits.

A Naafi mobile canteen was open. The pilots stood in line and bought hot sweet tea and sticky buns. They were sitting on the grass, eating and drinking and watching the flames flower over the billet roofs, when Flash Gordon arrived in the back of a truck. He was carrying a great bundle of white silk, which he dumped. "It works!" he shouted. "The bloody thing really works!" His eyes were bright with excitement, he was grinning like a child, words bubbled from him. "I never thought it would, never trusted it, to be honest, just kept on falling and waited for something to turn up but it never did, bloody hell is that tea, smashing, love one, sell my soul for . . . Anyway, kept on falling, hit this cloud a terrific smack, nearly broke all my bones, I'll have a bun as well, inside this cloud, thought come on Flash time to get your finger out, see if it works, nobody watching, bags of privacy, as per manufacturer's instructions I pulled the bit of string and by God it works! Works like a dream! Loud bang, nasty kick in the crotch, poor old Flash never the same man again but lots of lovely parachute overhead to keep out the nasty rain and tracer and bombs and birdshit and thank you, Fitz." He took the mug in both hands and drank deeply.

"Nice to have you back, Flash," Moran said. "You made damn good time."

"Had to, Flip. Had to. All the way down I kept saying to myself if I can still walk when this is over, first thing I'm going to do is find that stupid fucker Rex and screw his head off."

"What: right off?"

"Right off. Like a light bulb."

"Have you been on the booze, Flash?" Fitz asked.

"Certainly not. Tell you what, though. Met this awfully nice doctor. RAMC. Made me take some pills. Big fat stripey pills. Pep pills. Pep you up. Give you lots of zip. So he said. Why? Got any booze? Wouldn't mind a swig."

"Better not, Flash," Cox said.

"Why not? Miserable lot of buggers. Where's bloody Rex? Screw his fucking head off. Where's everyone? Where's Lloyd? Where's Moke?"

"Lloyd bought it," Barton said. "Moke nearly bought it. He's—"

"Silly bastards, should've done what I did, pulled the bit of string, it really works you know, really does, not a trick, it—"

"I know, Flash," Barton said. "I've done it myself."

It was very similar to the earlier briefing, with Sedan substituted for Maastricht. Rex announced that the Battle groundcrews were working flat-out on the damaged Hurricanes and expected to have them operational within the hour. The other piece of good news was that, in recognition of the great courage and resolution shown during the Maastricht show, the squadron had again been awarded the place of honour in another vital mission. Air Commodore Bletchley would explain.

"Fritz," Bletchley said, "has outsmarted himself. He's managed to do the impossible: he's got an armoured column through the Ardennes and over the river Meuse into France. The French are not amused, and intend to biff Fritz extremely hard. Fritz, of course, is very pleased with himself. However, his attack is an arrowhead without an arrow behind it. The Ardennes is appalling terrain. No supply column could cross it in less than a week. So Fritz has cut himself off from his support. Fritz is on his own. Our bombers will now isolate him completely by pulverising his bridgehead at Sedan. Every available Battle and Blenheim is joining in the attack. Your job: keep the sky clear for them. With a little bit of luck, Fritz will shortly discover that he has stuck his neck out too far and cut his own head off."

"Any questions?" Rex said.

"Flak," Moran said.

"Minimal," Bletchley said; and was surprised when they laughed.

"Fighters?" Cox said.

"Bound to be a few, but nothing you can't handle." They laughed again, more coarsely, and Bletchley looked at Rex for explanation; but Rex just smiled.

"The usual formation, I suppose," Fitzgerald said.

"Of course," Rex said.

Nobody laughed at that. Bletchley was puzzled, and slightly offended.

The pilots had left the mess, and Rex was talking to Bletchley, when Rex began to feel faint. His legs were rubbery and his face was clammy with sweat. Bletchley took his arm and, with the help of a mess servant, walked him to the sickbay.

"I'm not surprised," the doctor said. Rex's face was the colour of wet cement.

"Look, old boy," Bletchley said, "you've done one show today. Let somebody else lead this one."

Rex shook his head: a single, feeble movement.

"Is he fit to fly?" Bletchley asked.

"He isn't fit to breathe."

Rex touched his fingers to his mouth. "Pills," he whispered. Bletchley looked at the doctor. "Too slow," the doctor said. "Besides, they don't make a Lazarus pill; not yet, anyway. I can give him a shot of something to get him on his feet. The only question is: how strong is his heart?" Bletchley shrugged.

The doctor loaded the syringe and prepared Rex's arm. Five minutes later Rex was on his feet. The doctor put on a stethoscope and listened to his heart. "Ravel's *Bolero,*" he told Bletchley.

"By the way, how's Miller?" Rex asked.

The doctor went to a washbasin in the corner of the room and began soaping his hands. "Miller," he said. "I wouldn't worry about Miller if I were you."

Bletchley had a job to keep up with Rex when they got outside. He was glad to see Jacky Bellamy approaching: conversation might slow him down. "Greetings, scribe!" Rex shouted. "How goes the battle?"

"Isn't that my question?" She looked depressed and discouraged, which made Rex feel even brighter by contrast. "What did you think of Maastricht?" she asked him.

"Lively. Quite lively. We enjoyed good sport."

"How thrilling." The sarcasm escaped before she could stop it. "What was the score?"

"Ah, now that's asking!" Rex chuckled. He felt slightly drunk. Pain hovered around him like an aura: it was there but it couldn't touch him. Not yet, anyway. "Trade secret, old girl. I can tell you that we definitely drew blood."

"At a price, I gather."

"Oh well . . . You can't make sauerkraut without chopping cabbage, can you?"

She looked from Rex to Bletchley.

"Figure of speech," Bletchley said. "Got all you need now?"

"Yes. No. How is morale?"

"Oh, morale's fine," Rex said. "Top-class. All the chaps are itching to do battle."

"But didn't Maastricht—"

"Maastricht was a wizard show," Rex said cheerfully. "Piece of cake. Everything is absolutely tickety-boo."

"Tickety-*what*?"

"Boo," Bletchley said. "Or as you would say, 'Hubba-hubba.'"

The pilots were standing around, watching the final checks being made to their machines, when Skull arrived. "Got something for you," Fitzgerald said, and handed him a letter. "For Mary. In case I turn into a pumpkin."

Skull nodded. "I've been doing some telephoning. Apparently this isn't the first raid on Sedan. Several squadrons have had a go already. Bombers *and* fighters."

"You're becoming a terrible sceptic," Moran said. "You weren't like this when you joined us. War has depraved you, so it has."

"They say the flak's worse than Maastricht and the 109's are thick as tarts at Piccadilly Circus."

"You see?" Moran said. "Thoroughly corrupted."

"What's the time?" Barton asked. "I ought to go and see how Moke is, before we go."

"No need," Skull said.

There were a few seconds of bitter silence. "Oh well," Cox said. "Probably for the best, in the circs."

Rex came striding towards them, waving his gloves. "All set, everyone?"

"Heard about Moke, sir?" Barton asked.

"Yes indeed. The doctors say there's nothing to worry about . . . Right, let's go." He headed for his Hurricane.

"God give me strength," Fitzgerald said.

"Remember the punch-line," Cattermole said. "Close ranks."

It was about a hundred miles from Amifontaine to Sedan. The wind that had blown the bank of cloud across their path as they flew north now began to blow it away. Soon Rex was able to pick up landmarks: Cambrai, Le Cateau, St. Quentin off to his right, Maubeuge away to his left: names that had a comfortingly familiar ring: he had known them well as a boy, sticking little paper flags in a big map of the Western Front. There had been no official maps at Amifontaine so he had borrowed somebody's Michelin guide. From Cambrai and Le Cateau route nationale 39 led to La Capelle which had a six-way crossroads. It showed up clearly. He was dead on course.

The squadron was in vic formation, sections astern. Rex was at the point, flanked by Cox and Cattermole. Barton flew behind him with Patterson and Gordon on either side. Moran led the arse-end section with Fitzgerald and CH3, but in fact CH3 kept his distance and flew a continuous weaving pattern. That was a little bit sloppy, but Rex didn't really care. Whatever magic juice the doc had pumped into his arm made him feel remarkably happy. He was alert and keen and very ready for action. Some of the dressings on his back seemed to have come adrift when he heaved himself into the cockpit, and his shirt felt strangely slippery with what must be blood, but that only strengthened his sense of accomplishment and well-being. It was a fine late afternoon. Everywhere he looked he saw colours of an extraordinary beauty and brilliance. It was going to be a splendid evening.

Rex set a keen pace. After thirty minutes they crossed the first hills and forests of the Ardennes, and they picked up the Meuse, looking as looped and twisted as a fallen strand of wool. Rex turned south and followed it. In the distance he could see black smoke and the faint flicker of shellfire. Sedan. "Close up," he said. "Nice and tight."

He saw the bombs burst before he saw the bombers. Hornet squadron was at eight thousand feet, and the bombers were at least a mile below. The sudden fountains of earth caught the setting sun and stood briefly golden on one side. He hunted for the bombers and found them, ten or a dozen, looked like Blenheims. At once he looked up and searched for enemy fighters, and he found them too, a great pack of 109's arriving from the east at about fifteen thousand feet; had to be 109's, there weren't that many Hurricanes in Europe. By God, what a scrap this was going to be! He checked the bombers again, hoping they had finished and were going home, and they were, but as they banked the sun lit up the crosses on their wings and they were Junkers 88's, not Blenheims. So those bombs had fallen on the wrong side of the bridgehead. Those bombs had killed Allied troops. What evil. What savagery. What filth.

Rex felt the clear, pure rage of a Crusader knight. He was washed clean of fear or pain or worry. He was indestructible. They could kill him but they could not destroy him! "Bandits below," he called. "Eighty-eights at three o'clock. Going down, chaps. Let's get 'em." It was the first time he had said *chaps* in an order. He thrust the stick forward and fell on the enemy. "Close ranks," someone said, and Rex fell alone.

"Close ranks!" the voice repeated. Immediately the formation tightened. Cox and Cattermole edged in to fill Rex's space. Patterson slid half-under Barton's left wing, Gordon eased over Barton's right wingtip. Fitzgerald crowded Moran. The flight commanders were so boxed-in they could do nothing but fly straight and level.

Rex, plunging through the mile of empty sky, heard none of their shouts and curses: someone else's transmission switch was open, blocking the channel. If he had heard, he would not have turned away. This was his mission, his crusade: to smite the ungodly! To biff Fritz extremely hard! *Bring me my bow of burning gold,* he sang to himself. *God, who made thee mighty, make thee mightier yet!*

The bombers saw him coming. Their gunners raked his path with crossfire. At four hundred yards he squeezed his gun-button and experienced a jolt of exultation as the Hurricane kicked and trembled and his shining streams of death raged across the formation, ceasing only when he sliced between a pair of 88's.

Rex hauled the stick back and opened the throttle to climb and attack again, tracer still hounding him, and he wondered where everyone was. Then a pair of 109's appeared above. For a second they hung like trapeze artists at the height of their swing. They turned and dropped, and he climbed into their fire. Cannonshells ripped through his tank and the fuel gushed over his legs. Rex never saw that, never even heard the walloping impact. Before the stench of fuel could reach his nostrils, a bullet smashed into his oxygen bottle. It exploded. The Hurricane blew up like a bomb. Pure oxygen mixed with high-octane fuel made a furnace-heat that incinerated Rex, literally in a flash. His clothing turned to ash in a second, and his body was boiled in its own fluids. The cockpit melted around him. The fighter separated into a hundred parts which blew away like a handful of dust. Looking down, Pip Patterson saw only the flash of white, as stark as lightning. And then nothing.

For a moment what remained of the squadron cruised on. They were so bunched-up that Patterson and Gordon and Fitzgerald could see Moran and Barton gesturing furiously, shouting silently. The unknown pilot's transmission switch was still open, swamping all the earphones with his cockpit roar. Then CH3 went past them, waggling his wings. He put his nose up and fired a warning burst towards the pack of 109's, now tumbling out of the sky. At once the switch was closed. The formation relaxed and spread itself. Barton found himself in front, leading. "Trouble at ten o'clock," he announced. "We'll go up and meet it." The Hurricanes turned and climbed.

It was a quick scrap. Not even a proper fight: just a sudden firestorm. Cattermole, watching the 109's get bigger, told himself: *I've got a ton of engine to protect me.* The enemy arrived in a rush. Both sides blasted away. Cattermole's head wobbled to the pounding of his guns. The enemy vanished. Cattermole still lived.

The 109's outnumbered them four to one but they did not return for another attack. A more attractive target lay below. Battle bombers, a dozen of them, came flying along the Meuse valley, their pattern heavily outlined by German flak. A couple of 109's, dribbling smoke, had broken off and were heading for home, but the rest went hunting the Battles. Barton's instinct was to dive and fight. "I'm hit in the engine, dammit," someone complained. "Who's that?" he asked.

It was Pip Patterson. His engine was coughing and missing and shaking the plane so hard that it frightened him. Then Moran called up: he was losing glycol. "Any more?" Barton asked. Cox reported that his guns had jammed. Barton looked about him and saw Flash Gordon gesture thumbs-down and tap his earphones. Radio dead. God knows what else damaged. "Let's try and get back to Amifontaine," Barton said. As they banked to head north he saw the 109's go slamming into the Battles. Before he had levelled out, two Battles were on fire and falling.

Moran had to switch off his engine when he lost all his coolant. They circled while he force-landed in a field, and saw him get out. Patterson's engine shook itself to death soon afterwards. He could see nothing but woodland beneath. He baled out. Six machines reached Amifontaine. Half of Flash Gordon's undercarriage collapsed on landing and the Hurricane made hectic circles across the grass, destroying its left wing.

Black Thursday
The Second Schweinfurt Attack, October 14, 1943

BY BRIAN D. O'NEILL

(Excerpted from the book *Half a Wing, Three Engines and a Prayer* by Brian D. O'Neill)

On paper, the plan must have looked as promising as it did audacious: Send the largest B-17 force ever assembled deep into Germany—far beyond the range of friendly fighter cover—to bomb Hitler's source of ball bearings at Schweinfurt. Without ball bearings, the Nazi war machine would grind to a halt.

In that summer of 1943, the U.S. Eighth Air Force in England was fighting desperately to prove that daylight precision bombing with its B-17s, when combined with the ongoing British nighttime bomber attacks, was worth the high losses of men and planes. For whatever casualties they endured, the Americans and the British pilots were the only forces taking the battle to the Germans in Germany itself.

The B-17 raid of August 17 and the follow-up attack of October 14, 1943, resulted in two of the most savage battles ever fought in the skies. The planning of the missions, the ultimate ways they were flown, and the damage inflicted are subjects that have been detailed and debated in several books and individual articles that make fascinating reading. One well-known novel set around the subject, *The War Lover* by the distinguished writer John Hersey, was even made into a film. Although it is a lesser-known novel than *The War Lover,* Jim Shepard's *Paper Doll* is, in my opinion, as riveting and unforgettable as any fiction ever written on B-17 action in World War Two. Shepard's novel reaches its climactic moments during the August 17 mission, and I recommend it highly.

197

For our presentation on the Schweinfurt campaign, we have chosen a distinguished work of nonfiction that will bring the action of the October 14 raid right into your reading room. Brian O'Neill's highly readable *Half a Wing, Three Engines and a Prayer* has been described by some authorities as, "The best collection of stories about a B-17 Bomb Group that has ever been published." Originally published in 1989 and republished in 1999 by McGraw-Hill in a revised special edition, the book focuses on the tour of duty of the 303rd Bomb Group, one of the key elements in both Schweinfurt attacks.

Here is the war as it was actually fought in the skies over Germany on that October day. If monuments could be placed in the skies, they would be there right now.

Fog covered England the morning after Münster, and all over East Anglia the Eighth's bomber bases were quiet—there was no mission on. The losses the Eighth had suffered during its last three raids—88 heavy bombers, plus eight Category Es—far exceeded those of any comparable period, but Hap Arnold was still focusing on what had been achieved rather than its cost. On October 11th he sent General Eaker a telegram which confidently predicted a turn of the tide:

"The employment of larger bombing forces on successive days is encouraging proof that you are putting an increasing proportion of your bombers where they will hurt the enemy. Good work. As you turn your effort away from ship-building cities and towards crippling the sources of the still-growing German fighter forces the air war is clearly moving toward our supremacy in the air. Carry on."

This was exactly what Eaker and Anderson intended to do. The weather remained bad on October 12th, but that afternoon General Anderson and VIII Bomber Command put the finishing touches to a new plan and then briefed General Eaker on it. On the next clear day over the Continent the Eighth would return to Schweinfurt, the most heavily defended target in Hitler's Reich.

The attack appeared to be a clear military necessity. General Eaker was aware that the Germans were putting maximum pressure on neutral Sweden for its total ball bearing output, and that they were scouring the Continent for other sources of supply. He knew too, that the damage done to Schweinfurt on August 17th had not been that great, and photo-reconnaissance flights confirmed that Germany would soon have its most important ball bearing factories back up to full production.

Thus, if the Eighth was to destroy the Luftwaffe by cutting this strategic jugular, the time to do so was now, regardless of cost.

October 13th offered no opportunity. Fog continued to blanket England, and the still cloudy skies over Europe gave VIII Bomber Command only two options: Waste another day or try a PFF mission. General Anderson opted for the latter, but the effort was stillborn. Elmer Brown wrote that "we started out to bomb Emden, but it was recalled before we left England."

That afternoon, however, the picture started to change. There was no reason to think the fog over England would lift, but the 1600 weather briefing at Pinetree showed encouraging signs: It appeared the sky over Germany would be clear the next day. And so General Anderson put out the fateful order—there was a mission tomorrow and the target was Schweinfurt.

The word went through the chain of command as on any other operation, finally being passed by teletype to the bomb groups, where the plan would be completed down to the last detail. At Molesworth, Captain Mel Schulstad was one of those playing his part as Assistant Group Operations Officer.

"I was working with the squadron operations officers, telling them how many planes we needed from each squadron, and figuring out where they would be assigned. These discussions became fierce fights and got pretty personal because so much was at stake. I'd tell one guy we needed so many airplanes and crews from his squadron, he'd say it wasn't possible because of battle damage or whatever, and we'd go around and around till things were worked out."

This day Major Ed Snyder was slated to lead both the 427th Squadron and the Group, flying Mr. Five by Five as part of Jake James's crew—James, now a Captain, had taken over for Strickland's crew when they finished their missions and went home. Lt. Carl Hokans was the lead navigator and the lead bombardier was none other than Mac McCormick, who had traded places with James's regular bombardier, Lt. Walter Witt, for this all-important effort. Leading the Squadron's second element was David Shelhamer (he was a Captain now, too), while Hullar's crew flew the No. 6 position off Shelhamer's left wing in Luscious Lady.

The 360th was high squadron, but ships and crews from the 358th helped complete its formation. Among them were Lt. Roy Sanders's crew filling the No. 5 position in Joan of Arc, B-17F 42–29477, and Lt. Bill Fort's crew (with Lt. Calder Wise as Instructor Pilot again) flying the No. 6 slot in Yankee

Doodle Dandy. The 359th filled six of the seven low squadron slots. The No. 5 position was taken by a crew on their fifth raid, Lt. Ambrose Grant's in the Cat O' 9 Tails, B-17F 42–5482, while the No. 6 position was filled by Flight Officer T. J. Quinn's crew in Wallaroo. Completing the formation in the tail-end Charlie position was Lt. Jack Hendry's crew in their favorite Fortress, the War Bride, B-17F 42–5360.

At the time these assignments were made, most of the men had no idea what lay ahead; only lead crews were put in the know as part of the standard mission routine. But Molesworth was alive with men in the early morning hours beginning to get the planes ready to go. It was impossible to see them going to work in the darkness and the fog, but as he toiled through the night, Captain Mel Schulstad kept an ear cocked for their first stirrings.

"When the ground crews got out to the hardstands, one of the first things they did was to start a portable gasoline engine. They used it to provide electric power to the airplane and lighting to work. You'd hear the noise as a simple 'putt-putt-putt' when the first engine was started, and then soon it would be joined by others and grow until you heard this 'hummmmm' coming in from the hardstands. To me there was always a terrible drama about it greater than the opening of Beethoven's Fifth, or the beginning of any symphony. When you heard that sound, you knew for sure that today men were going to die."

On no day of the Eighth's air war was this truer than "Black Thursday," October 14, 1943.

The bomber crews were roused at 0630 for breakfast at 0700 and then the crews filed into the officers' and enlisted briefing rooms to get the word at the 0800 briefings. Mac McCormick already knew what the objective was, but there was no way to soften the blow which hit the other officers in Hullar's crew when the curtain was raised.

"Schweinfurt again!" is how Hullar began his notebook entry, and Elmer Brown started his diary for the day with exactly the same words.

Klint's reaction was visceral: "When they uncovered the map, and showed us going back to Schweinfurt again, it was like getting hit in the face with a baseball bat. Again, the keynote of the CO's introductory remarks was: 'Today's raid, if successful, can shorten the war by six months.'"

At the enlisted briefing, Miller remembers "sitting down with the rest of the fellows and wondering what that map was going to look like when they took the cover off. A couple of officers came in and someone yelled 'Attention.' We stood up, then sat back down, and my heart started beating a little faster as the suspense of waiting to see where that red line went up.

"Then they peeled the cover off the map, and we were a bit startled to see the line go deep into southern Germany again—to Schweinfurt. I knew then that we were in for a bad day."

George Hoyt recalls "being told at the briefing that the ball bearing factory at Schweinfurt, which we had knocked out on August 17th, was now back to 65 percent of production, and we were going back to destroy it again. This was not agreeable news."

Norman Sampson remembers feeling that "this mission was going to be the roughest of them all. They knew we would be back, and would be ready for us with more planes than they had the first time."

Intelligence estimated that the bombers would be facing 1100 German fighters, and since U.S. fighter escort couldn't penetrate deeper than Aachen, all the B-17s had to oppose this strength was their own numbers. The plan called for a mass assault on Schweinfurt with every bomber the Eighth's three divisions could muster: 360 B-17s and 60 B-24s.

The First Division was designated the first air task force and was to send its three combat wings—the 40th Wing leading, followed by the 1st and 41st Wings—on a more-or-less direct route to the target running just north of Aachen. The Third Division's three wings formed a second air task force. They would follow the First to Aachen at a 10-minute interval, taking a parallel track 30 miles to the south. During this time the two forces were to fly with division formations as near to line abreast as possible, creating a gigantic wedge going into Germany.

Beyond Aachen the First Division was to continue along its original track while the Third cut sharply south, flying along the Belgian-German border. The First Division was to draw off most enemy fighters as it flew to a point north of Frankfurt am Main, where it would take a southeasterly course to its IP west of Würtzburg and southwest of Schweinfurt. The Third Division would continue south just inside the German border halfway down Luxembourg, where it would swing on a course due west to its IP northwest of Würtzburg, following the First Division into Schweinfurt.

Meanwhile, the 60 B-24s of the Second Division, constituting a third air task force, would use their faster cruising speed and greater fuel endurance on a more extended route to the south. They were to arrive over Schweinfurt shortly after the Third Division dropped its bombs.

After the plan was outlined, John Doherty remembers that "the officer who gave us the briefing said, 'Gentlemen, look to your right. That man won't be coming back.'"

George Hoyt "did not see anybody flinching," but in the officers' briefing, David Shelhamer's keen eye took in a correlation between the length of the tape going out to the target and another, human line.

"Our Catholic Chaplain, Father Ed Skoner, would wait at the end of the briefing hut to pray over any crewmen who wanted a blessing before a mission. The longer the ribbon, the more crew were back there kneeling in front of the priest. On the day of Second Schweinfurt, the men who had been unfortunate enough to go on the first Schweinfurt mission lined right up like a British queue. It was really something to see."

The bomber crews arrived at their aircraft around 0935, an hour and ten minutes before takeoff. Once there, Hullar, Klint, and Rice conferred with the *Lady's* ground crew and went through their preflight routines. Miller and the other gunners "checked all our equipment, our guns, and everything we could so we could feel a little bit better, a little bit safer."

As Norman Sampson explains, "Each gun on our plane meant the life of our crew."

Extra ammunition was especially important. The *Lady* had two extra ammunition boxes installed in the tail gunner's compartment, which were filled to overflowing, but more vital still was the extra ammo secreted aboard in portable boxes. On First Schweinfurt the *Lady* had about 12,000 .50-caliber rounds aboard; today she was crammed with over 14,000 rounds.

Other crews followed suit. David Shelhamer remembers that "On Second Schweinfurt I loaded about half a ton of extra ammunition aboard because, having had the pleasure of going on the first mission, I knew quite well what to expect."

John Doherty likewise explains that "We had been on the first Schweinfurt mission and most of us took new gun barrels and extra ammunition, a considerable amount more than we normally carried, which was a real lucky thing because we had used it all up by the time we got back."

Ed Snyder would have intervened, had he known. Weight loading for this mission was especially tight. Elmer Brown wrote that "This time we had a bomb bay tank on the left side. The bomb load was three 1000-lb. bombs and three 100-lb. incendiaries."

And yet, as Snyder recalls: "The gunners, everyone who fired a gun for that matter, tried to take additional ammo. It was done in an emotional, unscientific way. They didn't ask, 'Will this airplane fly with this much weight on board?' The airplanes were overloaded with fuel and bombs as it was, so we had to go out and take ammunition off because many guys would have creamed themselves otherwise. It made a lot of them real unhappy but they would have stalled out on takeoff and never gotten off the ground. The old B-17 did all kinds of things they said it couldn't, so they figured it always would. But there was a limit."

Hullar's and Shelhamer's crews escaped their Squadron Commander's notice, and there finally came a point where all the preparations, good and bad, ended in the foggy stillness. As on August 17th, the crews were left to wait for the word—to go or not to go.

In *Luscious Lady,* Bud Klint put his thoughts down on paper: "The fog still hangs on, and we're still trying to get a raid off. Today we're scheduled to go back to Schweinfurt. With 1100 fighters in range of our course, and 56 flak guns over the target alone, it promises to be rough.★ If we get it off. Our first trip to that target was the roughest one I have been on to date. Maybe we'll hit a new high today. My only prayer is that we come through as well as we did then."

In *Yankee Doodle Dandy,* Lt. Bill Fort was similarly preoccupied: "We were scheduled to go to Schweinfurt, which had a reputation for being a real tough target. The weather was terrible, and I never thought we'd get off the ground, figuring they would scrub it most any time. But they put us in the air anyway."

This decision was made after a weather report was passed to VIII Bomber Command from a British Mosquito reconnoitering over Germany. The sky was crystal clear. General Anderson put out the fateful, final order: "Let the bombers take off."

All over East Anglia the Eighth's bomber bases now rang with the noise of radial engines as Wright Cyclones and Pratt & Whitney Twin Wasps

---

★ There were, in fact, far more flak guns than Klint knew about. On First Schweinfurt, August 17, 1943, the Germans had sixty-six 8.8 cm flak guns in eleven, six-gun, single batteries defending the City. By October 14, 1943 the Germans had increased the defenses to the equivalent of 22 single batteries of 8.8 cm guns—broken down into four single batteries, four double batteries, and three triple *Grossbatterien*—for a total of 126 8.8 cm flak guns. In addition, there were two, four-gun batteries of the larger caliber and even more dangerous 10.5 cm flak guns. Thus, there were 134 rather than 56 flak guns defending Schweinfurt on Black Thursday. According to a German historian, at this time Schweinfurt was "the best defended town in Western Europe" in relation to its size. Friedhelm Golücke, *Schweinfurt und die strategische Luftkrieg 1943.* Schöningh, Paderborn, 1980, p. 173.

Manning these batteries were approximately 250 regular *Luftwaffe* personnel and another 1500 schoolboys, fifteen to seventeen years old, euphemistically called *Luftwaffenhelfer* or *Flakhelfer* ("Flak Helpers"). Five hundred of the *Luftwaffenhelfer* served the 8.8 cm guns directly while the remaining 1000 worked their associated radar, optical tracking, and battery control switchboard equipment.

coughed and came to life on more than 400 B-17s and B-24s. They taxied from their hardstands and waddled to their runways, turning onto them and disappearing into the foggy air one by one.

At Molesworth, Major Ed Snyder and Captain Jake James's crew was first, speeding down the main east–west runway in *Mr. Five by Five* and climbing aloft right on time at 1045. Three bombers later, Captain David Shelhamer got *Vicious Virgin* up at 1047. The timing of these takeoffs is all the more remarkable when one learns how they were done. David Shelhamer describes the procedure:

"The airdrome was fogbound, and the ceiling was zero. When we taxied to the takeoff end of the runway there was a jeep there that would get ahead of the aircraft about 100 to 150 feet. It would then start right down the center of the runway at about 15 to 20 mph, while the aircraft followed directly behind it. We established a direct course down the center of the runway, locked the tailwheel, and set the gyrocompass to 'zero it out.' The gyrocompass was extremely sensitive and would react immediately to any change in course. The jeep then pulled away from the aircraft and we went on full power."

At 1048 Bob Hullar got a signal that the runway was clear and *Luscious Lady* began her run. As the overloaded *Lady* lifted off, the crew encountered the first of the mission's many hazards. George Hoyt remembers that "we were violently buffeted about by the propwash of the plane that took off seconds before us. I always sweated this out, since it took place only 30 to 50 feet off the end of the runway, with treetops and rooftops whipping by just under our wing."

The Group takeoffs continued until 1057, when Lt. Jack Hendry got the War Bride in the air, the Group's 20th and last ship. Now the 303rd faced the difficult task of assembling above the cloud cover and joining up with the 379th and 384th Groups to complete the Wing. Captain James's Group Leader's Report describes this phase:

"We climbed to 6,500 feet before breaking out of the lower overcast. Then we continued our climb through a cloud-free space of about 700 feet, at which time we began to climb through another cloud layer, which we broke out of at 9,000 feet. Our assembly was made at 11,500 feet and we left the base on course for Eyebrook on time. On leaving the base we had spotted the other two Groups of the Combat Wing. We got into Combat Wing Formation slowly en route to Eyebrook. We experienced a lot of high cloud around the English Coast and gained and lost altitude intermittently to avoid this cloud. We left the English Coast about seven minutes late, probably due to turning south to go around a very heavy cloud layer at the intended point of departure from the English Coast."

Though assembly went reasonably well for the 41st Wing, the same cannot be said for the First Division's other two wings. The 40th CBW was to head the Division, but when its three groups and those of the 1st CBW sorted themselves out in mid-Channel, the latter was in the lead with the 305th Group of the 40th CBW flying an unorthodox "double high" position in its formation—the two remaining groups of the 40th Wing, the 92nd and 306th, were trailing behind. The First Division was flying into battle in disarray, its lead wings in poor condition to meet the German onslaught.

The Third Division's assembly went more smoothly; its seven groups crossed the English coast in proper order on track 10 miles to the south of the First Division. In contrast, the B-24s of the Second Division never assembled at all—of the 60 Liberators scheduled for the raid, only 29 came together above the cloud cover. This paltry force was ordered to abandon the main effort, and instead was fated to conduct an ineffectual feint towards Emden.

Aborts reduced the weight of American numbers further as the B-17s crossed the Channel. The Third Division lost 18 out of 160 aircraft. The First Division's ranks were cut from 164 to 149. In the 41st Wing, the 384th lost two of 19 Forts, the 379th pressed on with 17 ships, and the 303rd had two of 20 B-17s return.

Of the cross-Channel flight, Elmer Brown wrote: "We switched the gasoline over and dropped the [bomb bay] tank before we reached Holland. We had P-38 escorts at first, and then P-47s."

George Hoyt and Merlin Miller also saw the twin-boomed fighters, which were from a new unit in the ETO, the 55th Fighter Group. The Lightnings had been ordered to escort the Fortresses as far as Flushing, Holland, but their mission was canceled at the last moment because of problems the big fighters still had operating the ETO's frigid air. From what Hullar's crew observed, however, some of these eager fighter pilots either "didn't get the word" or chose to ignore it. Unhappily, George Hoyt recalls that "little if any enemy fighters were spotted during the P-38 escort time."

The primary defense would remain the Fortresses' firepower, and readying this defense was the final prebattle ritual. In the *Lady,* Norman Sampson had long since gotten into the ball turret.

"When we got off the ground, the turret was locked into place with the guns pointed towards the tail. As soon as possible I entered the turret through the door, put the safety belt on my back, and locked the door from the inside with two handles. I decided to wear a chest-type parachute. It made for close quarters, but I felt more comfortable with it. If something happened I could just roll back, unfasten the safety belt, open the door, and roll out. I put the power on, and was free to move the turret all around to protect the bottom on the plane."

Dale Rice got into the top turret, perched on its "bicycle seat," turned on its electrohydraulic power drives, and was soon rotating his twin .50s to protect the top of *Luscious Lady* fore and aft. Merlin Miller also took to his bicycle seat in the tail, and prepared himself to protect her rear with his flexible twin guns:

"Crossing the Channel we had fighter escort, but I knew they couldn't stay very long, and it was going to be a long fight in and out if the last Schweinfurt mission was any indication. So I sat there, anticipating things, wondering what was going to happen, how the mission was going to go, and I got a little tense.

"Then all of a sudden I realized the battle was going to start pretty soon, and I said to myself, 'Merlin, get mad, get angry.' I started thinking about the Germans, what they were trying to do to us, what I was going to do to them, anything to stir up my blood and get anger coursing through my system, anything to take away from that feeling of fear I otherwise would have had. By this time I had figured out that it was difficult for a person to feel two emotions at the same time. In combat I thought I was better off being angry than afraid.

"Pretty soon I heard a call on the interphone. It was Hullar saying, 'Crew, check your guns.' So I hand-charged each machine gun and fired a short burst from each one, and said, 'Tail guns okay.'"

The process was repeated all the way up to the *Lady*'s nose as each man at a machine gun tested it and reported in. Hullar's crew was as ready as they would ever be for the fight which lay ahead.

At the head of the First Division, the 1st CBW crossed the Dutch coast at 1250 in the company of P-47 escorts from the 353rd Fighter Group. At 1311, the 41st CBW also crossed the enemy coast, and Elmer Brown wrote that "P-47s took us into just south of 'Happy Valley,' or rather the town of Bonn. We saw several dogfights."

The Thunderbolts stayed with the Fortresses approximately 25 minutes, but even as they did, B-17s were leaving the 41st Wing from the 384th Group. Four of their Fortresses turned back due to various problems over enemy territory. Their loss left the 384th terribly exposed. It now had only three B-17s in its high squadron, four in its lead, and six in its low.

While this was occurring, Captain David Shelhamer had a few brief moments to observe events in the wings ahead.

"The visibility was unbelievably beautiful. You could see the Swiss Alps about 200 miles to the south. I was personally aware of at least four B-17s swinging out of formations from other groups heading right straight down to the Alps. With the fighter opposition that was only moments away from us, it is very questionable whether they were able to make it to Switzerland."

A short time later, Lt. Bill Fort saw what Shelhamer was talking about: "There was a wing ahead of us about a couple of miles. They seemed to be hit real hard. We saw a number of planes explode, and others going down."

Meanwhile, George Hoyt saw "a lone Me-109 paralleling our course on our starboard side about four o'clock high. I swung my flexible .50-caliber to frame him in my ring sight. He appeared to be 800 yards or better out, and I hesitated to see if he would turn to attack. In a split second he did, and I hosed him with two good short bursts, giving a good allowance for the trajectory drop at this range. Much to my surprise, he appeared to be hit. He abruptly broke off his attack by throwing one wing up and made a long dive out of my field of vision. I breathed a sigh of relief."

This single encounter was like a raindrop before a cloudburst; after the P-47s left, the storm struck with indescribable fury, progressively engulfing each wing of the First Division as it flew to meet the enemy.

The first two combat wings were decimated. The 1st CBW and its tagalong, the 305th Group, lost 13 of 55 aircraft. Twelve of these came from the 305th, leaving it with three ships. The 40th CBW fared even worse. On the way to Schweinfurt its 92nd Group lost five of 19 and the 306th Group 10 of 18.

The story of the 41st CBW's ordeal is less well known. What happened to the Hell's Angels and to the other groups in the Wing is a tale worth telling, for on Black Thursday no part of the sky was still.

Elmer Brown's diary sets the tone for what took place around 1345, just east of Aachen: "We had fighter attacks from that time to the target and deep in France. We were fighting for over two and a half hours constantly." They came, quite literally, in the hundreds: FW-190s, Me-109s, Me-110s, Me-410s, Ju-88s, Do-217s, and "at least 3 FW-189s," a twin-engine, twin-tailed reconnaissance plane whose presence shows how completely the Germans were pulling out the stops.

Official 303rd reports describe it this way: "At least 300 E/A were seen. Generally speaking the E/A seemed to have tried just about everything ... The attacks by E/A were made from all around the clock generally from the high and level positions. Most of the T/E [twin-engine] fighters attacked from the side—chiefly on the nine o'clock side. FW-190s attacked from the nose. The E/A carrying rockets made their attacks from the tail. Several T/E fighters slow-rolled through the formation to break it up. The rocket carrying E/A attacked in trail usually three at a time . . . The attacks are reported as particularly vicious and the pilots gave every indication of being experienced and clever. The Yellow Nose FW-190s were particularly effective in their attacks."

Up to this point in the Eighth's air war, Major Ed Snyder felt that he had pretty much seen it all. Though he missed First Schweinfurt, he had flown so many other missions that they had almost become routine. Now he realized he was in the center of something truly exceptional.

"There was a tremendous amount of aerial combat activity going on the whole time. On all the missions I had been on before, there always seemed to be goodly periods when nothing was really happening. You just anticipated the enemy being there. On this mission there was always something going on, flak or fighters all the way in, and all the way out."

The impression was the same from front and rear. In the *War Bride's* tail, John Doherty remembers that "going in, we'd be picked up by a group of fighters. When they got to where their fuel supply was getting low, they'd go down and you could see another bunch coming up. It was kind of a comical situation in one way; the one bunch would leave you, and the other bunch would start coming up at about 10,000 feet. That's the way they kept us all the way into the target.

"You could also see these flak guns shooting, because you were looking right down into the barrels when they'd shoot and you could see the fire. And then you would sit there and say to yourself, 'Well, I wonder where that one's gonna hit? I wonder how close that one's gonna be?' You could see them all along in different places like that."

Ed Snyder continues: "You could see fires all over the place, and parachutes dropping all over the place. You can't believe the number of aircraft that were being shot down, the number attacked, and the different kinds of aircraft that were involved. The Germans came at us with just about everything in their inventory. On one occasion I saw a German fighter come through our formation that was badly hit. The pilot must have been wounded or killed because he jammed the controls. As he went past under our wing his left wing was where his nose should have been. He just skidded through the formation sideways."

Just behind Snyder in *Vicious Virgin,* David Shelhamer had a different kind of head-on encounter: "I very vividly recall this one FW-190 who obviously was attempting a head-on ram on my aircraft. He was not firing at all when he was well within his own range. Our rate of closure had to be about 400 to 450 mph, and I eased the aircraft down, realizing full well that he was attempting a ram. It required split-second timing, but my timing was good. I pulled the aircraft up and he went about 50 feet under me. I will never forget the green scarf that German pilot had around his neck."

Frightening as these experiences were, many veterans, particularly the pilots, say that they were simply "too busy to be scared." As Ed Snyder explains, "I was concerned about my own safety, but that had to take its place at the back of my mind as long as we were not on fire, or incapable of flying. I was too busy trying to make everything in the formation work properly to get too concerned. I don't mean this to sound macho or anything. That's just the way it was. There was just too damn much going on."

Every crew was now locked in the ultimate test of teamwork. Hullar's fought with hardly a moment's pause. Lt. Witt was at the *Lady's* single .50-caliber in the nose; while he fired at fighters making head-on attacks, Elmer Brown was "crouching behind him looking to see which way they would break. I would jump to either the left or right cheek gun and try to get a few rounds off." In his diary, Brown also wrote that "Jerry would cue up at about one-thirty [o'clock] time and again and attack our lead group."

In the cockpit, Hullar and Klint were consumed by the twin tasks of staying in formation and taking evasive action. At debriefing they complained of "too much variation in airspeed over enemy territory. Varied from 130 to 170 mph." Their job was made doubly hard by the need to take evasive action, as Bud Klint explains:

"We were tucked away in the No. 6 position of the lead squadron, and while this offered a definite advantage from the standpoint of enemy fighters, we were hemmed in on all sides by other B-17s. The Forts were packed in a tight defensive formation, and everyone was using violent evasive action, which was an added hazard. Bob and I were trying to get the most evasive action possible without ramming one of our own planes.

"Evasive action was largely a psychological thing. Probably our chances of kicking the airplane into the path of enemy shells were almost as good as kicking it out of their path. In addition, those violent gyrations certainly made an unstable platform for our gunners, and disrupted their aim. In spite of this, it was a tremendous psychological boost for everybody on the airplane, and it was virtually impossible for Bob and me to sit in the cockpit and hold the airplane straight and level while we were under attack by enemy fighters. We just felt that we had to do *something*.

"We both were on the controls almost constantly doing evasive action. The sky was literally filled with aircraft, and we were trying to hold as tight a defensive formation as possible. Bob was responsible for aircraft to our left and I was responsible for those to our right. When he would kick the plane to the right it was up to me to kick it back when we got too close to some other aircraft. There was no method or plan to the evasive action we took. We just did whatever we felt was necessary—kick the rudder, drop a wing, pull the nose down, anything that occurred to us that we could do in the limited airspace we occupied.

"The B-17 had no boosters on the controls. It was a very stable airplane and easy to fly on a straight-and-level path. But when you did strenuous evasive action it really required a lot of muscle and a lot of perspiration. Evasive action would wear one pilot out in a very short period of time."

Bob Hullar offered a more succinct explanation of these efforts in his notebook, where he wrote that "Bill and worked like Hell on evasive action."

Amazingly, however, Hullar found time for humor during the worst of all this. In his diary, Elmer Brown noted that "They were firing a lot of rockets at us. We sure took plenty of evasive action. In the midst of the battle the flying got a little rough, and on the interphone Hullar said, 'Oh, my aching back,' just as innocent as he could be. Rice said, 'Is someone calling the engineer?'"

When he wasn't trading quips from the top turret, Dale Rice was firing nonstop at enemy fighters, and using his position's superior visibility to call them out to the other gunners. Merlin Miller vividly recalls Rice's help with some Me-110s making head-on attacks:

"I heard Rice call out, 'Fighter over the top!' Then a 110 came over the top of our tail upside-down. He seemed to hang there for a moment and I could see the pilot so clearly that if he hadn't been wearing an oxygen mask I could have recognized him on the street. But the plane was going too fast for me to react.

"So I called to Dale and said, 'Rice, one of those fighters comes over the top again, tell me where and when and maybe I can take a crack at him too.' Rice said okay and the next time an Me-110 came over I was ready. My guns were pointed up, and when I saw the German I held the triggers down. He flew through the bullets, and pieces flew off of his airplane."

It wasn't all give for Rice, however, as George Hoyt explains: "At one point Dale ducked down out of his Plexiglas dome in the top turret to get some reload belts of ammo on the floor. While he was down a couple of slugs went through the Plexiglas right where his head would have been." Rice fought the rest of the mission with the wind screaming through the splintered dome.

For Hoyt, the big shock was the large number of rocket attacks from the rear:

"I kept thinking, 'How can the Jerries have so many rocket-firing Me-110s?!' They would hang out beyond the effective range of our .50-calibers and fire their rockets into our formation.

"As soon as I saw the flashes when they fired, I would call to Hullar to 'Kick it around!' You could track the rockets as they came in. It always seemed to me that Hullar would dive out of the way just as the rockets zoomed over the top of the radio room.

"After rocketing us the Jerries would bear in with guns blazing, and that was when Dale, Merlin, and I really zeroed in on these boys. Through all of this it was very hard to hold one's fire to short bursts to keep from burning out the machine gun barrel."

The first encounter with rocket-firing fighters also took Merlin Miller by surprise: "I looked up and saw two Ju-88s sitting in the back wingtip-to-wingtip. I wondered 'What are those bastards up to?' because they didn't seem to be closing in. They just sat there. All of a sudden I saw big smoke puffs from under the wings of both planes, like they'd been hit. Then, to my dismay, I saw four 'gizmos' coming at us which looked like black softballs. I pushed down on my mike button and yelled to Hullar, 'Kick it, kick it!' He bounced the airplane around and the four rockets exploded above us.

"I got used to it after this, and would tell Hullar to get ready to either climb or dive when the rocket ships lined up. When I said 'Kick it!' he would abruptly maneuver up or down, depending on which way he was heading at the time. As things turned out, none of the rockets actually exploded too close to us."

While all this was going on, George Hoyt "could feel through the soles of my flying boots Norman down in the ball turret firing away at Jerry. He had a lonely and detached station down there with his Sperry computing gunsight." The *Lady*'s evasive action didn't make his job any easier, as Sampson himself explains:

"The gunsight was a frame which had two lines, with one on each end of a box. You turned handles to 'frame' an enemy fighter between the lines as it came in. When you had the fighter framed, you shot at him by pushing the firing buttons, which were on top of the handles. I searched mostly to the sides and the tail because there was a lot of blocked-out area up front from the body of the plane and the propellers. The turret had stops that wouldn't let you fire there. Also, the head-on attacks came too fast for the sight.

"I didn't have any trouble spotting the enemy fighters. You could see them way off like vultures before they came in. But shooting them down was another thing. Just about the time I got a good frame on one of them, someone would call out for evasive action."

Back in the waist, Marson and Fullem were busy at their hand-held guns, and the image of them in action is something George Hoyt will always carry with him:

"Through the open radio room door I caught glimpses of Chuck and Pete firing away at their right and left flexible guns. Chuck Marson was a bit superstitious about the door, and insisted that I keep it open despite the slip-stream that blasted back through my open upper hatch. But seeing him and Pete at their guns gave me a great feeling of confidence. Our tracer ammunition had been eliminated before this mission, since the powers that be had decided that it threw our aim off in aerial combat. So all of us at the flexible guns were on our toes to use our sights to the best of our ability. You had to be good to hit anything with these guns, and both of them were.

"Marson was a real pro in every way. The .50-caliber had a recoil plunger and spring device which impacted against the metal discs that were on the backplate. Marson had taken the gun's backplate off and he added some coins to the discs to make the plunger and spring go faster. This upped the rate of fire, but it also increased the danger of the gun 'blowing' or the barrel burning out unless you stayed down to very short bursts. Marson knew just how to handle it. With him at the gun I knew I had nothing to worry about.

"Pete didn't have the familiarity with guns that Chuck had, but he was a quiet, tenacious gunner who never complained or backed down in any way. Bob Hullar always showed complete confidence in him, and that was good enough for me. Pete was a great guy."

In the *Lady's* tail, Merlin Miller was performing his one and only real duty: "My job was to protect the airplane. The fun really started after our fighter escort left. I looked back and saw maybe six to eight enemy fighters behind us, so I called to the crew, 'Fighters, six o'clock.' I watched them as they strung out one behind the other, and started in. Then I called, 'Bandits, six o'clock.'

"When they got to within 500 to 700 yards I fired a couple of short bursts to discourage them from coming in closer. But they came in hard, and fast, and that's when I started to get really angry, because the closer they got, the more dangerous they were. From here on in it was just a matter of shooting and calling the fighters out to the other gunners as they came in and went past.

"We all had to work together. I relied on Sammy to see below and behind us, and on Rice to see above and behind us. Marson and Fullem would keep me posted too, and I would call out any fighters I saw, particularly when more than one was coming in at the tail so that Bob Hullar could skid the plane sideways a bit and maybe give one of the other gunners another shot. We worked together that way all the time. It was the only way we could survive."

So it went all during the flight from east of Aachen to the IP between 1345 and 1437, each crew doing their utmost to stay alive. In the low squadron, Lt. Jack Hendry moved to the No. 6 slot from tail-end Charlie after another bomber aborted, but the Group's records show him waltzing the *War Bride* with *Wallaroo* all through the fighter-filled sky that day. A formation diagram note says: "Ship #029 and #360 flew Nos. 3 and 6 position alternately due to intense enemy opposition."

It was near the IP that enemy opposition reached its greatest intensity, for the Germans knew this was their last chance to stop the 41st Wing. Their all-out assault finally bore fruit against the low 384th Group. Three B-17s went down—at 1425, at 1430, and at 1431. Shortly thereafter, Bud Klint witnessed other losses from German rockets.

"As we began our bomb run, I saw three '17s from the Wing ahead completely disintegrate and fall earthward in flaming shreds. This served as my introduction to the 'rocket gun,' which played an ever-increasing part in the European air war."

This happened as the Wing's three groups shifted out of the combat box into a line-astern formation in order to begin individual runs on the target—the Kugelfischer ball bearing plant in the center of Schweinfurt. The 379th went first, the 384th second, and the 303rd last.

Captain James wrote in his Group Leader's Narrative that "At the IP we made a left hand turn and made our run on the target on a magnetic heading of 40 degrees at 24,000 feet using AFCE . . . Flak in the target area was moderate and accurate. Enemy fighter attacks began on us before we reached the IP and continued to the target." In another report, he added: "I have never seen the like of those fighters in my life."

Ed Snyder recalls these moments well: "As we got into the target area and got onto the IP, we were attacked very heavily by twin-engine German fighters. One of them put a cannon shell or two into our radio room and blew a good portion of the radios up. But it didn't disturb the bomb run.

"Back in the tail I had a pilot, Lt. John Barker, riding as an observer to help me with the formation. He was going crazy because there were these 109s, 110s, 210s, and 410s back there shooting at us. He kept screaming, and I finally did have to cut him off. I can remember Mac very calmly saying, 'Please tell him we can't do anything now. I'm on the bomb run.' There was only one thing on Mac's mind, and that was to put those bombs on target."

Lt. Barker later offered this assessment: "It was a hell of a day. I'll bet there were over 300 enemy planes."

The twin-engine fighter that hit *Mr. Five by Five*'s radio room may well have been an Me-410 that attacked at 1438. It came in from about four o'clock level to about 400 yards and the right waist gunner, Sgt. Daniel Harmes, fired about 100 rounds in two bursts. He told the interrogator: "He went down out of control—parachuted out about 1000 feet below, and plane was observed to crash." Sgt. Harmes got credit for a kill.

The 303rd's mission file contains so many combat reports from this phase of the fight that a chronological account of what took place would be meaningless. But the experiences of individual crews, as reflected in the records and their own words, do provide one way of sensing what Black Thursday was really like. What follows is a selected chronology, set against the larger backdrop of events in the Wing, showing *part* of what happened to six of the 303rd's crews: Bob Hullar's, Lt. Bill Fort's, Lt. Ambrose Grant's, Lt. Jack Hendry's, Captain Lake James's, and Captain David Shelhamer's.

At 1442, a Do-217 went after *Luscious Lady*. Dale Rice told the interrogator: "He came in from seven o'clock high. He was about 1200 yards away. I kept firing as he came in. At approximately 700 yards he began to smoke. He burst into flames—two men bailed out of E/A. Plane went down." Rice got a "probable."

At 1445, the 379th dropped its bombs on Schweinfurt. Just after they went, two of its Forts were lost. Moments later the 384th Group dropped its bombs. Three of its B-17s were shot down, but the action was so hectic they weren't seen leaving the formation.

Just before the 303rd dropped at 1446, calamity struck Lt. Ambrose Grant's crew in the *Cat O' 9 Tails*. The crew's engineer, Robert Jaouen, tells what happened:

"For some reason that I do not remember, I was a waist gunner that day. Just before the target, I turned to yell something to Woodrow [Woody] Greenlee, who was on the other waist gun, and there was a bright flash between us. Woody was hit in the right side of his face by an exploding 20mm shell. Ed Sexton, the Radio Operator, came back and bandaged Woody's face and eye. We didn't have much time for Woody, as we had to keep shooting in hopes the fighters couldn't tell one of our crew had been hit.

"They threw everything at us, and the fighters swarmed down in wave after wave. It was a chaotic time for the entire crew, both in trying to protect the plane and in keeping it flying. The *Cat O' 9 Tails* was like a sieve. Someone stated there were around 200 holes of various sizes. Ed Sexton, being a devout

Catholic, often signed himself. I had seen him do this before, and it hadn't bothered me. However, things being as they were this day, it scared the hell out of me. I thought that maybe he knew something that I didn't, and that things were even worse than they appeared."

In the same seconds before the 303rd's bombs dropped, a Ju-88 went after *Luscious Lady.* From the left waist, Pete Fullem saw it coming in "from about seven-thirty o'clock, slightly high, but almost level."

He said later: "I started firing at him at 600 to 700 yards. Fired several bursts at him as he was coming in. At 200 yards he began smoking, burst into flames, and went down."

Fullem was credited with a damaged fighter, but the *Lady* was hit, too. George Hoyt recalls that "Pete threw four .50-caliber slugs from his left waist gun into the leading edge of our left horizontal stabilizer. The armor-piercing bullets ripped through the inner tail structure and came out through the rear, tearing away a section of the elevator some three feet in diameter, and leaving a gaping hole with shreds flapping in the slipstream. I can remember Merlin calling Bob Hullar, saying 'Hey, we got a hole in the tail right beside me big enough to crawl through.'"

Miller remembers the incident as well and says, "It didn't seem funny at the time, though we joked with Pete about it later. It was an easy thing to do in the heat of battle."

Meanwhile, it was Mac McCormick's moment at *Mr. Five by Five*'s bombsight. As Ed Snyder recalls, "Mac laid the bombs right on the target."

Bud Klint later learned that "The bomb run was perfect. Strike photos showed our bombs completely blanketed the target area."* The 303rd then made a sharp right turn off of the target, returning to its high slot in the wing formation.

"As we came out to the south," Ed Snyder continues, "you could see the fires from the aircraft that had been downed, and the smoke plumes going up. And, of course, we were still under fire."

The next minute was Merlin Miller's. At 1447 an Me-109 with a 30mm cannon slung under each wing came in from five o'clock high. These large-caliber guns were truly lethal; three hits, on average, were all it took to kill a B-17.

---

*The 303rd's target was the same Kugelfischer ball bearing plant that the Group had bombed on August 17th. According to George Schäfer, whose father owned the factory, the October 14th bombing could not have come at a worse time. The night before the factory had received a precious shipment of fuel oil which completely filled a large tank on the plant premises. The bombs exploded the tank, creating a massive fire that could be seen for miles.

Not surprisingly, the 109 was, Miller remembers, "one of those fighters you saw every once in a while who was determined to add a B-17 to his score. He came in straight and level with our tail. He throttled back, and I could see his prop slow down. I could see him fishtail as he started to aim at us.

"There wasn't much doubt who he was going to shoot at, but to put it bluntly, I sneezed first, before he could pull the trigger. I hit his plane just at the base of the left wing where it joins the fuselage, and it blew the wing completely off the airplane. He immediately flipped upside down and spun away. I didn't have time to see whether or not the pilot bailed out." Miller got credit for a kill.

At 1449 the Germans claimed the first 303rd Fortress. It was *Joan of Arc,* flown by Lt. Roy Sanders's crew. To her left in *Yankee Doodle Dandy,* Lt. Bill Fort saw the events leading up to *Joan of Arc's* demise.

"The sky was covered with quite a number of planes, and debris from planes that had exploded. It was real tough, close fighting. We were being hit from all sides. The slower German fighters were hitting us from the back with 20mm cannon shells and rockets. The Me-110s kept picking on this one plane off our right wing. They knocked all the fabric off its elevator and rudder, and eventually set it on fire, but apparently this didn't slow the plane down much, since they stayed with us another five to ten minutes.

"Then this German plane came back of us and under us, in a vertical bank that almost cut our wing off, and set the plane on fire again. It was mostly smoke, coming from the bomb bay, and then it flared up. We tried to move over to get out of the way of this plane, and shortly after I looked around and it was gone." Others in Lt. Fort's crew reported *Joan of Arc* being hit in the tail by an Me-110 rocket and going down on fire. Ten chutes were observed.

At 1450, the 379th lost another Fort. An Me-109 collided head-on with the second element leader of the Group's lead squadron, destroying the 109 and cutting 15 feet off the B-17's right wing. The Fort dropped out of formation; five chutes were seen.

While this was occurring, Lt. Grant's *Cat O' 9 Tails* was being assailed by FW-190s. Two with yellow noses came in at six o'clock level and Sgt. Francis Anderson opened up on them 600 to 700 yards away from the *Cat's* tail. One FW broke away but the other kept coming in, and at 300 to 400 yards Anderson's fire took effect. The 190 nosed up and seemed to stop in midair. It went into a spin and as it spiraled down, the pilot bailed out. Sgt. Anderson got credit for a kill.

Another yellow-nosed FW-190 attacked *Luscious Lady* at 1452. From the right waist, Chuck Marson took aim as it came in from five o'clock high. He later told the interrogator:

"I opened fire at 400 to 500 yards, bullets going into engine—Parts of engine began falling off—He began to smoke—His prop stopped completely—Then he rolled on his back and fell down."

Miller, Hoyt, and Rice were all certain the FW-190's engine was knocked out and Marson got a "damaged."

In the 14 minutes between 1453 and 1507 there were no less than 15 recorded attacks on the 303rd. One occurred at 1500, when an FW-190 came in to attack the *War Bride* from "six-thirty o'clock high to about 100 yards." In the top turret, Sgt. Loran Biddle "gave him 200 rounds. E/A started to smoke—rolled over on his back and pilot bailed out." Sgt. John Doherty confirmed from the tail and Biddle got a kill.

Germans weren't the only ones hitting the silk. The 379th Group lost a ship "at 1510 hrs, with No. 2 engine out and a large hole in the wing, going down under control. Five chutes seen." A short time later another B-17 was seen "with No. 3 engine feathered, dropping out of the formation."

And, in the most telling observation of all, the 379th's Group Leader, Lt. Colonel Louis M. "Rip" Rohr, stated: "There were reports of other B-17's going down, but action at that time was so intense crews could not keep track of them. As many as 10 to 15 B-17s were seen going down at one time."

It was some time during this phase of the fight that Merlin Miller accounted for another Me-109. Neither he nor George Hoyt reported it, but both remember the incident.

As Hoyt describes it: "An Me-109 flew up our tail so close that I was afraid his prop would chew it off. The vertical stabilizer was in my way, so I could not safely shoot at him. But Merlin let go with a burst from his twin tail guns that riveted the pilot back in his cockpit seat just as he was releasing his canopy and starting to push himself up to bail out."

Miller remembers it differently. "An Me-109 came in at six o'clock, at just about our altitude. I fired at him, and I think I hit him a couple of times. Pieces of his canopy flew off and he stopped shooting. But he kept coming in, closer and closer, nearer than any other fighter had ever come to us before. I just sat and watched, wondering what was going on. He was wobbling around, about 20 to 30 feet below us.

"Then all of a sudden I got this terrible sinking feeling. I realized all he had to do was pull back on his stick and hit his firing button, and I'd be a dead duck and the rest of my crew with me. So I put about 20 rounds into the cockpit, pounding it in on top of him. I remember the canopy flying off, but I don't remember the pilot trying to get out. My only thought was, 'He won't do it to us now!' After that he just continued on, wobbling away below us."

This fighter didn't get the *Lady's* tail gunner, but another almost did. Miller recounts that "I was back there with my head on a swivel, looking right to left, left to right, up and down, all the time. My head was never still for a moment watching for fighters. The tail position had windows to the left and the right, and it had a thick, flat bulletproof window in the back I looked through to see my post and ring sight. You could see quite well.

"Suddenly, out of the corner of my right eye, I caught a twinkling, like a flickering neon sign. I leaned back and turned my head to see what it was, and the next thing I knew I was lying on one of the ammo boxes, half on it and half on the floor wondering what the hell happened. My face didn't feel right. I rubbed it and realized that my oxygen mask was half off, so I put the mask back on. My right shoulder tingled and I had a sore spot above my right ear.

"I was still trying to figure it all out when I saw that my side windows were gone. The twinkling had to be a fighter sliding in at nine o'clock. He blew out my windows, and almost blew my head off. The lump on my head had to be from bouncing off the ammo box. I worked my right arm a bit, checked my shoulder, and found no holes in my uniform, so I must have been hit by flying Plexiglas from the windows.

"I really got angry at this point, because I was now getting a blast of cold air. I wanted that fighter to make another run so I could take a crack at him. It was a futile wish, but I hoped someone else would set him on fire."

Other crews in the Group continued to score. One of the victors was Sgt. Howard Zeitner in *Yankee Doodle Dandy's* ball turret. At 1515 an FW-190 passed to the rear of the high squadron 600 yards out. Zeitner started firing at six-thirty o'clock, and at five thirty o'clock the 190 went into a spin and exploded, with the tail and rear fuselage breaking off.

Sgt. Zeitner said afterwards, "He turned his back to me as he banked around and made a perfect target. I let him have a long burst and he tumbled down and blew up."

In the same minute an Me-110 made a tail attack on *Vicious Virgin*. The German came in at six o'clock level, firing rockets from 300 yards out. In the *Virgin's* tail, Sgt. Robert Humphries opened fire, hitting the 110 at the same time its rockets exploded under the B-17. The 110 went straight down with its left wing and engine on fire. Sgt. Humphries got credit for the crew's second "damaged" fighter.

It was also at 1515 that the *Cat O' 9 Tails* got still another Me-110. Sgt. Anthony Kujawa told the interrogator:

"I was in the right waist gun position firing the guns. My regular place is top turret, but it was out of commission and I had gone to the waist guns because [the] waist gunner had been wounded. I was firing his gun.

"An Me-110 was out around four o'clock. He was just standing there shooting at us when I came to the waist gun position. I turned the gun on him and began firing away. He was very near. I kept firing. Then he began smoking and he burst into flames. He began to spin towards the ground. He was disintegrating in midair." Kujawa got credit for a kill.

So it went, interminably. To Merlin Miller the return flight was "a constant battle, back and forth, fighters from all directions, lots of head-on attacks, and lots of attacks from the tail. I checked my ammo and realized that all I had left was in my auxiliary boxes behind the main boxes. So I called Marson, and told him to bring me back some ammo.

"When he came back, I told him to put the ammo in the boxes with the bullet points out, because that's how the belts feed into both guns. He made an appropriate remark, because that was like telling someone to put sod down with the green side up."

Chuck Marson came back at least two more times with extra boxes of ammo, and George Hoyt did too. "Later during this grueling dogfight, Merlin ran low on ammunition. I grabbed a wooden box full of .50-caliber belts, clipped on a portable oxygen bottle, and headed back to the tail with it. After I crawled past the tailwheel, he greeted me with a big wave of his hand. I gave Merlin the spare ammo, and felt relieved, as I knew he was the most capable defender of our most attacked point on the plane, the tail."

Hullar's crew stuck it out, joking to ease the tension, but finding themselves slowly worn down by the sights, sounds, and emotions of endless combat. Miller's other memories evoke "the way it was" better than any work of fiction:

"I remember hearing from Rice in the top turret. He was getting a blast of cold air just like I was. It didn't help his temper any more than mine, but he stayed pretty calm, and joked with the fellows.

"During a crew check I said, 'Everything's okay except the sons of bitches are trying to freeze me back here. They shot my windows out.'

"Rice said, 'Yes, I got a hole up here. They're trying to freeze me, too.'

"Marson piped up with, 'I don't know what the hell you fellows are griping about. Me and Pete don't have any windows at all—never did.'

"It was quiet for a moment, then Rice said, 'Don't pay any attention to him, Merlin. He's just mad because we're having all the fun.' And we were all still shooting at fighters as they came in."

"It was really cold up there. I was uncomfortable. All of a sudden I wanted a drink—drink of water, drink of beer, drink of whiskey, a drink of anything, I didn't care, I was thirsty. Why did I get so thirsty when I couldn't drink? Then I thought about smoking a cigar, but I knew I couldn't do that either. I tried it once at altitude, but when Hullar found out how I lit the thing with my oxygen mask on he threatened to kill me if I ever tried it again. He could smell the smoke through the oxygen system all the way up in the nose. All I could do was shoot.

"Off to my right I saw a B-17 flying alongside us. He got hit. His outboard engine on his left wing, No. 1, caught fire, burning back through the wing. I started yelling to myself, 'Come on, you guys, get out of there, get out of that plane!' It started down in a shallow dive, burning more all the time, and I counted the chutes as they came out. I felt a big relief when I saw the tenth chute pop out, because now at least they'd have a chance. A few seconds after that, the burning wing folded over the top, the plane went into a steep dive, and blew up. But I'd seen that happen before, so it wasn't that much of a shock.

"I was getting more and more uncomfortable. My nose itched and I couldn't scratch it. My oxygen mask felt like a dull knife gouging at my cheek-bones. My shoulder still ached a bit, and I felt generally bad all over. I'd been in this one position too long, and I was hoping this would get over soon.

"I was still shooting at German fighters, which were coming in with pretty fair regularity. They were shooting down a lot of bombers. They weren't hurting our Group so much for some reason, but I could see others all around going down.

"Off to my right I noticed a Ju-88 pull up under a bomber and hang there, shooting into the bottom of the bomber. The ball turret gunner was shooting back. I could see both his guns flashing. They got each other. I saw the German burning, and cannon shells exploding inside the ball turret, flashes inside, so I know the fighter got him.

"The bomber peeled off, burning badly. As it peeled away below me I could see a glow behind the pilot and copilot from flames inside, and I could also see flames coming out the top of the radio room. The plane was burning from behind the cockpit clear through the bomb bay and out through the top. I started thinking, 'Get the hell out of there,' like I did on the one a while ear-lier, but before I could even think it the plane cleared the formation and blew all to pieces. Outside of what looked like the four engines dropping, there was nothing left.

"I took a deep breath, was glad it wasn't me, and continued watching for fighters, shooting at them when they came in, worrying about my ammu-nition, and wanting to get this raid over with. It began to seem like it was last-ing forever when I heard Brownie say we were about 20 minutes from the French coast."

Each bomber crew has a similar story of courage and teamwork, a tale that must be told. The *War Bride* lost her No. 3 engine and Lt. Hendry could not feather the prop. She fell out of formation and was trailing below the Group, a prime target for fighters. John Doherty offers this account of her fight:

"Of that second Schweinfurt mission, what can be said? That raid will stand out in my mind for as long as I live. By all dimensions we should have gone down, but we made it back with a tremendous effort by every man on the crew.

"You couldn't single anybody out, but the pilot that day went far beyond what most pilots would do. I remember one time the call came over the intercom, 'Prepare to bail out!'

"I called back and I said, 'Don't give up!' That, I remember. The plane was vibrating so badly, at any minute a wing could have fallen off. But it didn't. Nobody did more for any group of people than Hendry did that day. That's why we all put him in for the Silver Star when we got back.

"I started to run out of ammunition. I called for the waist gunner to bring me back more. And the action kept going but it never seemed to get there. So I called on the intercom again and I told them I was leaving the tail. I was out of ammunition.

"Jim Brown, our radioman, called back and said, 'Stay where you are, I'm coming.' So Brown got me ammunition back there. But I had one gun, the left one, that was completely out [of ammunition]. The ammunition was all used up in the belts. The other gun had eight or ten or twelve rounds left. When we put a new box of ammunition in, we used to like to string it into the old bandolier. We took one shell out and put one from the new one in and it was just like a chain, it kept going.

"But I had this left gun completely out, so I had to get down on my back and crawl down in there where the gun was to get this new ammunition in. It didn't take too long, and as I was crawling back out I thought I would kind of roll over to the other side and link up my other box while I was at it. But something told me I had better check up and see what's happening, and as I looked up there was a plane coming in on us.

"This guy had got in there in our blind spot, and nobody had seen him. He had a kill—he knew he had a kill. He was getting right up to where he was going to give it to us with 'both barrels,' and there wouldn't be a question. He'd have us dead to rights.

"He was close enough to where I could see the silhouette of his face. I grabbed both them guns and shot right into the windshield. Never aimed, just grabbed them and shot. The windshield splattered, and I must have hit the pilot right in the face. The plane tipped, went into a dive, the wing broke off or something, and I forget the rest of it."

"The rest of it" is revealed in a 1602 combat report. The fighter was an Me-109 that had closed to 50 yards. Sgt. Doherty "gave him 200 rounds" and the 109 "went into a spin and then went over and over. Pieces of plane came off and it seemed to break. John Doherty got a "probable."

By this time, the *War Bride*'s No. 3 propeller had broken off its shaft and she had fallen completely away from the formation.

"We were using the clouds as much as possible for cover from the fighters," Doherty remembers. "And our navigator, Lt. McNamara, done a tremendous job of getting us *out of there*. We were by ourselves, and he got us out of enemy territory as best he could, directly."

The remainder of the 41st Wing crossed the French coast at 1655, south of Bologne. Now the bomber crews had to get back to England and land safely with damaged aircraft in weather conditions that rivaled the morning's. At 1720 the Wing crossed the English coast between Beachy Head and Dover flying at about 18,000 feet.

Captain James reported that "From the English coast to the base, we worked our way through the clouds and landed at the base at 1758 hours."

In his diary, Elmer Brown described this return with unintended poignancy: "This was another foggy day when even the birds were not flying, but we made another instrument takeoff and landing."

Bill Fort remembers that "Eventually we got back to England in about the same kind of weather that we left in. Luckily, we got down OK, along with a number of other planes. We were very fortunate not to get a bullet hole. Others were landing in any field they could find that was open."

There was one 303rd bomber that didn't land at all—Lt. Grant's *Cat O' 9 Tails*. On the return flight, the *Cat* had also strayed from the Group due to damaged engines, and she was little more than a flying wreck by the time she arrived over England. Robert Jaouen describes her last moments:

"We were struggling along, trying to keep the engines going. Several passes were made over the field, but we couldn't see the landing lights. Either the fog obliterated them or they weren't turned on, as there were reports the Germans might try to follow planes in.

"During these landing attempts, as we were circling, a break came in the clouds. We saw we were passing by a church steeple, and the pilot struggled to gain more altitude. Besides the crippled engines, we were running low on fuel, our control cables had been damaged, and keeping up altitude was near impossible, as was landing. Finally, after once again struggling to gain altitude, Lt. Grant ordered the crew to bail out."

"First out was Woody. Being semi-conscious, he was ejected with a static line, and I was ordered to follow him down. After leaving the chaotic conditions of the plane and the mission, I never have experienced such silence, before or since, as floating down through the clouds.

"I watched Woody disappear into the fog and was certain I knew where he was. After landing in a cow pasture, I started looking for him in the opposite direction from where he landed and I never did find him. He landed, apparently revitalized by the cold air, not knowing if he was in Germany or England. He was found by a farmer trying to read his compass. The farmer took him to a nearby air base on the back of a tractor. He was sent to an English Hospital and when well enough was sent to the States, where he underwent extensive plastic surgery.

"My landing was my first good break of the day, as I hit the mud near a gate where the cows had been standing. Thus, I couldn't have had a softer landing, except for the inconvenience of a little manure on my clothes. When I arrived at a farmhouse, Chester Petrosky [the ball turret gunner] had entered a few minutes earlier. The people were very nice, but stayed a long ways away from me. I finally deduced it was from the 'cow perfume' I'd picked up in landing. Eventually a constable picked us up and took us to a Canadian Air Force base, where we were hospitalized for the night, given sedatives, etc."

Lt. Grant's crew all landed within a four-mile radius, not far from the *Cat*'s point of impact near Riseley, a small English village 10 miles south of Molesworth. Her end made an indelible impression on a young English boy named John Gell, for the *Cat* came down in his family's backyard:

"I was a young lad nearly six years old, just old enough to be aware of what was going on. We lived in one of a gathering of cottages, my parents, myself, my two baby brothers, and my grandma. We had been watching the Forts come home that miserable evening, when we heard this particular B-17 making a lot of popping noises. I remember going to the front door as it passed over our group of cottages. A very loud crash was heard soon after it passed over, and on looking through a back window a cloud of dust was visible, with a large metal object nearly 80 yards away.

"The *Cat* came down in our back garden. She broke up on hitting an oak tree, which ripped off the starboard wing. The port wing separated from the fuselage when it chopped an elm tree off eight feet above the ground. The fuselage finished in two pieces 150 yards away in an adjoining field.

"My father was a special policeman who was involved with many plane crashes in the area. He and the wife of a neighbor went—holding hands—to the wrecked fuselage, not knowing what they would find inside. To their amazement, nobody was in it. My father put out a small electrical fire, and gathered up some used dressings. Soon the MPs arrived and darkness fell.

"The wreck was taken away two days later, on October 16th. Some days after, my father noticed something lodged in the crook of the oak. It was one of the *Cat's* propeller blades. I still live in that cottage today, and I still have that prop blade."

The *Cat* was not the only 41st Wing bomber abandoned over England. The 384th Group lost three in this way, plus six shot down over enemy territory. Three of its bombers landed away from its home base, and only the Group's lead ship returned to Grafton Underwood! The event had an enormous impact on the 384th's ground echelon that is well recalled by John F. Bell, a Lieutenant then serving as Assistant Squadron Engineering Officer in the Group's 547th Squadron:

"Our ground crews were waiting and waiting at their hardstands. And *nothing* was coming back. It was a pretty sad looking sight. These people had worked their butts off preparing these aircraft over a long series of missions that summer. They really thought of themselves as part of a team with the flight crews. And when the other half of the team didn't come back, it was a hard thing to take."

The 379th Group paid a high price, too. Six of its ships were lost and one landed away. Lt. Colonel Rohr summed up the day's events as well as anyone when he reported: "This, in my opinion, was an extremely rough mission."

At Molesworth, the mood that evening was far more upbeat. The Hell's Angels had come through the roughest mission ever with minimal losses and much to be proud about. Thanks to McCormick, the Group's bombing had been superb, and the prolonged aerial combat yielded a bumper crop of fighter claims. When all the crew interrogations were completed, the 303rd sought credit for 20 fighters destroyed, four probables, and 13 damaged.

There were other, more personal reasons for celebration as well. Lt. Carl Hokans and Lt. Walter Witt had both beaten the long odds against completing a 25-mission tour. Elmer Brown wrote that "The boys took Hokans's pants off and he was running around without any at interrogation." Two other Group crews also marked this mission as their last. There were, of course, others in the Group affected by the loss of the one 303rd crew. That evening Lt. Ralph Coburn wrote, "Sanders went down. Good boys, we'll miss them."★

For Hullar's crew, the feelings were mixed. McCormick took satisfaction in his accomplishment, saying afterwards: "I think we did the job right today."

Norman Sampson summed his opinion up in a word—"Rough!"—and Bob Hullar was equally taciturn when he wrote: "Rough do this trip."

Bud Klint thought about the losses: "I saw what seemed to be a fabulous number of planes go down, and later I found out how right I was."

Elmer Brown remembers that he was "very impressed by the number of planes going down. It was just terrible to see all those B-17s on fire and to know our people were being killed. It really hit home that the high command thought we were expendable. All we could do was fight the Germans as best we could." In his diary he noted: "Our plane shot down five enemy fighters and we shot up more ammo than any other plane in our squadron."

Merlin Miller recalls thinking at the time, " 'The raid's over. We've done it again. We're all still alive.' It was something I wouldn't have bet on a few hours earlier. We shot up all our ammo. Even the spare boxes in the radio room were empty. There were less than 150 rounds left in the whole plane. I must have fired at least 1000 rounds from each tail gun, and Rice reloaded his guns a number of times too. When I came out of the tail to the waist the brass was knee-deep. We figured we shot down at least five German fighters, and though we really didn't keep count, some of the other crews tried to give us credit for a few more. We used up more ammo than any other plane in the Group."

Pete Fullem took some ribbing about the holes he shot in the *Lady's* tail and was rather sheepish about the incident. He felt even worse about the duel between the Ju-88 and the bail turret gunner, which he had seen just as Merlin Miller had. Miller remembers that "Pete was sitting on his bunk comparing notes with me. He talked about wanting to shoot at that Ju-88, but not being able to. I know he felt bad about it, but that was just the way it went sometimes."

Hoyt didn't want to think about any of it. "I wasn't shaky, but I was badly fatigued, thoroughly zapped. I wanted to forget it all, to wash it out completely. So when someone said we should go to the 'NAAFI,' I was all for it. NAAFI stood for the British 'Navy, Army, Air Force Institute,' but it really

---

★Lt. Sanders's right waist gunner and tail gunner were killed in action; the rest of the crew became POW's.

meant the enlisted bar on our base, the 'NAAFI Club,' which the British kept open for us. As I recollect, all the enlisted men went, and we had quite a time. They had a kind of potato beer there—you had to watch for the sediment in it—and they later broke out some Canadian ale. By the time the evening was done I had washed a lot of the day out."

The "someone" who suggested the NAAFI was Merlin Miller. "I went to eat my evening meal with the rest of the crew, and a bit later on in the barracks I said to Rice, 'Let's go over to the NAAFI and have a couple of beers.'

"He said, 'That's a damn good idea. Only I'm going to have more than a couple! He really liked the mild and bitters that they served warm. So we wandered over there and drank a few beers, and we had some cookies too. And after a while we all headed back to the barracks.

"Rice looked at me and kind of grinned, and he said: 'Merlin, you better get a good night's sleep, 'cause you know, we might have to do this all over again tomorrow.' "

The following days brought a sober reassessment at all levels of the Eighth. The losses had simply been too great. Second Schweinfurt was to prove a true turning point in the daylight bombing campaign.

It was not long before the Eighth's bomber crews discovered the full extent of Black Thursday's bad news.

On October 15th, Elmer Brown recorded that "Sixty B-17s were lost yesterday. We were in the first task force composed of nine groups; 45 B-17s were lost by this task force. Our Group lost one."

These numbers established that the First Division had indeed fulfilled one part of the October 14th plan. It had drawn the greatest number of enemy fighters, though there were still plenty left over for the bombers that followed.

The Third Division bombed Schweinfurt's factories five minutes after the 41st Wing left the target area; the balance of 15 B-17s lost had come from its ranks. Total losses amounted to a stunning 19 percent—of the 291 B-17s that pressed on to Schweinfurt, nearly one in five failed to return.

Nor was this all. Seven bombers were written off as Category Es and many of the planes that *did* come back were out of commission for two to seven days undergoing repairs.

A total of 139 B-17s suffered battle damage, including *Luscious Lady*. Bud Klint recorded that she "suffered a broken wing spar, a shattered top turret, and a wide assortment of holes which sent her off the field to a Unit Repair Depot for a major overhaul. The *Lady* even had to be patched up before she could be flown to the Depot."

While the ground crews and repair units tackled the huge task of getting the Eighth's aircraft ready for the next raid, the combat crews tried to unwind. The day he recorded the Eighth's losses, Elmer Brown also wrote: "We have a 48-hour pass but we officers didn't even leave the post. Just laid around and relaxed."

In contrast, the crew's enlisted men did spend their time off-base, as George Hoyt recounts: "The next day we enlisted men set out on a 'special mission' to London, a two-day pass. These passes, with their escapades, provided very important relief for the stress that built up from high-pitched missions like Second Schweinfurt."

At the same time, America's leaders were trying to manage the public relations crisis that had grown from the October 14th losses. The toll, when combined with he aircraft lost on the preceding three missions, amounted to 155 B-17s. It was a serious military setback, for no combat organization could long endure losses on such a scale, and the military problems were made to appear worse by a series of gaffes President Roosevelt committed during a press conference on October 15th.

The President inadvertently confirmed the loss of 60 bombers on the mission and then stated that the Eighth could not afford to lose that number on every operation. He confused the issue further by denying that it was losing that many aircraft on every mission, and then tried to offset the losses by mentioning the 100 German fighters the Eighth had claimed were destroyed on the raid.

The President also mentioned the severe damage inflicted on Schweinfurt's ball bearing factories, but he was unable to convey the impression that a true knockout blow had been delivered. And, in fact, this objective was not achieved, despite the excellent bombing by both the First and Third Divisions and the fears of German leaders such as Albert Speer, Nazi Minister of Armaments and Munitions.

Three days later it was General Arnold's turn to fumble during a press conference. He was greatly concerned about the impact the loss of 60 B-17s would have on the public's attitude toward the air war, but his comments were poorly conceived. He opened by stating that high losses were inevitable against an objective such as Schweinfurt, and then implied that even a loss ratio as high as 25 percent might be acceptable. His closing, exaggerated comment—"Now we have got Schweinfurt"—offered neither solace nor reassurance.

Back in England, Eaker and Anderson also sought to quell public unrest, and to this end VIII Bomber Command called upon a surprising source: Major Ed Snyder. He recalls the occasion very well:

"I was called down to Bomber Command Headquarters after Second Schweinfurt to make a propaganda-type broadcast to the folks back home. I understood that Bomber Command was interested in getting one of the group leaders who participated in the mission to give a personal account. It was done on the radio, and was sent on to the broadcasting systems in the United States.

"I think the point they were trying to get across was that this was a very important mission and, yes, we had all those people go out on it, and it was quite hairy, but even for the guys that didn't get back there was some hope. I believe I was put on mainly as an eyewitness who saw a lot of parachutes come out, who could say that just because we lost all those airplanes, it didn't mean no one was going to make it back.

"I had no reservations about making the broadcast. I realized the gravity of the situation with us losing that many airplanes, but felt this was something that just had to be. You couldn't win the war without taking some losses, and this happened to be one of the tough ones. I always had the feeling that the losses were justifiable some way or the other. I had faith in the people who were planning these missions and I believed they wouldn't have sent us there unless they really thought it was necessary, and that they could get us back with reasonable safety. And you have to remember that we were practically the only Americans fighting the Germans at the time. The Invasion wasn't until June of 1944."

Despite the determination of men like Snyder, it was now absolutely clear that the Eighth needed new resources to make deep penetrations with acceptable losses. Until new escort aircraft—the P-38 and P-51—became available, the Eighth would be restricted by weather and limited fighter range to shallower raids into Occupied Europe and Germany.

In the week after October 14th, the Eighth was incapable of even this kind of effort. On October 18th Elmer Brown wrote that the Group "started out for Düren and was called back at the English Coast. Weather."

But this explanation merely heightened doubts among the press and the people back home and the crisis of confidence continued.

Along the Yalu

BY JAMES SALTER

(Excerpted from the novel *The Hunters*)

Even though he has published only a half-dozen novels and books of short stories, the prose of James Salter has earned the loftiest international literary accolades and the praise of his peers. Distinguished by its terseness, every word seeming to be so perfect and vivid, like the diamond sparkles striking your eyes from a sun-struck field of snow, James Salter's pages of description, action, and dialogue make the average novelist's efforts seem bloated and turgid.

A veteran filmwriter in addition to producing novels and short stories, Salter's works range through such diverse subjects as mountain climbing, the air war over Korea, and life in Paris. Although everything he writes has the veracity of personal experience, Salter's novel of air combat in Korea, *The Hunters,* is particularly striking in this regard. In the early 1950s, when American Sabre jets flew against North Korean and Chinese pilots in Russian-made MIGs over North Korea, Salter was there. Although fiction, *The Hunters* is based heavily on Salter's personal experiences.

The lead character is the pilot Cleve Connell, an experienced jet jockey but new to the Korean action and anxious to prove himself against the enemy. As Salter points out in his preface to the 1997 edition of *The Hunters* (originally published in 1956), the dream of every pilot was to score a kill. "A small red star painted on the side of a pilot's plane below the cockpit was the symbol of a kill. Discreet, almost invisible in the air, a row of five was a mark of highest honor, greater than any trophy or prize."

Korea has been called "America's forgotten war." The veterans who fought there have strong feelings about what they endured and accomplished. James Salter writes in his preface:

"It was said of Lord Byron that he was more proud of his Norman ancestors who had accompanied William the Conqueror in the invasion of England than of having written famed works. The name de Burun, not yet Anglicized, was inscribed in the Domesday book. Looking back, I feel a pride akin to that in having flown and fought along the Yalu."

At 5:15 in the morning it was piercingly cold, with an icy moon still bright in the sky. The windows of the barracks were dark as Cleve walked down the road to where a truck waited in front of the mess, its parking lights on and an asthmatic smoke wreathing from its exhaust. The mud beneath his feet was frozen into hard ridges and swirls. The cold bit at the tips of his fingers through his gloves. He had given up eating breakfast on mornings like these. The result was an insistent hunger later on, but he preferred the extra sleep. He had finished all the training flights and indoctrination. During the week past he had started flying missions. There had been four of them, all uneventful. This was to be his fifth. He was scheduled on Desmond's wing.

After the briefing, they dressed in a locker room grim with the light from the early day and a single unfrosted bulb. Desmond always loaded himself with equipment in excess of that which everybody carried in standard seat packs. His pistol he wore at his waist, the holster tied down against his thigh with leather thongs. On the other hip was a heavy hunting knife and a canvas packet of extra ammunition clips. Besides that, he filled his flying-suit pockets with plastic boxes of jellied candy, cigarettes, and hand warmers, wrapping friction tape on the outside of the pockets to hold them firmly. Everything had to be secured or it would be lost upon bailing out, ripping right through cloth at the shock of the parachute opening.

There was some erratic humor. Robey, one of the flight leaders, read an imaginary telegram he had received from Big Stan Stalenkowicz—"You all remember him as tackle on last year's team." Stan was going to be at the game today, and he wanted they should get out there and really fight. In reply, there were some pledges to win this one for Big Stan.

Robey was credited with four victories. He was the leading man in the squadron, and he did not look the part at all. He had a small, pale mustache, which seemed to have been pasted as an afterthought onto the face as bland as a piece of fruit. His complexion was bad. The one thing that distinguished him was the self-assurance of an heir. One more aircraft destroyed, and he would have his title. Because of this, he was treated with deference. In return, he was patronizing. He moved among them as if they were, even unknowingly, his flock.

"Going to open up a hardware store, Des?"

"Very funny."

Cleve dressed himself slowly to reduce the time he would have to spend standing around and taking little part in the talk. He was not fully at ease. It was still like being a guest at a family reunion, with all the unfamiliar references. He felt relieved when finally they rode out to their ships.

Then it was intoxicating. The smooth takeoff, and the free feeling of having the world drop away. Soon after leaving the ground, they were crossing patches of stratus that lay in the valleys as heavy and white as glaciers. North for the fifth time. It was still all adventure, as exciting as love, as frightening. Cleve rejoiced in it.

They climbed higher and higher, along the coast. It became difficult to distinguish earth from water where they met. The frozen river mouths blended into white land areas. The rice paddies south of Pyongyang looked like cracked icing on pale French pastry. He saw the knotted string of smoke go back as Desmond test-fired his guns. He checked his own. The sound of them was reassuring.

They climbed onto the contrail level. Long, solid wakes of white began flowing behind them. Formations left multiple ribbons of this, streaming sky pennants. Frost formed on the rear of Cleve's canopy. He was chilly, but not uncomfortable. They were north, and he was busy, looking hard, clearing himself, Desmond, and the two other ships in his flight. The sky seemed calm but hostile, like an empty arena. There was little talking.

In half an hour they had reached the Yalu, an unreal boundary winding far below. The sun was higher now. The sky was absolutely clear. His sunglasses made it a deeper blue, like deep ocean. He could see a hundred miles into a China that ended only with a vast horizon, beyond the lives of ten million rooted people. At forty thousand feet they patrolled north and south, turning each time in great, shallow sweeps.

They had been doing this for about ten minutes when somebody called out contrails north of the river. Cleve looked. He could not see them. Then he heard,

"They're MIGs."

He heard Desmond: "All right, drop them."

He dropped his tanks. They tumbled away. He looked north. Still he saw nothing. He was leaning forward in his seat, intently. He stared across the sky with care, inch by inch.

"How many of them are there?" somebody asked.

"They're MIGs!"

"How many?"

"Many, many."

He looked frantically. He knew they must be there. He began to suffer moments of complete unreality. He felt he was staring holes in the sky.

"Where are they crossing?" somebody called.

"Just east of Antung."

Then at last he saw them, more than he could count. It seemed unbelievable that he had been unable to locate them only seconds before. He could not make out the airplanes, but the contrails were nosing south unevenly, like a great school of fish. They were coming across the river. They were going to fight.

Soon they were near enough to distinguish: flight after flight of from four to six ships, the flights in a long, tenuous stream, all above them, at forty-five thousand, he guessed. The van of this column was approaching fast. Suddenly, he understood why these formations were called trains. He expected the fight of his life momentarily.

"Let's take it around to the right," he heard Desmond say.

They started a turn toward a position beneath the MIGs, with unbelievable lassitude it seemed, and began traveling south with them. Cleve felt very alone in the cockpit. He was acutely aware then of being far into enemy territory. He squirmed in his seat. His mouth and throat were dry. It burned to inhale. Still they went south, the MIGs staying above. It was like watching a fuse burn.

At that altitude they could not climb the five thousand feet up to the MIGs without losing speed and falling behind or else leaving themselves almost motionless in the air to be attacked, so they continued underneath and a little to one side, watching the ships and contrails floating high above like the surface after a deep dive. Cleve was shocked by the number of them. He could count more than fifty. At that moment he had only one friendly flight besides his own in sight. There were sixteen friendly ships altogether, four flights.

Suddenly, the radio exploded with voices. The fight had started somewhere. He felt his nerves twitching. Then there were four of them, Desmond called them out, turning down for a pass. They did not come all the way, however. They swept overhead, going at an angle. Cleve saw them closely for the first time. He watched the nearest one sail across, silver and abrupt, with speed fences on the wings, as soundless as a great fish. Then they were gone.

Two others started down in a high side pass. They turned into them, and the MIGs pulled up and continued on. It was all sparring. Desmond was cautious. He kept them out of trouble, but constantly turning so that there was little chance for him to make a pass himself. He flew like a boxer who keeps moving away, waiting for an opening.

Even though Cleve could see the MIGs easily now with the contrails marking them plainly at great distances, he still had a pressing sensation that they might be coming in from all sides, unseen. He sweated, twisting in the cockpit, straining to look everywhere. They turned indecisively among the MIGs for about ten minutes. Once he saw one firing at him from a long way off. The cannon shot firm, heavy tracers that arced through the air like Roman candles.

Finally, he and Desmond were chasing four of them north, unable to close; and when they broke off, it was all over. The MIGs were gone, vanished, as characteristically as they had appeared. The sky was empty except for the fading traces of contrails, left like ski tracks in blowing snow.

They turned toward home. Cleve felt tired. As he listened to the talk of the withdrawal over the radio on the way back, he realized that he could not remember having heard anybody except Desmond after they were once in it, he had been so absorbed.

"It looks like they came up early in the morning for a change," Desmond said when they had landed and were waiting for the truck to throw their equipment on and ride back to operations, "but it wasn't much of a fight."

"No, it wasn't," Cleve agreed, although he felt very spent.

"They were too cagey today. It's usually like that when the fight is in the cons. They can see you too easily, and you can't get close to them. Not only that, but they just didn't seem to want to mix it up this time."

"I thought they were doing their share."

"What do you mean?"

"It seemed we were playing it pretty safe," Cleve said.

"You got back, didn't you?" Desmond said flatly.

"So did they."

There was a silence. Cleve regretted having said it.

"You did a good job, Cleve," Desmond said simply.

"Thanks." He thought with despair that it had not been as he had anticipated, easy fight or not. He was going to have to push himself beyond what he had expected. A sense of inadequacy made him feel exhausted and as fragile as a dry stalk.

There came a morning like autumn or the long marble corridors of some museum. The sunlight seemed preserved as it gleamed from sleeping surfaces, and the air was still. They were on the second mission of the day.

"Great, Billy," Cleve told Hunter as they left the briefing and walked through the mild winter noon. "You've got No Go, and I have the Guzzler."

They were scheduled in one of the last flights and had been assigned the oldest, most troublesome ships. Hunter's was notoriously slow, and Cleve's drank fuel. Hunter laughed a little.

"This is the day that we're bound to run into a hundred of them," he said.

The previous mission had been in a running fight along the Yalu, but with no conclusive results. It was the first time that the MIGs had been up in several days, and there was a chance that they would be flying again this time, Cleve hoped.

"I'll be surprised if we just make it up there and back in those dogs," he said.

The locker room was never pleasant for Cleve. As they dressed, he felt the usual pre-mission discomfort. He was glib, but there was a looseness in his knees and the insistent uneasiness of what was he doing involved in all this? There was plenty of time to dress, too much time he had always felt. He talked with DeLeo and Pell, briefing them additionally. At last they all went out to the ships. Two members of a flight that was not going stood near the door as everybody left.

"Get one for me," they burlesqued.

The time before takeoff was always difficult, too. The mind could occupy itself, but the dumb, quavering heart could do nothing. Cleve sat in the cockpit, checking the second hand on his wrist watch. He drummed his fingers on the tight metal skin of the ship. Finally, it was time to start engines. He passed gratefully into the realm of function.

Once they were into it, the sky was clear, and bright sunny blue. It was a sky, Cleve thought, you could see tomorrow in. He looked over toward Hunter on his wing. DeLeo was flying number three, wide on the opposite side, just then moving into position with Pell number four out beyond him. They climbed north, over the quiet Haeju Peninsula and then across the edge of the Yellow Sea, heading the shortest way for Antung.

A fight seemed to have started already. They could hear the loud, excited transmissions of one flight among the MIGs. They were late, Cleve thought angrily. He pushed the nose of his ship down slightly, lowering the rate of climb and increasing the forward speed. He wanted to get to the Yalu as soon as possible.

At thirty-four thousand feet they began to leave smooth, persistent trails in the air. Cleve stopped climbing and dropped down several thousand feet to remain below the contrail level where they would be less visible. The river seemed deserted when they reached it. They could not locate the fight. Cleve asked several times where it was, but was unable to get any clear answer over the radio jammed with voices. He heard the ground control radar calling out train number four. There were many MIGs in the air, he knew, somewhere.

"Bandit train number five leaving Antung." In the tunneled voice of a stationmaster another one was announced. "Train number five leaving Antung, heading three five zero."

"Drop tanks," Cleve ordered.

He felt a buoyancy that was both fear and expectation. From here on, he was working against time to find them. He headed up the river, passing occasional elements, all friendly. He scanned the wide sky meticulously, high and low. There was a speck of dirt on the plexiglass canopy that looked like a distant airplane every time his eye passed over it. Despite himself, it tricked him again and again. Aside from that, there was nothing. As he turned to go down toward the mouth of the river he saw four ships chasing two MIGs far below, flashes of silver against the snowy ground. The radio was cluttered increasingly with cries of battle.

"Bandit train number six leaving Antung."

It seemed impossible to be traveling through so big a fight without finding anything. A desperate sensation of futility seized him. He was certain he was heading in the wrong direction, but he had turned less than a minute before. He could not cover ground fast enough. He felt as if he were merely hanging in air.

Someone called out sixteen MIGs heading south.

"Where?" Cleve asked.

No answer.

"Where are the sixteen MIGs?"

"Heading south! Sixteen crossing the river!"

"For Christ's sake, where?"

There was no answer.

Suddenly Pell called out something at three o'clock. Cleve looked. He could not tell what it was at first. Far out, a strange, dreamy rain was falling, silver and wavering. It was a group of drop tanks, tumbling down from above, the fuel and vapor streaming from them. Cleve counted them at a glance. There were a dozen or more, going down like thin cries fading in silence. That many tanks meant MIGs. He searched the sky above, but saw nothing. They were somewhere in that deep blue, though; they had to be. At great distances planes could appear and vanish before the eye as they turned or rolled out, depending upon the surface they presented, but these must be close. He had to see them. Segment by segment he checked and discarded the sky above. Then, from nowhere, there were two MIGs sailing past, headed the other way.

"There's two on the left!" Cleve called. "Let's go."

He turned—a delicate, speed-killing proposition at altitude—and fell in far behind them. It was another chase, long and useless, but they were going south. Sooner or later the MIGs would have to turn back. Cleve took a chance on that. He would never catch them otherwise, in the extended straightaway.

He looked back to check Hunter. DeLeo and Pell were not following. DeLeo called that he was breaking off to go down after some others. Cleve could not see them. He looked forward again. A moment later the MIGs he was trailing started a wide, climbing turn. It was sooner than he had hoped for. He cut to the inside, gaining on them.

"You're clear! You're clear!" he heard Hunter calling.

They continued to turn. He drew nearer. It all seemed childishly simple. He wondered if they had seen him yet. He was almost in range, closing on the second MIG steadily. He ducked his head to see the gunsight reflection on the armor glass. The MIG was growing bigger and bigger in the bright reticle.

"Keep me cleared."

"You're all right."

Before he could fire, the MIG banked steeply and tightened the turn. He's seen us, Cleve thought. The limber sight computed itself off the glass screen as Cleve turned hard after him. The MIG began to climb. The sight swam back into view. Everything seemed to be going at a sleepy pace. They were not moving. They were all completely motionless in a glacier of space. The leader had disappeared. There was just this one. He fired a brief burst. The tracers lined out and fell short, like a bad cast. He pulled the pipper forward a little as the MIG turned, still climbing. He squeezed off another burst. It fell around the wing. He could see a few flashes there and the minute debris of glancing hits. He managed to move the pipper forward again, leading more.

"There's one coming in on us," Hunter shouted. "We'll have to break."

"OK," Cleve said, "tell me when."

"It's two of them."

With just a few grains of time he could do it. He had no thoughts but those that traveled out on a line of sight to the plane ahead of him. He needed only seconds. He fought the impulse to look behind. The pipper refused to stay in the right place. He kept calmly adjusting, holding his fire. It was like standing on the tracks with his back to an express already making the earth tremble.

He fired again. A solid burst in the fuselage. The silver lit up in great flashes of white. He was playing a machine in a penny arcade. Suddenly he saw something fly off the MIG. It was the canopy, tumbling away. A second later the compact bundle of a man shot out.

"Did you see that, Billy?" he shouted.

"Break left!"

Cleve turned hard, straining to look back. Two MIGs, firing, sat close behind. Their noses were alight. He was turning as hard as he could, not gaining, not yet feeling himself hit, thinking no, no, when at the last moment they were gone, climbing away, in the direction of the river.

Cleve saw nothing more of the fight. He headed north for a while, but it had all ended. There was only the meager conversation of flights withdrawing from the area. It was over. The fight had dissipated. The MIGs were gone.

Cleve had never felt so fine as when finally they headed back through the quiet sky. This was the real joy of it all. He understood at last. He looked across at Hunter. His ship, far out, was like a silver, predaceous minnow with an abrupt, featherish tail. It seemed to be fixed against the azure blue of altitude. At that moment, Cleve could not remember ever having doubted that he would know this heady, sweet surfeit. Instead, it was just as he had always felt it would be. He knew then that he would never lose.

He was unprepared for what happened soon after they had landed. He thought he heard a crew chief say it, and then they told him as they walked to debriefing: Pell had gotten one, too. Cleve saw DeLeo waiting for him outside the sandbagged operations building. He appeared angry, tight with fury.

"What happened, Bert?" Cleve asked.

"Haven't you heard?"

"They tell me that Pell got a MIG."

"That's right. The son of a bitch went off alone and got one."

"Alone? By himself?"

"Sure, by himself," DeLeo said.

"He didn't say anything to you?"

"Not about leaving me. I was going after a flight of four of them. It was after we left you, later, and he called that he had some more of them out to the side of us. I said OK, and the first thing I knew he was gone, and I had two right in back of me that I damned near never got away from."

Pell came up, his face circumspect, but subduing a grin.

"How'd it go?" he sad to Cleve casually. "I understand you got a MIG."

"That's right. I hear you got one, too."

"I did," Pell said happily. "I guess I was pretty lucky. I got hits all over him, though."

"Where did you get the idea that you could take off alone in the middle of a fight?"

Pell's expression was innocent.

"I didn't know I was alone," he protested, "until I was just about to start firing on this MIG, and then it was too late to do anything else. I lined up behind him . . ."

"What do you mean you didn't know you were alone?" Cleve interrupted. "What made you think you could go off and leave your leader?"

"He said it was OK. I asked him."

"Listen, you son of a bitch," DeLeo began, "you never asked me a thing."

"Yes, I did. I called out two MIGs to the right of us, and you said it was OK to go after them. I thought you were with me all the time."

"I didn't tell you to go after anything," DeLeo said flatly.

"I thought you did. Well, that's probably what caused us to become separated."

"I don't care what caused what, Pell. You never said a word to me, and even if you did, I didn't tell you anything about going after them. When you're flying wing, your job is to cover me, and you stay there and do that no matter what you see or think. You almost got me killed today."

Pell did not reply.

Cleve was tempted to let it go as a misunderstanding. Things like that could happen easily enough in the excitement of fighting, he reasoned. Meanwhile, it seemed as if a dozen people were crowding around him, offering handshakes and asking how he had done it. He found it difficult to sustain any displeasure. He was swept along in a flurry of rejoicing. There were two MIGs in his flight.

"Cleve," Imil said, punching him on the flat of the shoulder, "I knew you'd do it. It took a while, but I knew you would."

"He bailed out," Cleve grinned. "I could have kissed him."

"You should have given him a squirt."

"Oh, no. That one's my friend. He may be back tomorrow with another MIG for me."

Imil laughed.

"It's only the beginning," he said. "You're on the way now. I hear a wingman in your flight got one, too."

"That's right."

"Who was it?"

"Pell. He's a second lieutenant."

"Pell, eh? They tell me it was only his seventh mission at that. Well, that's good work."

Everybody was saying nice going. Nolan came by, and Desmond. The debriefing was continually interrupted. A sergeant was standing by to take pictures for press releases. Cleve felt the full warmth of exhilaration devouring him. So this was what it was like to win. Already he could no longer recall the hunger and despair of days past.

DeLeo stood in the background silently. Cleve took the opportunity to talk to him as soon as he could. He wanted to smooth it over.

"It won't happen again," he said.

"He's going to get shot down," DeLeo swore. "They'll get him up there alone and murder him. He's a smart one, but I don't care how smart he thinks he is or how good he thinks he is. If he's alone, he can't cover himself, and they'll get him. I don't give a damn if they do. He's asking for it. He'll never leave me again, though. I won't fly with him."

"He's all right," Cleve argued, feeling the words awkward in his mouth. "It was probably a misunderstanding, that's all. Give him the benefit of the doubt."

"It was no misunderstanding."

"It might have been. Those things happen."

"Who do you believe anyway?" DeLeo asked. "Me or him? It has to be one of us."

"It's not a question of that."

"It's not, eh?"

"I'm just trying to bring out that it could have been an honest mistake."

"Honest?" DeLeo said. "He knew what he was doing."

"We'll see."

They stood there for a while close to one of the flat, interior walls of the quonset, not talking. Pilots still thronged about the map-covered tables, explaining what they had done and seen, and the room was filled with voices. Cleve caught sight of Colonel Imil talking to Pell near the doorway. The colonel seemed happy. Pell must have been elated, too, but his expression was composed, a smile both modest and assured.

Hunter came by. He was brimming with words and excitement.

"You should have seen it," he told DeLeo. "MIGs in front. MIGs in back. It was a circus."

He turned to Cleve. "I still don't know how we got away with it," he said.

"You played it just right, Billy."

"Oh, no," Hunter cried. "You were the cool one in there. I was scared. I admit it. I was keeping my eye on the ones behind, though. I was trying to judge it just right. You know, the last second, like you said."

"You were perfect. I mean it."

"It worked out, didn't it? Just right. We'll get them again, too."

"You bet we will, Billy." Cleve was grinning.

Carrier Qual

BY DOUGLAS C. WALLER

(Excerpted from the book Air Warriors:
The Inside Story of the Making of a Navy Pilot)

For anyone interested in aviation, one question is inevitable: What's it really like to fly a jet fighter like the F-14 Tomcat onto the deck of an aircraft carrier? The task—or feat, if you want to call it that—seems absolutely daunting, an impression verified again and again by movies or television documentaries. One can't help wondering: Could I be trained to do that? Would I choke? Bust my butt?

There's no easy answer to those last questions, but some excellent insight into the first—what's it really like?—is coming right up in these pages.

Douglas Waller's book *Air Warriors* brings naval fighter pilot training to the printed page with a force of engaging realism few books have captured so well. You want to fly an F-14 Tomcat? Then read Waller's book.

For now, we'll cut to the chase, as they say. Before Uncle Sam lets you get your hands on an F-14, you've got to prove you are a carrier-qualified jet pilot. You'll be using the T-45 Goshawk, in the most difficult aviation tests you've ever faced.

The view was spectacular. Behind him and to the west only wisps of clouds danced across the light blue sky. In front, billowing dark thunder clouds stacked in stairsteps to the morning sun. But they were far to the south, no threat to the dangerous landings Rob Dunn would make today. He felt so serene. If there was a heaven, it had to be off Key West with the rippling dark blue Florida Straits below him. Streaks of aquamarine from crosscurrents interrupted the dark blue, along with the white speckles of cigarette boats and trawlers sailing near the coast. He tried to make out Cuba, which lay fifty miles to the south. Dunn had naively asked how he would know if his jet accidentally strayed into Cuban airspace. "The missile coming off the rail of a MiG ought to be the first clue," Mango had said with a laugh.

Dunn had been circling his Goshawk at almost a mile high. He was with a group of four jets, led by an instructor piloting one of the aircraft. Below him, the aircraft carrier looked so peaceful as it steamed sixty-nine miles from Key West toward the east. From his altitude, Dunn could just make out the trickle of a churning light blue wake the behemoth left behind. Everyone had told him the ship would look like a postage stamp from this altitude. To Dunn, it seemed larger. But he was amazed at how narrow the vessel appeared, a thick dark tube on the sea almost like a submarine. How the hell could anybody land on that pencil, he wondered.

"It's show time!" a voice on the radio broke his thoughts. The carrier's air traffic controllers had taken command of his flight and ordered him to drop his altitude to enter the landing pattern.

Dunn banked the jet to the left and down, then lined up behind the carrier at 1,200 feet. He could feel his heart thumping in his chest. For every other new maneuver, the student had an instructor sitting in the plane's back seat the first time. But not for the first carrier landing. The student landed on his own. Dunn was flying solo. No instructor would dare sit in the back seat. Too nerve-racking. The instructor would be too tempted to grab the controls and pilot the aircraft himself. Landing on a carrier was considered one of the most dangerous things a student did. It had to be performed alone.

Mango and Wolfie stood on the landing signal officer platform rubbing sunscreen lotion all over their faces. It would be a hot one today. Almost two o'clock and the Caribbean sun beat down on the carrier deck. The steel platform jutted out from the right side of the carrier at the rear where the jets landed. Two large gray consoles sat on the platform with television screens that played videotapes of the jets coming in. Large white dials on the consoles gave readings on the wind speed across the deck and on the rocking of the ship. (Today the ship sailed smoothly on glassy calm seas and a headwind blew an ideal twenty-two miles per hour from the bow to stern.) Four black phones hung from the console to patch the landing signal officers—LSOs—to the planes above or to the ship's captain perched in a padded chair on the bridge. Mango tried to avoid talking to the skipper, who never called the platform except to complain.

Stretched out to the left of the platform was a large padded basket made of black vinyl that the LSOs could jump into if a landing plane veered off course and was about to crash into them. Only once in his Navy career had Mango edged close to the basket when he thought a jet would hit him.

Mango adjusted his black Blues Brothers sunglasses and straightened the white canvas vest LSOs were required to wear to identify them on the flight deck. Some wiseass had written "I like boys" on the back of Mango's vest. His wavy black hair blew back in the wind. Mango and Wolfie refused to wear the cranial helmets with their Mickey Mouse ears that were required for other deckhands. No LSO would be caught dead in those goofy things. Fuck the carrier's "safety Nazis." A landing signal officer had to look cool.

Lieutenant Mike "Mango" Carr and Lieutenant Rusty "Wolfie" Wolfard were a breed of aviators becoming extinct in the Navy. Mango—he earned the call sign in a politically incorrect era because his mother was Filipina—was a civil servant's son from Washington, D.C. He was proud of the fact that he was nothing like these obsessive yuppies coming into flight school today. Mango picked a major in college that would graduate him the quickest, joined the Marines briefly on a lark, transferred to the Navy, then didn't give a damn about grades in Navy flight school. Wolfie was a tall country boy from Sheffield, Alabama, who managed country music songwriters before deciding he wanted to become a Naval pilot.

Mango and Wolfie met in flight school and had been close friends for eleven years. They both served in Desert Storm. Mango flew E-2C Hawkeye planes, the airborne command posts directing the carrier's air war. Wolfie piloted A-6E Intruder attack bombers.

For Rusty Wolfard, flying combat missions over Iraq was the most intense experience he had ever been through in his life. Carrier deckhands would rush up to him with tears in their eyes before closing the canopy on his cockpit, yelling at him to "Kick ass!" Nights bombing oil complexes in Iraq became surreal with antiaircraft fire and SAMs whizzing over, under, and around his jet. Like the Fourth of July. He'd catch himself becoming mesmerized by the light show for brief seconds, then terrified by the reality that he could die in an instant. It brought out the best and worst in the men around him. Wolfie saw some pilots so stricken with fear they turned in their wings and refused to fly. Other friends paid with their lives. After the third night of low-level bombing, the carrier crew presented the Intruder squadron with a barrel of brass balls because they flew so close to the gunfire. Before a combat mission, the pilots would play Wagner's "The Valkyrie" full blast in the carrier ready room. Then one aviator would pull out his penis and crank it like a

Model T as the others cheered. It would take Wolfie six hours in the ship's gym to unwind from the fear he felt after each mission.

The Navy was now all fucked up, Mango thought. He was getting out. Wolfie already was out. The only reason Wolfie stood on the LSO platform was because he loved flying and had stayed in the Naval Reserves after quitting. When the Navy retired the A-6, it tried to stick Wolfie on a ship. He refused and took up training students in the Goshawks as a reservist so he could stay in the air.

It was all this political correctness that had finally gotten to them, all the ass covering, all the goddamn pencil-pushing admirals in Washington who let Congress lead them around by the nose. The warriors were leaving. The good aviators were fleeing in droves to the airlines. A generation of fighters lost, and left behind were the careerists more interested in punching tickets than flying.

Mango had been passed over for lieutenant commander. Because of Tailhook, he was sure of it. He had been at the 1991 convention. What he saw was one big drunken frat party, not the assaults. But those Gestapo agents from the Pentagon's inspector general's office interrogated him as though he was a prisoner of war. Even if he had seen anything he wasn't about to rat on his brothers to these guys. Mango was labeled "uncooperative" with the investigation. It drove a stake into his Naval career. The brass wouldn't admit it, but Mango was sure there was a list of officers who attended that infamous convention and anyone on that list would not be promoted.

Mango and Wolfie were the last of the hell-raisers, the last of the old dogs. They were outgrowing it anyway. Both were in their thirties and raising families. Mango had left his wife, Lisa, with two sick kids so he could play on this carrier off Key West. Lisa kept telling him he would have to grow up one day. Mango knew it was true. But he would dearly miss the camaraderie, the brotherhood of warriors he had been allowed to join for a brief part of his life. The kids coming into flight school today were a bunch of neurotic engineers. They analyzed things too much, Mango thought, put too much pressure on themselves. They were like sponges soaking up every word he and Wolfie uttered, asking a million questions. Flight school would be the best years of their lives. Enjoy it, Mango would tell the students. "The most fun you can possibly have with your clothes on and they're paying you," Mango liked to say. "All you have to do while you're here is eat, sleep, fuck, and fly."

But these kids wouldn't listen. The ten chicks Mango and Wolfie had brought to Key West for their first carrier landings were coiled tighter than steel springs. Wolfie had ordered them all to speak over their plane radios in low, manly voices to try to make them feel like warriors. Before leaving their home base in Kingsville, Texas, he had passed around a Styrofoam cup filled with malted milk balls. "Take one," he ordered each student. "It's good juju and it lasts for a week." The ten nervous chicks carefully swallowed the candy as if it were a magic potion to bring them good luck.

Dunn had descended in his jet to 800 feet. He was now three miles from the carrier, just off its right side. The training squadron's students would be landing on the USS *John C. Stennis,* a brand-new, 100,000-ton, nuclear-powered supercarrier named after a dead Mississippi senator whose only claim to fame had been larding the Navy with billions of dollars of pork when he chaired the Armed Services Committee. Dunn glanced down and to his left. The ship still looked damn skinny even from this lower altitude, he thought. Dunn could feel his heart now racing. Pablo and Baby Killer followed him in their three-plane formation.

Pablo was Paul Rasmussen, a twenty-five-year-old lieutenant junior grade who got his call sign because he had learned to speak Spanish fluently working summers with Mexican-American dishwashers in a steak house. Pablo was like a character out of *Fast Times at Ridgemont High,* Wolfie thought. Short blond hair spiked in every direction. High-strung like he just stuck his toe in a light socket.

Mango had given Lieutenant Junior Grade Brian Burke the call sign "Baby Killer" to make him feel like a warrior. Burke, also twenty-five, was one of those thoughtful silent types, Mango said. A preppy who read *U.S. News & World Report* every week, Burke kept everything churning inside.

None of the ten chicks qualifying on the *Stennis* this weekend were female. The instructors let their guard down more when no women were in the detachment. The language became saltier. The Goshawk no longer was a jet but a "pointy nose pussy getter." There were three rules to live by when landing on carriers: one, when your jet hit the deck keep it at full power to take off again in case the tailhook didn't catch the arresting wire; two, follow the deck crewmen when they directed your plane after landing; and three, the landing signal officer was never wrong. The instructors gave the students one other rule when they went on liberty in Key West: "Never get caught in bed with a dead woman or a live boy."

Dunn and the nine other students all grew mustaches to show manly solidarity for the carrier qualification—or at least tried to grow them. They decided to order T-shirts commemorating the event; printed on them, a cartoon of a woman performing oral sex on a pilot as he landed his jet on the carrier. The instructors were a little edgy about that one. Don't buy anything you couldn't wear in front of your mother, they warned. The ten chicks deliberated. What the hell, they decided. The prudes in the Navy would probably disapprove of any shirt, so let's go for the crudest and keep it hidden from mom.

Dunn raced past the right side of the carrier at 250 miles per hour. So many things suddenly seemed different from the hundreds of landings he had made on the field back at Kingsville. This airfield moved constantly. On land there was always *land* beyond the airstrip. At the end of this 350-foot airstrip was a sixty-foot cliff that dropped into water.

Dunn was surprised at the smooth ride. The ground had different temperatures, which created bumpy thermal pockets for the pilot flying above. But the water temperature was the same around the ship so there were no pockets.

Dunn tried not to let the new experiences overwhelm him. He had grown comfortable in the Goshawk over land. But this was all new, like when he slept over at a friend's house the first time as a kid, listening to unfamiliar noises, lying wide awake, keyed up.

Dunn banked his Goshawk sharply to the left so it flew in front of the bow of the carrier. Pilots called this the "break," the beginning of the final landing pattern before touching down on the ship. He started to descend the plane another 200 feet so it now flew at 600 feet. Dunn willed himself to pay attention to his flying. Students became so mentally overloaded by their maiden trip to the carrier that the first couple of passes could become a blur and minds went blank. Some students forgot their plane numbers when the carrier tower radioed them, even their names.

Now flying downwind in the opposite direction that the ship sailed, Dunn quickly lowered the aircraft's landing gear, flipped out the speed brakes to slow down the jet, then checked his seat harness to make sure it had him locked tightly to the backrest. He didn't want to fly forward from the violent stop he would make.

Dunn's jet reached a point perpendicular to the rear half of the ship. It was called the "abeam point." He had hit it perfectly—600 feet altitude and one mile to the left of the ship just parallel to the LSO platform where Mango and Wolfie stood. Dunn keyed the transmission switch on his throttle. He felt like he had a sock stuffed down his throat. He could barely choke out the words. Forget the manly talk.

"Echo four abeam," Dunn said, feeling almost out of breath. Echo four was his call sign for this flight.

Dunn had accumulated a long list of personal call signs. The other students had first tagged the twenty-five-year-old lieutenant junior grade with

"Muddy" because once after a raucous party in Pensacola he had raced a four-wheel-drive Jeep over a sand dune and into a muddy bog. It remained stuck there for two days until low tide and his buddies could help him dig it out. Mango alternated between calling him "Under Dunn" or "Dunn Deal." Dunn could be his own worst enemy, Mango believed, another one of those wound too tightly who thought too much. It took Wolfie an hour of shooting hoops with him before Dunn stopped calling him Lieutenant Wolfard.

Mango didn't know how well Dunn would do today. He could land on the boat if his head was screwed on tight. But he could also unravel. Mango and Wolfie made their own private bets on who would succeed or fail at carrier landings. They were right only about half the time. Dunn was a toss-up, they thought.

For Rob Dunn, fear of failure now overwhelmed any anxiety he felt about the danger of carrier landings. A student could excel at dropping bombs and dogfighting in the sky, but if he couldn't land his jet on the boat, he was useless to the Navy as a combat pilot. If Rob failed to qualify today, he would be given a second chance. If he failed again, he was out of jet school, perhaps even out of aviation entirely. Sometimes students who failed would be transferred to another school training the service's land-based pilots, but often they refused to take the jet school's rejects. A career could be destroyed in the next few seconds.

If he qualified, the grades he made landing this weekend would be used to decide what kind of jet he would fly. Score poorly and it hurt your chances of flying the highly coveted F/A-18 Hornet. Poor landers would also be steered away from the F-14 Tomcat fighter, which was difficult to fly onto the carrier, or the EA-6B Prowler, which, crammed with sophisticated electric countermeasures gear, was too expensive to crash. An instructor had chalked the day's date in large letters on a board in the squadron's temporary ready room at Key West. For the ten chicks it would be a day as seminal as their wedding or the birth of a child, the instructors told them.

Rob Dunn couldn't stand the thought of failure. He had been clawing his way through flight school from the beginning. Short and wiry, Dunn seemed born with a wild streak. A Chicago boy, he spent his last two years in college tending bar in the evenings and sleeping in a flat he rented from the bar owner just across from Wrigley Field. He stayed on the go every waking minute, from riding a bike fifteen miles each day to attend classes at the University of Illinois at Chicago, to pouring beer for drunken Cubs fans, to catching glimpses of the game from his bedroom window. Dunn studied hard and played hard.

It was the playing hard that always seemed to land him in trouble. After one wild party at his apartment, Dunn and several buddies sneaked into Wrigley Field to run the bases waving flashlights. Neighbors telephoned the police. Dunn barely escaped. If he had been caught and arrested, the Navy would have taken away his scholarship to be an officer and packed him off to the fleet as an enlisted sailor. The scare was sobering.

But Dunn couldn't seem to stay out of bad scrapes. He had scored in the top of his class during primary flight training. But at Meridian, Mississippi, where he began jet training in the T-2 Buckeye, every student came from the top rank. Dunn's grades were average. When he finally soloed in the jet, Dunn was so excited he began flying loops and victory rolls, then roared into the landing pattern 100 miles an hour faster than he was supposed to. Nobody would notice the hotdogging, Dunn thought.

He was wrong, of course. An instructor pilot flying nearby took down the tail number of his jet and spent an hour on the ground later chewing him out. From that day on, Dunn had a reputation as a pilot who might not be trusted in the air. The instructors started tightening the chain on him, watching anything that might prove the rap.

The chain finally got yanked after a weekend Dunn spent drinking and chasing girls in Pensacola. He rolled into work Monday morning hung over and dehydrated and flew miserably the next two flights. When an instructor asked what was wrong, Dunn made the fatal mistake of telling the truth.

Within a day, he was standing at attention in his crisply pressed khaki uniform before a Performance Review Board. The PRB was the Spanish Inquisition for student pilots, a panel of instructors who reviewed the cases of problem students and decided whether they should be thrown out of flight school. Dunn managed to convince the board he was salvageable. He escaped with another tongue-lashing that left him watery-eyed.

But the next stop was the squadron commander's office. Waving Dunn's training file in front of him, the CO began to carve out his piece of Dunn's rear.

Dunn then made his fourth mistake. "Sir, have you looked at my file?" he asked defiantly. "I couldn't be as bad as they're saying I am. Just to get here you had to have good grades."

The skipper turned beet red and flung the file across the room. Its papers showered down around Dunn. "Now you listen to me!" the commander bellowed. "I'll take a pilot who follows the rules any day over a hotshot with a stellar file!"

Dunn scooped up the papers and practically crawled out of the office. He became a model student afterward.

But back luck continued to dog him. He cracked a joint in his back one weekend during a rough game of touch football. Dunn visited the flight surgeon the next day. His back was broken, the doctor warned. He might not be able to fly in any plane with an ejection seat. The blast out would be too incapacitating.

Dunn was distraught. His dream to become a fighter pilot could be destroyed by a silly sandlot game. He flew to Pensacola for a second opinion from a Navy specialist.

If his back healed correctly, Dunn might be able to survive an ejection seat launch, the specialist concluded. Dunn waited. His back finally did heal correctly. He was allowed to continue jet training. Of course, he might still have to quit if a back problem flared up, the doctors warned. We'll be watching you.

Dunn transferred to Kingsville to finish the final phase of his jet training in the T-45 Goshawk. He was glad to be out of Meridian. The problems in Mississippi hopefully wouldn't follow him. Kingsville would be a fresh start. The flight surgeons there wouldn't be aware of all the back problems he had had unless they dug into old records.

So far, Dunn had had no more severe pains. The back ached a little when he pulled Gs, but if anyone noticed him rubbing it, they wouldn't immediately associate it with an old injury. He just had to make it through flight school. Then the flight surgeons would stop watching him like hawks. The doctors were always pickier with students, or at least that's what the students thought. Once he pinned on his wings and became a pilot, Dunn figured the Navy would have too much invested in him and the flight surgeons would be more tolerant of minor ailments.

For students in training, the flight surgeon became the enemy. Navy regulations prohibited pilots from self-medicating or visiting private physicians. Flight surgeons were the only doctors allowed to see them so the service could constantly monitor the aviators' fitness to fly. But the rule was routinely violated. Too much was at stake. A flight surgeon might find a minor ailment that could keep a pilot out of the cockpit, end his career. Students instead would secretly visit private doctors for treatment. Or they would see a private physician first for a diagnosis and, if the ailment wouldn't disqualify them from flying, then go to the flight surgeon to be treated. In the weeks before they flew to Key West for their first carrier landing, the instructors privately advised

Dunn and the other students that if anything was wrong, treat it themselves so they wouldn't be disqualified from this critical test.

Dunn banked the jet gently to the left to begin lining up behind the *Stennis.* He adjusted the power and the position of the jet's nose ever so slightly so the Goshawk began descending at a rate of about 250 feet per minute. He glanced at the vertical speed indicator in his cockpit to make sure the plane was descending at the proper rate. The VSI said he was. It was critical that Dunn line up at the proper altitude and distance behind the carrier; otherwise, he would have to make too many last-second corrections to touch down on the carrier deck at the right point. During the practice landings over ground, Dunn had had a problem with reaching the final lineup point too high and with the aircraft traveling too fast.

As his jet continued to bank in a wide turn to the left, Dunn's eyes darted quickly to four points on the cockpit control panel in front of him. His brain raced to instantly absorb the readings of each gauge and send a signal to his hands to make slight adjustments to the stick and throttle he gripped. He glanced first at the angle-of-attack indicator, a tiny box with colored arrows and a circle perched on top of the control panel that registered the angle of the wind relative to his aircraft. It looked fine. Next the vertical speed indicator. Fine also. The Goshawk was falling faster—500 feet per minute—exactly as it should. Next the gyroscope that measured the angle his aircraft was banking to the right. Okay as well. Twenty-six degrees. Then finally to the radar altimeter.

Damn! he cursed himself. He was high again. The same problem he'd had at the field. At the end of his left turn to line up directly behind the *Stennis,* Dunn's altitude should have been no more than 375 feet. He was slightly higher.

There was another problem. When he leveled his wings to start his final descent flying to the carrier deck, Dunn was not only too high, he was also too close to the carrier.

Ever so slightly, Dunn pulled the throttle back with his left hand so the plane would drop because it had less power. All the throttle and stick movements had to be finesse ones at this point. This wasn't dive-bombing or dogfighting where the pilot yanked the stick like reins on a horse. He had to have a surgeon's touch now.

Dunn was now at the start of the "groove," which began about a half mile behind the carrier. The groove represented the final flight path his jet had to travel on in order to touch down on the carrier. He looked outside at the ship in front of him.

You're kidding me! he thought to himself. I'm supposed to land this six-ton jet on that skinny runway? The carrier's landing strip was angled to the left so jets could land and take off at the same time and so that an aircraft could get airborne again if it missed the arresting wire. Dunn felt like he was swinging a station wagon into a tiny parking space. His throat felt like sandpaper.

The next fifteen to eighteen seconds—the time it took from the start of the groove to touchdown—would determine his future in the Navy. Of the two days Dunn would spend with his "carrier qual," no more than four minutes of it would be consumed by the critical time in the groove, the time that the instructors would scrutinize the closest to see if he could land on the boat.

Dunn, in effect, now had to thread the giant Goshawk through a long thin pipe. He had to be nimble with the stick and throttle. In order to touch down on the carrier deck at the right spot, he had to fly the jet down the groove at a precise angle. His "glideslope" for the Goshawk was a gentle 3.25 degrees down.

To help Dunn stay on his glideslope, the carrier had the Fresnel lens perched on the left side of the deck. Down the center of the light box were the five lenses, each of which was angled up to project a light. The top four lenses would shine an amber light. The fifth and bottom lens would shine a red light. From his angle flying down the groove, Dunn saw only one light, the "meatball," at any time. The lenses were set so that if the amber light that Dunn saw was in the middle of the box, it meant his jet was on the proper glideslope. He was "on the ball." If the amber light appeared at the top part of the box, it meant Dunn was above the correct glideslope and had to drop his altitude. If the ball was low, he was below the glideslope and had to increase altitude.

But staying on the proper glideslope wasn't enough. As Dunn's jet drifted down toward the carrier deck, the aircraft's nose also had to be tilted up to catch the wind at just the right angle. Pilots called this the "angle of attack." It was important for a simple reason. What mattered to Dunn was not where the Goshawk's wheels hit the carrier deck but where the six-foot-long tailhook sticking out the back of his plane struck. The tailhook, after all, was what grabbed one of the two-inch-thick arresting cables strung across the carrier deck and brought the jet to a stop. Dunn had to worry most about landing the tailhook in the right place and that could be tricky. Like a seesaw, if he tilted the jet's nose up, the tailhook in the back dropped down and struck the carrier deck too early. If he tilted the nose down, the hook went up and struck the deck too late to catch one of the wires.

To keep the plane at the proper angle so the tailhook was positioned correctly, Dunn had to maneuver his stick and throttle the opposite from what pilots who landed on the ground normally did. Landing on a regular airstrip, an aviator held his throttle at a steady speed and gained or lost altitude by pointing his plane's nose up or down with his stick. But to land on a carrier, Dunn had to keep his plane tilted at a steady angle with his stick in generally one position, then gain or lose altitude by pushing his throttle back and forth to add or take back power. Delicately pushing or pulling the stick and throttle correctly could be nerve-racking. No pilot ever ended up doing it perfectly.

Dunn saw the amber light on the Fresnel lens. The meatball was one step too high. He was slightly above the correct glideslope.

Holy shit! Dunn thought to himself. I've arrived! Years of hard work and here I am.

He felt proud. But only for an instant. He still had to put this plane on that tiny steel deck.

"Two three six Goshawk, ball, three point oh, echo four," he radioed in a nervous and high voice to Mango at the LSO platform. "Two three six Goshawk" identified his aircraft and its tail number. "Three point oh" was the fuel left in his tanks, 3,000 pounds. "Echo four" was his radio call sign. "Ball" meant he could see the meatball.

"Roger ball," Mango radioed back in a casual and almost cocky voice. Dunn, Rasmussen, and Burke were Mango's students. Wolfie had other students such as Schroder and Sobkowski. The two LSOs were the chicks' lifeline to the carrier, their umbilical cord, the calm and confident voice who would talk their nervous souls down to the deck. It was the same relationship an athlete had with a coach. Mango and Wolfie believed they could talk any chick down.

Mango gazed into the light blue sky. He didn't really need the radio call from echo four to tell it was Dunn. He could pick out Dunn just by the way the plane flew. Mango and Wolfie had guided down tens of thousands of jets to the carrier deck. Each pilot had his own way of landing, generally made the same mistakes each time. It was like a signature or fingerprint the LSOs came to recognize.

"Here comes Dunn Deal," Mango shouted to Wolfie. Dunn always started the groove too high.

A landing signal officer's job was to guide the aircraft during the critical few minutes from the time it started the groove until it touched down on the carrier deck. In the early days of carrier aviation the LSO waved colored paddles to keep the pilot on course or order him not to land because other planes on the deck were in the way. Today, Mango and Wolfie signaled the pilot over radios and pressed a red button on a pickle switch, which flashed red lights on the ship's Fresnel lens to signal the pilot that the deck wasn't ready for landing.

Mango and Wolfie had been "waving planes" for almost a decade. It was one of the more curious professions in the Navy. A seasoned LSO developed a razor-sharp eye for distances, angles, and altitudes to the point that standing on the carrier a mile away he knew better where a pilot was on his glideslope than the pilot himself.

An aviator learned to be a landing signal officer much like craftsmen of the Middle Ages. In the four years it took to become one, an LSO would spend only one week in formal classes. The rest of the time the skill was learned on the job through a hierarchical apprenticeship. It began at jet training. A student with an aptitude for landing on carriers usually would tell an LSO he wanted to learn the craft. LSOs often were the best in the squadron at landing on carriers. The student was first given the job of secretary, standing with the LSO on the platform and writing down the complicated scores he dictated for each pilot's landing. He then climbed up three more levels of apprenticeship until certified to guide down all types of carrier planes. By then he was a seasoned lieutenant with more than a thousand hours of flying under his belt.

Dunn didn't have his tailhook down. For his first approach he would perform a touch-and-go. A pilot flew down the groove as if he planned to land, but instead of hooking one of the arresting wires his jet bounced on the carrier deck and took off without slowing down. Touch-and-go's were good practice for bolters, the times when the hook was down but failed to catch an arresting wire and the pilot had to take off quickly to try again.

Dunn first had to deal with the problem of being too high at the start of the groove. He pulled the throttle back slightly to reduce power and drop the plane. But the groove the pilot flew in was so narrow that every correction he made had to be followed quickly by a countercorrection or he'd fall out of the groove again. No sooner had Dunn dropped the plane than in the next four seconds it had dipped below the correct glideslope. As he approached the middle of the flight path down the groove, Dunn had to add back power. But though he flicked the throttle forward only slightly, Dunn had added too much power again.

A quirk in the wind also didn't help him. A breeze did funny things blowing behind a moving carrier. As the wind spilled over the back of the carrier deck when the ship steamed forward, it swooped down, then up like a rooster tail. The effect was called a "burble," which tended to push the jet up when it reached the middle of the groove, then back down when the aircraft was close to the back end of the carrier. Pilots had to adjust their altitude even more for the burble effect. But by the time Dunn had reached the middle of the groove he was too high again. What's more, he had lowered the nose of his jet when he added power so he was flying flat to the carrier with the tail too high.

Mango could see Dunn bobbing up and down in the groove ever so slightly. It didn't surprise him. Dunn Deal always seemed to be high in the middle of the groove as well. But what worried Mango more was that Dunn was flying his jet just to the left of the white centerline that ran down the middle of the carrier deck's landing strip. Most landing mishaps occurred because the pilot didn't line up his jet directly over that center line and plowed into another aircraft. Students were taught to keep the jet positioned so the white line in effect ran between their legs in the cockpit. No deviations. But keeping on the centerline could be tricky. The landing strip and the line were angled ten degrees to the left. The carrier was steaming forward. That meant the angled line constantly moved slightly to the right.

"Watch your lineup," Mango ordered over the radio with an edge to his voice.

Dunn quickly tapped his stick slightly to the right. Students were taught to respond instantly and without question to any LSO command. The right wing tilted down. The aircraft shifted to the right.

Engine roaring, Dunn's Goshawk raced across the back of the carrier no more than fifteen feet from the ramp. The rear part of the steel landing strip had thin, rusted red streaks, scars from the tailhooks of poorly flown jets that had come perilously close to the ship's stern. Dunn willed his eyes not to guide his plane down by looking directly at the carrier deck. Pilots called it "spotting the deck." Before the advent of the Fresnel lens, pilots watched the deck as they landed. Some older aviators still did, refusing to trust the meatball. But the Fresnel lens ball was electronically calibrated to guide the pilot's tailhook to just the right point on the deck in order to catch an arresting wire. If the pilot looked at the deck instead of the meatball when he landed, his plane tended to strike the deck early because from his vantage point the deck appeared about ten feet further away than it actually was. A deck spotter could kill himself crashing into the stern.

Thump! The Goshawk's tires, filled with extra air to make them rock hard, banged down on the carrier deck. Dunn felt as if he was in a giant-screen theater with the scene rushing at him. In a flash he was surrounded by the steel gray of the ship. The carrier's tower whizzed past on the right. Fleeting snapshots of deck crewmen and parked jets. Dunn shoved the throttle forward to full power and at the same time flipped a switch on the throttle forward to retract the speed brakes. He pulled back on the stick so the Goshawk banked up. In a flash, the gray below him was replaced by the bright blue of ocean water.

As the jet climbed, Dunn banked it slightly to the right, then just as quickly back to the left so his flight path wouldn't cross that of other jets being launched from the carrier's bow.

"Wow!" Dunn shouted to himself. He couldn't believe he was in a $16 million plane bouncing off a $5 billion carrier. He felt like a kid who had just been on the best amusement park ride of his life.

Dunn snapped his head to the side to shake off the wonderment. He wiped the first shock of touching down on the carrier out of his mind. It was now time to go to work. Concentrate, he ordered himself.

On the LSO platform, Mango turned to the young apprentice serving as his secretary for these flights. He thought for a brief moment and dictated, shouting into the apprentice's ear over the roar of jet engines on the carrier deck. "High start to in the middle. Too much power in close. Flat and lineup at the ramp. Touch-and-go."

The apprentice scribbled down the critique in shorthand: "HXIM TMPIC B.LUAR"

"Four wire," Mango continued. Even though Dunn had performed a touch-and-go, Mango calculated that if his tailhook had been down and he had tried to land, the hook would have caught the fourth wire. There were four arresting cables strung across the carrier deck's landing strip, forty feet apart from each other. An ideal landing had the jet's tailhook hitting the deck between the two and three wires and catching the third wire. Catching the second wire was considered satisfactory, but not as good as number three. Hooking the four wire meant the pilot had overshot his landing slightly and was close to missing all the wires. Catching the one wire was the worst of all; the pilot had dropped altitude too early at the last second and was in danger of striking the stern.

Mango paused for another brief moment.

"Fair pass," he finally shouted. That was Dunn's score. Every landing a pilot made aboard a carrier was graded. A "perfect" pass, something akin to an A+, which pilots rarely made, was awarded five points. An "okay" pass, the equivalent of an A, was given when a pilot had minor deviations in the correct flight path but made good corrections to get back on course. A "fair" pass, in effect a B, meant the pilot had even more deviations but managed to correct them. A C was a "no grade" pass, below average as far as the landing signal officer was concerned. If the pilot hadn't set up his approach properly for a safe landing, the LSO would order him not to land and instead fly over the carrier for another try. This was called a "wave-off" and the pilot received only a one, or the equivalent of a D. There was one other grade between "no grade" and a "fair" pass. Sometimes the tailhook banged the deck and bounced over a wire, never catching it. This was called a "bolter" and the pilot usually received a low grade when it happened.

Dunn climbed to 600 feet as he banked left for another touch-and-go on the carrier deck. The students performed two of them first as warm-ups before dropping the tailhook for an actual landing. Rolling into the groove behind the carrier, Dunn this time began too fast and too high. Then he never dropped the altitude enough to stay on the correct glideslope.

"Too much power," Mango radioed Dunn. If it had been an actual landing, Mango guessed he would have missed all the wires because he had been too high. Mango gave him a "no grade" for that pass. Two points.

"Don't let yourself get overpowered," Mango lectured Dunn over the radio.

Dunn was a little irritated with himself. Too much power, too high at the start of the groove. The problems he had been having with many of his landings at the field. Bad habits were returning.

He circled the carrier for the third time. Now he was supposed to land. Dunn pushed the gray handle marked HOOK on the right side of the cockpit panel.

He radioed to Mango that he could see the meatball. He started the groove just fine. But by the middle, his rate of descent had slowed dramatically. Dunn's plane was supposed to fall at 500 feet per minute. By the middle he had almost stopped descending. Dunn had only seven seconds to make up the corrections. As he neared the ramp at the stern of the ship, his Goshawk still flew too high.

Dunn tried to make the jet sink like a stone in the last second. But too late. He felt the thump of the carrier wheels on the steel deck, then the banging of his tailhook on the metal. He shoved the throttle forward as he had been taught to do. Even though he had his tailhook down to catch one of the wires and bring his jet to a stop, Dunn had to keep the engine at full power. If the hook didn't catch a wire, Dunn had to be prepared to instantly fly off the deck to try again. If he had reduced power after touching down, his jet wouldn't have enough speed after a bolter to take off again. The Goshawk would simply roll off the deck and fall into the ocean.

It was good that Dunn had been at full power this time. He had boltered. The hook banged the deck ahead of the fourth wire. A bolter gave him two and a half points, only slightly better than a "no grade."

Now Rob Dunn was angry. To hell with the thrill of a first carrier landing. If he didn't stop making the same mistake every time—flying too high—he'd never land this damn beast on the boat.

"You blew it!" Dunn shouted to himself in the cockpit. He flew the Goshawk around again for another pass. Lined up at the start of the groove, Dunn was determined not to be too high.

He turned out to be too determined. So he wouldn't fly the groove too high, Dunn increased his rate of descent to a whopping 700 feet per minute, 200 more than he should have. By the time he closed in on the carrier's rear ramp, Dunn was dangerously low.

"Power!" Mango ordered menacingly over the radio as Dunn neared the carrier deck.

Dunn pushed the throttle forward but it was too late. He was still too low.

"Wave off, wave off!" Mango shouted over the radio.

As Dunn's Goshawk crossed the rear of the carrier deck he pulled back the stick and pushed the throttle forward even more. The jet flew twenty feet above the carrier deck, then began to rise. Dunn received one point for the pass. Below average.

Mango looked at Wolfie. They were both thinking the same thing. This was usually the time Under Dunn either pulled himself together or unraveled.

"They should calm down now," Wolfie said hopefully. Dunn, Rasmussen, and Burke all were flying a bit ragged, he thought. It usually took about three or four passes before the light bulb turned on, the anxiety went away, and they realized that landing on a carrier wasn't impossible. They could do it.

"I need you to settle down and start flying decent passes," Mango radioed Dunn soothingly as his jet banked right behind the carrier. "Easy with the power in the middle there."

Dunn set up for his fifth pass. He started at the groove slightly low so he added power to lift up the aircraft. Dunn took Mango's advice and moved the throttle forward no more than a half an inch, but it still turned out to be an overcorrection. He was slightly high again as the jet closed in on the rear ramp.

But not too high. Dunn made more subtle corrections this time for the glideslope deviations. As he was just about to cross the carrier ramp, he had the aircraft almost in perfect position.

"Right for lineup," Mango ordered quickly.

Dunn flicked the stick slightly to the right to get the jet back over the centerline.

Thump! The tailhook banged on the steel deck. Dunn shoved the throttle forward to power up in case he boltered again. He didn't this time.

Dunn felt as if the jet had hit a brick wall. He had been traveling at 120 miles per hour when the arresting cable yanked him to a violent stop in just 200 feet. Dunn felt totally out of control. The Goshawk jerked and skidded from side to side. Dunn felt himself being thrown forward, his arms flailing in front, the seat straps cutting into his shoulders and waist, his head snapped back from the whiplash. He sat breathless for a second.

"Throttle back two three six, we gotcha," a voice from the carrier's tower radioed to him impatiently. Dunn shook his head almost dazed and powered down the plane. The arresting cable retracted a little so it made the Goshawk roll back and Dunn could lift and disengage the tailhook. But Dunn unhooked from the cable too slowly. The tower had to wave off the next jet approaching the ship because he hadn't cleared the landing strip quickly enough.

Rasmussen had landed just ahead of Dunn. Pablo had been a bundle of nerves from the beginning. His right leg had been bouncing so much during the preflight briefing, Dunn thought he was going to jackhammer it through the floor. Gremlins made Pablo even more nervous. Walking to his jet at the Key West naval base, where all the chicks had taken off that morning for their first landings aboard the *Stennis,* Rasmussen realized he had forgotten his knee board with directions for the flight. He had to run back to the hangar to retrieve it. Back at the jet, out of breath, he couldn't get the damn radios in the plane to work. Frantically he motioned a maintenance man over to check it out. Pablo had forgotten to push a button in the empty rear cockpit so the system would operate.

He tried singing to himself during the flight from Key West to the *Stennis.* "The End" by the Doors. He sang it over and over again. He was still nervous.

It was eerie, Pablo thought after his plane came to a stop aboard the carrier. The ship seemed so gray and lifeless as he landed. But as he shook the cobwebs out of his head from the violent stop, the carrier deck suddenly became alive with people stepping out of clouds of steam, swarming around his aircraft. They appeared from nowhere. Crewmen in Mickey Mouse helmets, darkly shaded goggles, different colored vests—yellow, green, brown, blue, purple, white, red. It was like being in the land of Oz with munchkins suddenly coming out of the rocks, all of them waving at him silently through the din of roaring engines he could hear outside his cockpit.

An aircraft carrier deck was the most dangerous place in the Navy. The crewmen who moved planes around it in cramped quarters risked their lives every day. The work was grimy, hot, and noisy. One slipup and jets crashed into each other. Bodies too close could be ingested by jet engine intakes. Limbs could be sawed off from whiplashing cables. Accidents were frequent. If the regular brakes failed on a plane, the pilot frantically applied the parking brakes to keep the aircraft from rolling off the deck. Sometimes he had to crash into the tower to bring the jet to a stop.

The taxi directors used complex hand signals to guide the pilots in the planes. Pablo tried to pay attention. Hand signals above the taxi director's waist were for him. Below the waist, the hand signals were meant for other deck hands. Their hands moved fast or slow depending on how quickly they wanted the pilot to move his jet.

The *Stennis*'s deck crewmen were a surly lot. They had just fifteen seconds to disengage Pablo's jet from the arresting wire and move him out of the way before the next jet landed. Rasmussen saw two taxi directors in yellow vests who seemed to be signaling him at the same time. Their signals were unclear. The taxi directors, who had to rush planes off the landing area so other aircraft could land, became impatient. In the *Top Gun* movie, the hand signals were all exaggerated, he remembered, probably so the audience would understand them. The taxi directors here made more subtle gestures as they ordered him off the landing strip to the catapult. Pablo became confused on whom to follow.

Finally, an irritated taxi director ordered him to stop the jet. He marched up to the left side of Rasmussen's cockpit, looked up and pointed his two fingers at his eyes menacingly. It was the signal for "pay attention to me, stupid!" Pablo had never been cussed at in sign language. The taxi directors had other choice signals for the inattentive, such as pulling a fist out of the other hand. It meant get your head out of your ass. The deckhands had a long list of bonehead things students did in their planes, such as taxiing along with the parking brakes on, which blew the tires. That usually earned the student the call sign "Boom Boom."

Dunn refueled his jet, then slowly taxied it up to the left catapult, which would rocket him off the carrier. A square of the deck behind his jet was automatically angled up as the deflector to catch the hot exhaust from his engine. A catapult director in a yellow vest signaled him to flip down the launch bar switch at the bottom left of his control panel to the EXTEND setting.

Deck crewmen bent over and scampered underneath the Goshawk to hook the launch bar into the shuttle on the catapult track. The crewmen have been known to have fun with the catapult when a plane wasn't being sent off. A catapult officer's last day on the job was usually commemorated by tying his shoes to the catapult and launching them off the ship. As a prank, one carrier even hoisted a junk car aboard and rocketed it off the catapult.

Once hooked to the catapult, Dunn moved the throttle forward to full power. The Goshawk shuddered and roared. Dunn turned the stick in a circle and pushed the rudder pedals back and forth with his feet to make sure the steering controls operated smoothly. He made a quick scan of the cockpit panel's engine readings, the RPM and fuel flow gauges, and central warning system light. No problems. Dunn flipped the launch bar switch up to the RE-TRACT setting. Then he grabbed the catapult hand grip in the cockpit with his left hand, locked his elbow, released the wheel brakes with his feet, and placed his boot heels on the cockpit floor.

Dunn felt the nose of the jet squat as the catapult pulled the launch bar down tightly. The catapult officer, the man who would finally order Dunn's plane rocketed off the deck, had taken over.

There was nothing left to check. Dunn turned his head right to the officer, gave a crisp salute with his right hand—the signal that he was ready for launch.

The catapult would propel the jet to 120 miles per hour in just two seconds. Many things could go wrong in those two seconds and the pilot had to be prepared to react almost without thinking. The catapult or its fittings could malfunction. The jet could be sent rolling down the deck at a slow speed with the pilot frantically trying to brake it before the plane fell off the deck. Dunn had to have his eyes glued to the jet's speed gauges. If the catapult didn't generate enough power and Dunn's plane had not reached at least 100 miles per hour by the time it crossed the ship's bow, Dunn had to decide if he should yank the black-and-yellow-striped handle between his legs and eject. Gray detonating cords of plastic explosives were laced over his cockpit canopy to blast the Plexiglas away so his seat would rocket him up with a crushing force of sixteen Gs. Ejecting in front of a carrier was hazardous. The ship could run him over as he parachuted to the sea. Dunn could try to fly the aircraft with not enough speed. But he might have to ditch the jet into the water, then blow the canopy and wrestle out of the seat harnesses before the aircraft quickly filled with water and sank to the bottom of the ocean.

Two seconds was hardly enough time for all those choices. "Make your own decision on when to eject while you're comfortable at home with a scotch in your hand and a cheap date," Mango had told them. At the catapult, just react and grab the ejection seat handle.

Most students, however, were too disoriented on their first catapult to react at all. They had to take the shot on faith, just pray ahead of time that nothing went wrong.

After saluting, Dunn turned his head to the center quickly and leaned it back on the helmet rest. His right hand lightly cupped the stick. Dunn wouldn't have control of the jet until it crossed the bow.

Time now crept into slow motion.

The catapult officer, squatted outside on his haunches, raised an arm, swept it over his shoulder, patted the deck in front toward the bow, then raised his arm again to point forward. The signal to launch.

Dunn had one second.

He took a quick breath.

BAM!

Dunn tried to look at his speed gauge in the upper left hand corner of the control panel but his eyes rolled back in his head. His knees jerked up. His stomach rose to his throat. It felt as if someone had dropped a bowling ball on his chest. He couldn't breathe.

One second.

The speed gauges shot up instantly to 120 miles per hour. His cheeks fluttered from the two-second blast.

Two seconds.

The Goshawk shot across the bow.

The pressure instantly came off as if someone had just finished rear-ending him.

"Yeaaaaah!" Dunn screamed inside the cockpit. Wolfie had warned the students not to key their microphones during the "cat shots." They didn't want every sailor and officer with a ship's radio to hear them squealing over the first shot.

For a little more than a second, Dunn was frozen in a daze. He finally grabbed the stick to begin flying the plane. Quickly he banked the nose up and instinctively to the left. "Every time a bat flies out of his cave he turns to his left," Wolfie had lectured them. Dunn didn't know whether that was true, but the story served as a good trigger in his mind to bank slightly left after the cat shot.

The jet seemed to him to fly so smoothly after its violent launch off the bow. Dunn accelerated the Goshawk to climb to the higher altitude for circling the carrier.

Brian Burke's Goshawk came to a jarring stop for its third trap aboard the *Stennis.* He had caught the fourth wire. Not as good as the third, but still a decent landing. He scored a "fair" for the pass.

"Good boy," Mango had radioed him. He had bet that Baby Killer would be the surprise at Key West. Burke was "Joe average" at the field, a plodder who said little, whose landings were predictably sound. Those kinds of guys often excelled at the carrier, Mango said. Burke's passes so far had been "fairs," a solid B.

Burke had calmed down considerably from the first touch-and-go. Now he wanted to land and take off from the carrier forever.

He didn't mind admitting that he had been afraid in the beginning. Burke had always been a thinker at the Naval Academy. It made him enemies among the upperclassmen when he didn't jump like a dog to their commands. At Annapolis, he could never stop thinking about Steven Pontell, an academy graduate who crashed in a T-2 Buckeye jet trying to qualify at the carrier in 1989. Pontell had served in Burke's company. He graduated just before Burke had arrived at Annapolis. He had stalled the jet coming in too slow for a landing. It had flipped over at the last minute, cartwheeled across the deck of the USS *Lexington,* killing Pontell and four others. Although the student accident rate was less at the carrier than over the ground, landing on the boat the first time was still dangerous—for no other reason than it was the first time a student attempted it. The instructors had shown them a grisly film of Pontell's fiery crash. Before coming to Key West, Burke filled up his free time with movies or workouts at the gym so he wouldn't dwell on the dangers. If you thought about it too much, you could scare yourself into an accident, he worried. Burke's mother didn't want him to tell her when he flew to the carrier to make his first landing. The white-knuckle worry would be too much for her.

"Dunn Deal needs a good pass here," Mango said, turning to Wolfie. Mango's secretary kept a calculator in his vest pocket. He totaled up the score after the five passes the three chicks had made. They would attempt four touch-and-go's and ten arrested landings today and tomorrow. Sprinkled among them would inevitably be an assortment of bolters and wave-offs, graded as well. In order to qualify, the chicks had to average at least 2.4, or a C+. After the fifth pass, Dunn averaged 2.3 and Rasmussen little better.

Pablo also bounced up and down in the glideslope, flying high and chasing the center ball as he did back at the field. He was a good pilot, Mango and Wolfie thought, but he second-guessed himself so much it eroded his confidence in the cockpit. "Spicolli," Wolfie radioed to him. He had his own call sign for Pablo. "You gotta keep that ball working on the crest there. You're letting it get too far out of parameter and trying to recenter it in close. Make those corrections a lot sooner! Okay?"

For his sixth pass, Rob Dunn started the groove low, then bobbed up and down along the glideslope, catching the fourth wire when he landed. Mango gave him a "fair," three points. It raised his average to 2.41, just barely above the minimum.

Dunn became frustrated. This wasn't his typical landing. He was starting the groove low instead of high and never completely recovering. Maybe he hadn't recovered from the shock of his first landing. His flying began to unravel.

Rounding into the groove for his seventh pass, Dunn had the jet positioned for an adequate start. He was still a little high in the middle of the groove and as he closed in on the carrier ramp, but he managed to drop the plane down at the last minute. Mango only had to order him to shift to the right slightly.

Dunn thumped down on the runway and pushed the power forward. Something was wrong. I should be stopping, he thought to himself in an instant. But there was no yank from the tailhook catching the arresting cable. In fact, no noise from the tailhook banging the steel deck.

Dunn raced off the carrier deck and banked his jet up. What the hell happened, he asked himself? He looked to the right control panel.

"Ah shit!" he shouted. "You idiot!"

There was his answer. The tailhook lever. It was still up. He had forgotten to lower the tailhook.

"The hook goes down on the pass," the carrier's air boss in the tower radioed to him sarcastically. Dunn felt dopey.

He came around for his eighth pass. His start at the groove again was adequate. But immediately he added too much power. The Goshawk didn't descend in the glideslope as fast as it should.

"Easy with it!" Mango radioed to him. But by the time Dunn neared the edge of the rear ramp his jet still was too high.

"Wave off, wave off!" Mango ordered. Dunn wouldn't have been close to any of the wires.

Mango shook his head. His secretary whipped out his calculator and began punching. A 2.2 average after eight passes. Dunn Deal wasn't doing well.

"Echo four, this is paddles," Mango radioed Dunn. Paddles was the call sign LSOs always used because of the colored paddles their forefathers waved to guide down planes.

"Echo four," Dunn answered.

"Just settle down," Mango radioed in a calm, slow voice. "You're making a pretty good correction in the middle. But you're leaving the power on for way too long and driving the ball way to the top there."

"Roger," Dunn said quietly.

But the ninth pass wasn't much better. Dunn's Goshawk boltered. "My boy's nervous as a whore in church," Mango told Wolfie. They were both now sweating under the early afternoon sun.

Dunn felt crappy. Two lousy passes. Was this the beginning of the end, he wondered. Would he be able to recover? He took a deep breath. Forget the last four passes, he told himself. Pretend they never happened. Start all over.

Dunn began the ninth pass. This time, he was on the glideslope perfectly. No deviations. No bumping up and down.

But just as he neared the rear of the carrier, Mango was forced to wave him off. "Foul deck," he radioed. Burke, who had been flying in front of him, hadn't cleared his jet out of the landing area fast enough.

Dunn cursed quietly. When it rained it poured, he thought to himself.

"Way to go," Mango radioed, trying to buck up Dunn's sagging morale. "It looked like you had a pretty good pass there, Dunn Deal. Keep it up."

Dunn struggled on. But the next pass resulted in another bolter. This time he flew just a little too slow at the start of the groove, the nose of his jet pointed up too high. Dunn overcorrected again. Every reaction created a counterreaction. When he pointed his jet down to gain speed, it slowed his rate of descent and again he arrived too high at the ramp. Dunn's shoulders sagged as he pulled the jet up from the carrier deck to make another pass. His wrists ached from gripping the throttle and stick. Fatigue had begun to set in. The water bottle he had been snatching sips from as he made each sweep around to the back of the carrier was now empty.

But a glimmer of light began to flicker. On the eleventh pass, Dunn still flew high as he neared the carrier ramp, but he stopped overcorrecting. As the Goshawk crossed the ramp, Dunn settled it down to catch the third wire.

Mango gave him a "fair" pass.

Dunn Deal began to feel just a little more comfortable about his landings. He started to develop a feel for making corrections in the groove. He had to think of them in threes. When he began the groove high, he had to pull back power slightly so the jet would descend quicker. But in the next breath he had to add power to climb or he would descend too much and fall below the correct glideslope in the middle of the groove. Then he would have to pull back on power once more or he would bump up and be high at the ramp. Power back. Power forward. Power back. Think in threes, Dunn told himself. Think in threes. His hands could never be idle in the groove.

The twelfth and thirteenth passes went smoother. The corrections became more subtle. Mango gave him a "fair" for both.

Dunn had pulled himself out of his hole. The glideslope correction for his fourteenth and final pass of the day was even smaller than before. Dunn even began to anticipate the corrections he had to make in the groove. Mango didn't have to radio them. Dunn finally felt he'd mastered the fine art of carrier landings.

Mango's secretary began punching his calculator to total up the first day's average. Dunn Deal had a 2.46, not spectacular but still passing.

Pablo, however, had dug himself into a hole. His first-day average was 2.25, not enough to qualify if the score didn't go up tomorrow.

The chicks usually flew better the second day, Mango knew. They had a chance to mull over the first day's landings, put pieces of the puzzle together. Mango could count on Baby Killer to keep up his "fair" passes. Pablo needed a good day tomorrow. Dunn Deal had to keep his head screwed on.

It was after eleven o'clock the next morning before Dunn and Rasmussen climbed into their jets for the second day of carrier landings. The hot Caribbean sun baked the steel deck. The *Stennis* sailed at an easy fifteen miles per hour. The three chicks had been scheduled to launch earlier in the morning but maintenance delays had cut back the first flights. Dunn and Rasmussen had gone to the bathroom when one plane became available shortly before 8:00 A.M. Burke had grabbed his gear and hustled off to the flight deck. Burke owes us a beer for stealing the first jet, Dunn grumbled when he got back from the head. Pablo and Deal spent the next three hours waiting and fidgeting in one of the *Stennis*'s squadron ready rooms. By the time they launched, Burke had already finished his second round of flights and passed with a respectable score of 2.63.

The three lieutenant JGs had spent their previous night aboard the *Stennis*. They inhaled a heavy dinner, then explored the ship. They wandered through the two wardrooms all with pictures on the bulkheads of the old senator the vessel was named for. They leaned out the carrier tower's fourth-story balcony above the flight deck, which was called "vultures row," so they could watch the jets land in early evening. They trooped over to the flag bridge and when no one was looking, spun around in the heavily padded seat that the carrier group's admiral would sit on if he had been aboard. They hiked back to the fantail at the stern of the ship where legend had it that the carrier's homosexuals hung out.

Like football coaches at halftime, Mango and Wolfie had huddled with them in the ready room that evening to critique the first day's flights.

"Baby Killer, you're making huge power corrections, not flying on the numbers. You're always high when you touch down."

"Pablo, you're doing the right things adding power, but you're overcorrecting. Your scan is breaking down at times. You're not watching the ball."

"Dunn Deal, you're flying too far outside the parameters. Watch out, you're getting way, way overpowered in the groove. Don't do the same thing wrong over and over again. You were picking up in the end. Settle down. You're doing fine."

Mango and Wolfie were gentle now. This was a pep talk. The three chicks could land on the carrier. They had proven that today. Tomorrow, they had to prove that they could land consistently well. Tomorrow Mango and Wolfie would give them fewer "candy calls" over the radio. The instructors now wanted to see how the students landed on their own.

Burke promised to position his plane more precisely at the abeam point and approach turn. He was amazed at the constant power corrections he had to make during the flight. He never seemed to fly smoothly on the glideslope. But he was no longer afraid. He was excited. In fact, Sunday had been the most exciting day of his life, he thought. The catapults and traps had become addictive. There was a Zen to following the meatball, Burke believed. He couldn't wait to jump back into the cockpit and try again.

Go back to basics, Rasmussen had told himself. Sunday had turned out disastrous. He had been overawed by the ship at first. It had distracted him. His eyes didn't dart about the cockpit gauges or to the outside fast enough to absorb the stream of readings his brain needed in order to keep the jet on course in the groove. It had been weird, Rasmussen thought. He had been having so much fun with the landings, he didn't realize how poorly he had been flying. The first day's score now haunted him. Pablo had had a nightmare before flying to Key West. In his dream he failed to qualify at the carrier and was kicked out of flight school. It jarred him awake, his heart pounding. He had to pull up that grade.

The three chicks had spent the rest of Sunday night slouched in the ready room's cushioned chairs watching the movie *Waterworld* over the carrier's TV network, then they trudged off to their staterooms and fell exhausted into bed.

Catapulting off the carrier deck Monday morning, Dunn felt as though an alarm clock had jarred him awake and he was expected to work immediately. On day one, the flight to the carrier from Key West had at least given him twenty precious minutes to collect his thoughts and become comfortable with the jet before landing. Now he blasted into the sky, turned left, and within seconds had to hit precise points in the flight path for the touchdown.

Dunn's head wasn't in the game yet. Flying around to the back of the carrier, he overshot the point where he should have flown the groove to the deck.

"Keep your turn in," Mango warned him over the radio.

But Dunn had edged too far to the right at the start of the groove. Mango could see that he would never be able to shift left in time to line up the Goshawk along the correct glideslope.

"Wave off," Mango said in a tired voice before Dunn even reached a half mile from the carrier.

Mango looked at Wolfie skeptically and shook his head. Pablo's first two passes had been fair. He might be climbing out of his hole. But this was another bad beginning for Under Dunn.

Dunn shook it off. He finally felt warmed up by the time he banked left to line up behind the *Stennis* for the next pass, this one a touch-and-go off the carrier. His jet was high for much of the flight to the deck. But he managed to drop enough at the last second as the Goshawk crossed the rear ramp so that when his wheels banged on the deck his tailhook would likely have caught the third wire if he had been landing. Mango gave him a "fair." He gave him a "fair" as well for the next touch-and-go. Dunn seemed to be bouncing back.

For the next pass, Dunn's Goshawk flew high again as he neared the ramp but its tailhook still managed to catch the magic third wire. Mango gave him another "fair."

But Dunn had made another mistake this time. Rounding into the groove he had forgotten to radio Mango when he saw the meatball on the Fresnel lens. Mango turned to his secretary. "Put a dollar sign by that score," he said in amused irritation. "Fine him for not calling the ball." The writer dutifully made the notation.

"Deal, this is paddles," Mango radioed Dunn. "Just relax. Very smooth out there. But don't forget to call the ball, all right?"

Dunn didn't make the mistake the next time around. "Two three six, Goshawk, ball, echo four," he radioed slowly and precisely like a pupil reciting a story in class.

Mango chuckled. "Roger ball, Muddy," he radioed back.

"He wants to make sure I heard him," Mango told Wolfie.

But this time Dunn began the groove too low. He pushed the throttle forward to add power. There was always a maddening delay of a second or more before the jet responded to the throttle's command and rose. By the time his Goshawk reached the middle of the groove it again flew too high. Dunn decelerated, then the jet dropped below the correct glideslope once more. The result of this weaving up and down through the glideslope was that Dunn ended up too low crossing the ramp. His jet's tailhook grabbed the deck's second wire.

Dunn realized instantly that he had caught the second wire. The taxi director waving at him to disengage from the cable was much further down the deck because his plane had stopped short. Dunn could kick himself. That was a sloppy landing, he knew. Mango wouldn't like it.

Mango didn't. He gave Dunn a "no grade." Just two points. At least Rasmussen and Burke made generally the same types of mistakes every landing. They knew the problems to correct. Dunn was becoming so inconsistent he didn't know what to fix, Mango worried. He had his secretary punch out the averages on the calculator again. Rasmussen stood at 2.39, still in trouble. Dunn's grade: a 2.5, passing but the "no grade" scores would begin to hurt.

Rasmussen boltered during the next pass. Mango shook his head, worried, turned to his writer and dictated: "He lined up to the left, too much power in close, high at the ramp."

Dunn rolled into the groove. Again, he flew too low in the approach, plus slightly to the left. The jet slammed down on the carrier deck and jerked to a stop. Dunn looked ahead.

Uh oh, he thought to himself. The deck hand seemed to be even further away than before when he caught the second wire.

Mango, Wolfie, and the half dozen other LSOs now crowded on the platform began cheering and laughing and pumping their fists. Not because of Dunn's landing. It had been lousy. Mango gave him another "no grade."

Mango was cheering because he had just won a bottle of liquor. The first arresting cable was the worst a pilot could catch. It meant his aircraft had flown too low and near the back end of the ship. Pilots who "aced" the landing, as it was called, were kidded unmercifully by their comrades. Before leaving for Key West, Mango and Wolfie had made four booze bets with each student. A student could put up any priced liquor or beer for each bet. If he lost, the booze went to the instructor. If he won, the instructor bought him the drink. The first bet: score better than 2.75 for the passes, Mango's average when he qualified as a student. The second bet: 70 percent of the passes graded "fair." Bet three: no catching the first wire. And bet four: whether the wire numbers the student's tailhook caught add up to an odd or even number. An LSO's liquor cabinet usually became the best stocked in the training squadron. But several students from another group in the air had been nailing their landings, making this an increasingly expensive weekend for Mango. Fortunately for Mango, Dunn thought there was no chance he would snag a one wire and had wagered a twenty-dollar bottle of Irish whiskey.

Rasmussen came around for the start of the groove. He had two more passes left. They had to be good. Pablo was dangerously close to disqualifying. Any more shaky landings and he'd have to return to the carrier another time for a second and final try. If he missed then, he wouldn't fly jets.

The LSOs stood silent on the platform as Pablo's Goshawk—engines screaming, hook down, wings wobbling only slightly for the lineup—glided to the carrier deck. They all knew that Pablo's career might be on the line with this pass. The jet snagged the third wire.

Mango thought for a few seconds before turning to his secretary.

"A little too much power at the start," he began slowly. "High in the middle."

He paused another second. "Okay pass."

His writer smiled. "That *was* a sweet landing," he agreed. Pablo had earned an A, and a much needed four points.

Pablo scored a "fair" on his final landing. He passed with a final average of 2.53. Mango could see that Rasmussen had been making steady, if slow, progress in his landings. Leaning to trap on a carrier deck was like learning to read, Mango thought. Some guys picked it up quicker than others. Pablo was a late bloomer.

Dunn was another story. The light seemed to be flickering off and on for Deal.

"Despite his best efforts, Dunn is trying to snatch defeat from the jaws of victory," Mango said worriedly to Wolfie.

This was it, Dunn knew as he began the groove for his final carrier landing. He had no idea what his average was at this point, but he did know he had to touch down safely or he wouldn't qualify.

"Echo four, Goshawk ball," Dunn radioed for the last time. He glanced quickly at his vertical speed indicator. The jet wasn't dropping as fast as it should. He was also just left of where the aircraft should have been for the final approach.

He drifted just a little too much to the right in making the correction. The Goshawk bobbed up as it reached close to the rear of the carrier, then dropped down as it crossed the ramp.

Thump, thump! The tailhook and wheels slammed onto the carrier deck. The Goshawk skidded to an abrupt stop. It had caught the second wire, one short of the ideal third wire.

Mango took a deep breath. "Make it a 'no grade' and see what it adds up to," he ordered.

His secretary scribbled on a pad, then pecked at his calculator.

A landing signal officer's grades ended up being subjective. The scores a student received boiled down to how comfortable the LSO felt with the pilot's landings. Sometimes it was good for a student to be disqualified on the first trip to the carrier. The second try would give him more practice and ultimately make him a better lander. But left unspoken among the LSOs was the fact that if Mango felt a student with potential was just having a bad day, he could massage the scores.

Mango's gut told him that Dunn could be a good pilot. He had given Mango a few nervous moments these two days. Maybe it was a case of being lazy with the stick. Mango didn't know, but he thought Dunn Deal was worth keeping.

His writer finished punching out the calculations. He looked up.

"It's 2.45," the secretary said.

Mango didn't have to manipulate the grade. Dunn had passed, not with flying colors, but he'd passed.

Mango rubbed his chin and squinted into the sunlight. "Yeah, that's about what it's worth," he told Wolfie. Dunn had turned in a 2.45 performance these two days.

Wolfie grabbed a microphone off the platform's control panel.

"Yeah Deal," he radioed with a broad smile. "It's eleven-eleven."

"You are the man!" Dunn radioed back, excited.

The higher-ups prohibited the LSOs from telling the students over the radio that they had passed. There had been too many cases of excited students celebrating in the plane when they learned the news. One exuberant youngster buzzed the carrier and sent his jet into barrel rolls. He was booted out of the program within two weeks. But if the LSO kept the student in the dark, he would fly back to Key West distraught that he might have failed and be just as accident-prone. One such distracted soul had wheeled his jet into the terminal parking lot for cars after he landed.

The LSOs therefore used codes to let the students know they qualified. Back at Kingsville the week before, Dunn had been telling a bawdy college story in the squadron ready room. It had been late one night, the story went, and Dunn and his buddies were tired of entertaining three girls in their apartment. The girls wouldn't take the hints and leave.

Finally, Dunn popped out of his chair and announced: "It's eleven-eleven." He explained that every night at 11:11 the four of them stripped naked and danced around the apartment wearing only their socks and sunglasses.

The girls squealed and rushed into the bathroom. "We looked at each other dumbfounded and I said, 'Well, I guess this means we have to take off our clothes,'" Dunn recounted. They then stripped down to socks and sunglasses.

In a few minutes, the three girls marched out of the bathroom wearing nothing but silly grins.

Whether the story was true or not didn't matter. By the next day, it had spread like wildfire through the squadron and Dunn had become famous for the eleven-eleven tale.

Dunn screamed after Wolfie's message came over the radio. He tried to dance an Irish jig in the cramped cockpit. He felt like Superman. A taxi director walked up to the side of his jet cockpit and angrily signaled him to pay attention. Dunn still had to park the jet on the deck and climb out for another student to take it for later afternoon flights.

Walking down a passageway deep inside the carrier, Dunn bumped into Charles Nesby, a burly Navy captain who commanded the air wing over the students' training squadron.

"Hey, what's this eleven-eleven crap?" Nesby growled.

"Oh, nothing, sir," Dunn answered at attention.

Dunn spotted Mango further down the corridor. He ran up to him, the hoses and communications cords of the flight gear that he was still zipped up in clattering.

Dunn wrapped his arms around Mango, gave him a hug, then walked off.

Mango stood there embarrassed for a moment, not knowing what to say. Pilots didn't do much hugging. This was the first he'd ever had from a chick.

He thought for a moment.

"Yeah," he finally said with a shrug. "I guess that's what makes the job fun."

An Airman's Letter to His Mother

(ANONYMOUS)

(From the book *An Airman's Letter to His Mother*)

Since the events of September 11, like many another American and citizen of the free world, I have been compelled to reflect on the values that we all too often take for granted—to remember that freedom and peace itself are the direct descendants of the courage and sacrifices of brave men and women.

There is a tiny book (if you could be so lucky as to find a copy), published by E.P. Dutton & Co., Inc., in New York in 1940, when the war against Hitler was raging in Europe and England, that holds me hard every time I look at it. Called *An Airman's Letter to His Mother*, consisting of a scant seven pages of main text, with a Publisher's Note of two pages and an Appendix of five pages that reprints the original Introduction of the English edition, the book is a heart-wrenching document on the real price of freedom.

The Publisher's Note of the book states the following:

"This letter was found among the personal belongings of a young R.A.F. pilot in a Bomber Squadron who was recently reported "Missing, believed killed." It was addressed to his mother—to be sent to her if he were killed." The Publisher's Note goes on to say that after the letter was sent to the mother, permission was obtained to share the airman's thoughts with his countrymen.

Here, in its entirety, is the text of the letter the airman wrote to his mother.

Though I feel no premonition at all, events are moving rapidly, and I have instructed that this letter be forwarded to you should I fail to return from one of the raids which we shall shortly be called upon to undertake. You must hope on for a month, but at the end of that time you must accept the fact that I have handed my task over to the extremely capable hands of my comrades of the Royal Air Force, as so many splendid fellows have already done.

First, it will comfort you to know that my role in this war has been of the greatest importance. Our patrols far out over the North Sea have helped to keep the trade routes clear for our convoys and supply ships, and on one occasion our information was instrumental in saving the lives of the men in a crippled lighthouse relief ship. Though it will be difficult for you, you will disappoint me if you do not at least try to accept the facts dispassionately, for I shall have done my duty to the utmost of my ability. No man can do more, and no one calling himself a man could do less.

I have always admired your amazing courage in the face of continual setbacks; in the way you have given me as good an education and background as anyone in the country; and always kept up appearances without ever losing faith in the future. My death would not mean that your struggle has been in vain. Far from it. It means that your sacrifice is as great as mine. Those who serve England must expect nothing from her; we debase ourselves if we regard our country as merely a place in which to eat and sleep.

History resounds with illustrious names who have given all, yet their sacrifice has resulted in the British Empire, where there is a measure of peace, justice, and freedom for all, and where a higher standard of civilization has evolved, and is still evolving, than anywhere else. But this is not only concerning our own land. To-day we are faced with the greatest organized challenge to Christianity and civilization that the world has ever seen, and I count myself lucky and honoured to be the right age and fully trained to throw my full weight into the scale. For this I have to thank you. Yet there is more work for you to do. The home front will still have to stand united for years after the war is won. For all that can be said against it, I still maintain that this war is a very good thing; every individual is having the chance to give and dare all for his principle like the martyrs of old. However long time may be, one thing can never be altered—I shall have lived and died an Englishman. Nothing else matters one jot nor can anything ever change it.

You must not grieve for me, for if you really believe in religion and all that it entails that would be hypocrisy. I have no fear of death; only a queer elation. . . . I would have it no other way. The universe is so vast and so ageless that the life of one man can only be justified by the measure of his sacrifice. We are sent to this world to acquire a personality and a character to take with us that can never be taken from us. Those who just eat and sleep, prosper and procreate, are no better than animals if all their lives they are at peace.

I firmly and absolutely believe that evil things are sent into the world to try us; they are sent deliberately by our Creator to test our mettle because He knows what is good for us. The Bible is full of cases where the easy way out has been discarded for moral principles.

I count myself fortunate in that I have seen the whole country and known men of every calling. But with the final test of war I consider my character fully developed. Thus at my early age my earthly mission is already fulfilled and I am prepared to die with just one regret, and one only—that I could not devote myself to making your declining years more happy by being with you; but you will live in peace and freedom and I shall have directly contributed to that, so here again my life will not have been in vain.

PART FOUR

Sound of Impact

"Flying is an act of conquest, of defeating the most basic and powerful forces of nature. It unites the violent rage and brute power of jet engines with the infinitesimal tolerances of the cockpit. Airlines take their measurements from the ton to the milligram, from the mile to the millimeter, endowing any careless move—an engine setting, a flap position, a training failure—with the power to wipe out hundreds of lives."

—Thomas Petzinger, Jr.

(Excerpted from the Prologue to his book *Hard Landing*)

Fire & Rain

BY JEROME GREER CHANDLER

(Excerpted from the book
Fire & Rain: A Tragedy in American Aviation)

Wind shear. The words sound dramatic, strangely mysterious, suggesting some transmutation of nature. In aviation they are part of a familiar lexicon of alarm bells, along with thunderstorm, turbulence, and ice. Crashes related to wind shear have resulted in hundreds of lives lost, magnificent airplanes shattered into rubble.

The phenomenon occurs most often near thunderstorms when sudden bursts ("microbursts," they're called) of air explode either straight down or in the opposite direction from which the wind has been blowing. Since airplanes take off and land into the wind to gain the critical airspeed and lift they need so near the ground, a sudden reversal of wind can be fatal. For it is airspeed, the speed of the air flowing over the wings, that allows the miracle of flight, not ground speed, the speed of the aircraft over the ground. A famous aviation bromide goes, "Thou shall carefully monitor thy airspeed, lest the earth shall rise up and smite thee." Pilots love to joke. But they know wind shear is no joking matter.

By 1985, when the events in this story took place, wind shear had been identified as the culprit in two of the most devastating, and therefore famous, crashes in commercial aviation history. Both involved Boeing 727s, although the type of plane was not thought to be a factor that caused the accidents. Eastern Airlines Flight 66 crashed during wind shear conditions while attempting to land at JFK in 1975, killing 113 of 124 on board. Pan American Flight 759 went down during a takeoff at New Orleans in 1982. The crew had been warned of wind shear conditions and had even discussed the need for extra airspeed, but the airplane went down seconds after leaving the runway, killing 153.

With these two crashes already established as aviation training landmarks, for establishing awareness and procedures to avoid future such accidents, wind shear struck again. On August 2, 1985, Delta Airlines Flight 191, a Lockheed L1011 tri-jet (in aviation parlance pronounced *L. . . Ten. . . Eleven),* ended up on the front page of newspapers everywhere after going down in wind shear conditions at the Dallas-Fort Worth International Airport, killing 137.

Jerome Chandler's book *Fire & Rain* is one of the most detailed and readable accounts of an airplane disaster ever written. Meticulously researched, *Fire & Rain* lays bare the minute-by-minute details of the accident and its aftermath. This excerpt cannot possibly do justice to the entire book, but from it you will learn what happened to Delta Flight 191, its passengers and crew, and to the world of aviation as a whole.

There are more pleasant places than North Texas in August. The heat is a palpable, enervating presence. It shimmers off the ground, giving the illusion of moisture where none exists.

All through the day of August 2, 1985, prairie and parking lot alike absorbed the radiated energy of the sun. On towards evening, after the heaviest bombardment of mid-afternoon, the earth gave it back. Through conduction, the air immediately above the ground heated. Through convection, the atmosphere drank in the heat. It was an uneven process. Heat given off by a paved area, a city, or other vast expanse of concrete was higher than that produced by ground or greenery. This particular afternoon, from the patchwork of farmland and parking lot that is northwest Dallas County, a hot spot was spawned. Surrounded by a relatively cooler circle, the pocket of warmer air rose. As it did, atmospheric pressure lessened and the pocket began to expand. At the same time, it cooled, although more slowly than the air around it. The higher the pocket ascended, the cooler it became. And simultaneously, its ability to hold water vapor lessened. Eventually, it reached a point called the *convective condensation level,* where the relative humidity is 100 percent. The product was a cloud.

At this point, something very interesting happened. By its very nature, condensation is heat-generating. Wet your finger. Wave it in the air, and you can feel the cooling evaporation. Wet your finger and hold it still in a breezeless room, and your skin becomes imperceptibly warmer. This afternoon, condensation was operating on an enormous scale. It stoked the molecular embers of the mass, a mass that had begun life not long before as merely a benign polyp. On this summer's day the structure of the atmosphere was right. The pocket mutated into a cancerous column, accelerating vertically like a roiling, living thing. Seen from a distance, it resembled the mushroom cloud of a thermonuclear explosion. The mimicry was apt, for hidden within the gangrenous giant was a heat engine, a furnace of enormous power. The hotter the day grew, the more malignant the monster became. On August 2 the thermometer registered above 100 degrees.

Land and air gave birth to the beast in private. Labor was rapid. One of those charged with monitoring such gestations was Ruben Encinas, a radar specialist at the National Weather Service's facility in Stephenville, Texas. Stephenville (SEP in aviation parlance) was some eighty nautical miles south-west of the spawning ground.

About 17:40 p.m. Central Time, Encinas broke for a brief dinner. Ten minutes before, a check of his scope had disclosed nothing unusual. The meal was brief, taken in a nearby office area. Between bites, he periodically glanced at a special Kavouras monitor linked to his scope. While useful for obtaining a general picture of the weather, the Kavouras is something of a blunt instru-ment. Prominent geographical features and landmarks are omitted from the display—landmarks like airports.

Encinas, a competent, conscientious technician, finished eating within a quarter of an hour. At 17:48 he took an upper-air observation. By 18:00 he was back at the scope. Where half an hour earlier there had been nothing of note, the beast was now apparent.

At 17:25, fifteen minutes before Ruben Encinas took a break, Richard Douglass too decided it was time for dinner. Also an employee of NWS, he was stationed at the Federal Aviation Administration's Air Route Traffic Control Center in Fort Worth. His job was to tell controllers when Mother Nature is going to interfere with their job.

Douglass was equipped with an RRWDS, a Radar Remote Weather Display System. The unit has a two-minute delay between what SEP radar de-tects and what is actually depicted. Just before leaving to eat, Douglass checked his display. Nothing out of the ordinary, not near DFW. As was the case with Encinas, no one took Douglass's place when he left his position for a break. Unlike Encinas, he had no Kavouras in the lunchroom. Douglass returned be-tween 18:08 and 18:10. There was no requirement that he coordinate the break with Encinas.

As the weathermen ate, so did the cloud—rapaciously. No longer a puffy white pocket, it quickly turned cumulonimbus. Condensation was by now a runaway nuclear reaction, as water droplets grew and collided with one another. The higher the process propelled itself, the cooler the droplets became. Some became crystals of ice. Some of the remaining moisture adhered to them. The result was hail that danced about in the mounting maelstrom until it was heavy enough to overcome the dervish updrafts and fall to earth.

And so it was that accumulated moisture—frozen and liquid—overloaded the capacity of the creature to carry it. Every thunderstorm produces a downdraft, a rush of air preceding and concomitant with rain and hail. In perhaps one in a hundred cases, the downdraft itself mutates. Although researchers don't fully understand all the mechanisms at work, one theory is that dry air somehow mixes with the cold, descending moisture. Evaporation occurs. The downrushing column is chilled even further. The colder the mass, the faster it drops. When conditions are right, it can assume the velocity of a freight train.

When the storm child smashed into the ground, it dispersed in all directions, a crazed circle of out-rushing wind. As it did, its claws sheared the very fabric of the air itself. Trees were flattened, signs tumbled, and a trailer overturned. A pair of cows grazing in a nearby meadow died when they were struck by accompanying lightning.

Had the storm's fury ended there, the afternoon would have passed into historical oblivion, simply another dog day in August. But the creature's fury wasn't spent. Unwittingly, an airplane on final approach to the world's third-busiest airport entered its lair. In its desperate struggle to live, it almost managed a miraculous escape. Later, as those few who survived the savage encounter were evacuated from the remains of the crash, one of them called the beast by name. "It was wind shear," he said. "Wind shear."

The scene at the gate was relaxed as 152 passengers waited to board the giant TriStar. In the lounge sat a microcosm of America. Once, flying was for the few, the privileged, but no more. Among those who would shortly enter the sky tunnel leading to the craft were a family of five on its way to California to begin a new life, a widow on a once-in-a-lifetime jaunt to Las Vegas, and a babe in arms.

There were also the "frequent flyers," those as much at home on an airplane as urban commuters are on a bus. They represented the infrastructure of the Sunbelt: young, ambitious, accomplished. Six IBM employees and their families were returning home from a business trip. Among them was the man credited with developing the company's phenomenally successful personal computer. The vice-president of the San Antonio Mexican Chamber of Commerce also waited to board the plane, as did a professor, a lyricist, and an engineer.

They were there for Delta Airlines 191, a flight that would span the Sunbelt, from the South Florida flatlands to the Los Angeles basin. En route, there would be a short stop at a Manhattan-sized piece of prairie known simply as "DFW"—the Dallas-Fort Worth International Airport.

Commanding 191 this humid August day was Edward M. "Ted" Connors, one of Delta's senior captains. His first officer (copilot), Rudy Price, was one of the bright young men at Delta, a future four-striper. The second officer (flight engineer) was Nick Nassick. When not flying the line, he taught others the intricacies of managing the sophisticated trijet. Together, the three men had 42,300 hours of flying experience. They represented the elite of a demanding profession.

When the crew reported for duty at 15:10 (3:10 p.m.) Eastern Time, they were given a standard packet of forms that attested to the health of their airplane and the weather they should expect. There was only one mechanical glitch: the first-class cabin movie system didn't work. Otherwise, N726DA, a 302-passenger Lockheed L1011, was ready to go. The weather at DFW was reported as "good," with scattered clouds at 6,000 feet and 12 miles of visibility. Cumulus clouds were noted to the northeast and southwest and north of the airport. Although forecast conditions at DFW were classed as "NF" (No Factor), there was a mention of possible widely scattered thundershowers becoming isolated after 20:00 (8 p.m.) Central Time.

Flight 191 departed Fort Lauderdale on schedule at 16:10 Eastern Time and raced the sun westward.

The man in seat 41J was in a hurry to get back to his wife and kids. If there is a typical frequent flyer, thirty-five-year-old Johnny Meier was one. District manager for a wholesale grocery chain, Meier sometimes thought that he lived on airplanes. He was on his way home to Temple in Central Texas, having been in South Florida for two weeks helping set up distribution routes. An associate had suggested that they fly home together via Atlanta, but Meier said no; he wanted to take 191. It was virtually a straight shot home. There would be only one connection, at DFW with a Rio Airways commuter plane.

Every time he flew, the soft-spoken Meier tried to book a window seat. There was something about the view—something magnificent—such that even after tens of thousands of miles encapsulated in an aluminum cocoon at 35,000 feet, he still got a kick out of it. Even over the flat, alluvial plains of Louisiana, nature was an artist. Bayous became silver crescents glinting in the afternoon sun. Farmland was a rich swirl of earth tones. And there was always the sky: robin's-egg blue at cruising altitude, pale chalk below, and, at the borders of the imagination, cobalt above.

On this day, Meier alternated contemplation of the heavens with bites of roast beef sandwich and swigs of Sprite from a plastic cup. While fitting the profile of a frequent flyer in many ways, Meier differs in a couple of respects. He says a silent prayer before each takeoff and he doesn't drink 'til he's back on the ground. Altitude potentiates the effects of alcohol. Should a crash ever occur, he wanted to have his wits about him.

Scheduled arrival time for 191 at DFW was 5:52 p.m. (17:52) Central Time. For most of the route, the flight was smooth. "I guess three-fourths of the way there, we had to put on our seatbelts because it got a little bumpy," remembers Meier. "Nothing unusual. Guess it lasted maybe two or three minutes."

While the choppiness meant nothing to Meier, it disturbed a young businessman a few rows forward. Flight attendant Wendy Robinson leaned over, assuring him that everything would be all right. The snack trays had been collected by now as the TriStar streaked towards DFW. This was the good part of the job, a chance to get to know the passengers and use subtle skills. Nearby, a woman was traveling with two young boys. The youngest had to go to the restroom. Robinson smiled and kept the four-year-old company until they returned. The young boy told her he was on his way to Dallas to visit an aunt.

One ninety-one was almost precisely half full this Friday afternoon. This gave flight attendant Vicki Foster Chavis a chance to meet some of her charges, too. Seated near the bulkhead in C Zone (the third of four passenger sections in the TriStar) was a family of five. Nearby, two young girls talked animatedly about a sorority function. At a window seat around row 27 or 28, there was a youngster traveling alone. Chavis paused to say hello. Such was the easy ambience of an altogether unremarkable passage.

Up front, Connors and Price were watching the weather—not the supposedly good conditions at DFW itself, but a line of weather along the Texas-Louisiana Gulf Coast. It was growing in intensity. Connors decided to alter 191's course and head into DFW via a more northerly arrival route. It would mean a ten- to fifteen-minute delay as the L1011 circled Texarkana; so be it. He was not one to trifle with heavy weather.

Meier remembers Connors's message over the public-address system: "He said we were seventy miles east of Dallas, and we were going to be in this holding position. He said the traffic was backing up at DFW. He was doing a little sight-seeing for us." As the giant blue and white airliner lazily orbited

Northeast Texas, Connors pointed out landmarks, cities, and rivers. The passengers might be late, but he was going to do his best to make sure they enjoyed the wait. "He was trying to make a nice flight of it," says Meier. "And it was nice. A nice, enjoyable flight."

Meier eased his seat back, staring out the window at the serpentine Red River below. "I knew when I got to Dallas I had a two-hour layover, so I was going to grab something to eat. I was just hoping my wife would be there in Temple. We live about three miles from the airport. I always liked her to be there early with the kids so they could see me get off the plane. I was thinking about whether I had time to get up a golf game. I knew I wouldn't be flying out [again] until at least Sunday afternoon." Family. Friends. Golf. The tension of the past weeks melted into the haze of a summer afternoon.

At 17:43:45 Central Time, the Forth Worth Air Route Traffic Control Center called 191. In the center, shirtsleeved controllers worked in an elongated, darkened room. As they hunched over radar screens, clockwise sweeps of golden wands illuminated alphanumeric data blocks. Each tiny, creeping block represented an airplane, an aggregation of human beings held aloft by the miracle of lift. At a glance, a controller could determine the airline, type of plane, altitude, speed, and assigned altitude. Gone were the familiar "blips" of earlier, less sophisticated times.

Linked to the scopes were computers. They, in turn, were linked to dish-like radar transceivers. The transceivers activated box-like transponders on board airplanes. Each transponder transmitted its own discreet code, which was translated into the data block on the scope:

FORT WORTH CENTER: "One nine one. Descend and maintain one zero thousand [10,000 feet]. The altimeter [barometric pressure setting] [is] two niner niner one, and suggest now a heading of two five zero [250 degrees, a west-southwesterly course] to join the Blue Ridge zero one zero radial and inbound [an arrival heading from a Very-High-Frequency Omni-Directional Radio-range navigational station]. We have a good area to go through."

17:43:56—CONNORS: "Well, I'm looking at a [thunderstorm] cell at about a heading of ah, two five five [degrees], and it's a pretty good size cell, and I'd rather not go through it. I'd rather go around it one way or the other."

17:44:06—FORT WORTH CENTER: "I can't take you south. I gotta line of departures to the south. I've had about sixty aircraft go through this area out here ten to twelve miles wide. They're getting a good ride. No problems."

17:44:16—CONNORS: "Well, I still see a cell now about heading two four zero."

FORT WORTH CENTER: "Okay. Head [here there's an overlaying transmission which obscures the cockpit voice recorder] when I can I'll turn you into Blue Ridge. It'll be about the zero one zero radial [heading]."

17:44:33—COPILOT PRICE: "He must be going to turn us before we get to that area."

CONNORS: "Put the girls down [get the flight attendants to buckle up]."

Fort Worth Center "turns" 191 before the heavy weather. Rudy Price is flying the airplane. It's normal practice. Captains and first officers often alternate takeoffs and landings.

17:49:29—CONNORS TO PRICE: "You're in good shape. I'm glad we didn't have to go through that mess. I thought sure he was going to send us through it."

At 17:51:19, Flight 191 is heading southwest toward DFW. At just over 11,000 feet, Connors, Price, and flight engineer Nassick have a good view of the sprawling metroplex.

NASSICK: "Looks like it's raining over Fort Worth."

17:51:23—PRICE: "Yeah."

17:51:42—FORT WORTH CENTER: "Delta one ninety one heavy [the word heavy here denotes the fact the flight is a widebody jet] regional approach control one one niner zero five." [Connors has been told to contact DFW approach control on a specific frequency.]

17:51:46—CONNORS ACKNOWLEDGES THE FREQUENCY "HANDOFF": "One one nine zero five. One niner one. You all have a nice evening. We appreciate the help."

17:51:50: FORT WORTH CENTER: "Good day."

At 17:52:08, Connors contacts the first of three air traffic controllers who will guide 191 into a landing at DFW. Each controller works a specific leg of the landing approach. The first position, Feeder East, is manned by Robert S. Hubbert, a seventeen-year traffic control veteran. Keying his cockpit microphone, Connors gives him a call: "Regional approach. Delta one ninety one heavy going through eleven [thousand feet] with romeo."

One ninety-one has received a transmission from DFW called "ATIS Romeo." ATIS is short for Automated Terminal Information Service—a recorded message designed to tell pilots arriving and departing a particular terminal what kind of weather to expect. ATIS Romeo—designating a specific message—tells 191 that all is well: ten miles visibility, calm winds, and 101 degrees. There is no mention of rain.

17:52:15—FEEDER EAST: "Delta one ninety one heavy fly heading two thirty five."

17:52:16—CONNORS: "Two thirty five heading."

17:53:41—FEEDER EAST: "Delta 191 heavy descend to seven thousand."

17:53:44—CONNORS: "Delta one ninety one out of nine [thousand] for seven [thousand]."

17:55:46—FEEDER EAST: "Delta one ninety one heavy turn ten degrees left. Reduce speed to one eight zero [knots]."

17:55:50—CONNORS: "Delta one ninety one wilco."

Inside the cockpit, Price calls for 10 degrees of flaps. As the devices deploy from underneath the trailing edge of the wing, drag increases and the TriStar begins to slow. In the coach cabin, in sections C and D, passengers hear the process, feel the rhythmic rumble of slipstream against metal. The craft is "dirty" now, no longer aerodynamically clean. The landing regimen has begun.

At the same time, Nassick has begun to read out the approach check, a litany of switches to be thrown and instruments to be read. Within minutes, 191 will be on the ground.

17:56:28—FEEDER EAST: "Attention, all aircraft listening except for Delta twelve ninety one is going across the airport. There's a little rainshower just north of the airport, and they're starting to make ILS [instrument, rather than visual] approaches. Other than Delta twelve ninety one should tune up one oh nine one [the Instrument Landing System frequency] for [Runway] one seven Left."

The ATIS that Flight 191 received a few minutes before is no longer valid. There is rain in the area, just north of an active approach runway—nothing serious, just enough precipitation to change from visual approaches to 17L, where pilots literally eyeball the runway, to the more exacting instrument variety.

17:57:00—CONNORS TO FEEDER EAST: "One ninety one, out of seven [thousand] for five."

17:57:03—FEEDER EAST: "Okay, one ninety one."

In the cockpit, the approach checklist continues. At 17:57:19, Nassick notes that the "No Smoking" sign is on in the cabin. In the smoking sections, passengers draw in a final few hits of nicotine before crushing their cigarettes in armrest ash trays. Tiny spirals of smoke curl up toward the ceiling where they are caught by cool blasts of air from the overhead vents. In the passenger cabin, too, they're performing prelanding rituals.

When the sign goes on, Wendy Robinson rises from her seat near the right rearmost exit door and walks through Section D, checking to make sure that seatbacks and tray tables are upright, cigarettes are out, and carry-on luggage is stowed. The practice is more than rote. There's a reason. In the event of a crash, passengers have to get out of airplanes—fast. A seatback sticking in your chest, or a bag on the floor, can be a potential killer when there is no time to spare. After perusing the cabin, Robinson returns to her seat, fastens her belt as tightly as she can, and sits in a brace position. It's standard procedure. In the event of a crash, flight attendants are supposed to be the best-restrained people on the plane.

At 17:57:45, the ILS is tuned and the sound of the identifier is heard in the cockpit. Twelve seconds later, Delta Flight 1291, overflying DFW from south to north, spots a budding cumulonimbus formation just north of the approach to Runway 17 Left. Flight 1291's visual perspective, *looking* north, is decidedly different from that of flights approaching *from* the north. From its vantage point, 1291 sees a rapidly building cloud formation. On the other side of the buildup is an equally rapidly dissipating rain shower. Planes approaching DFW from north to south are screened from the bilious phenomenon that 1291 sees by the shrinking shower. Flight 191 will make its final approach north to south.

17:57:57—DELTA 1291 TO FEEDER EAST: "Delta twelve ninety one. We'd like to go around this buildup twelve o'clock to us [straight ahead]. Can we turn left a little bit and go around the other side of it?"

17:58:03—FEEDER EAST: "Twelve ninety one. Twenty [degrees] left or so is approved. Call approach [control] [on frequency] one twenty five eight."

One ninety-one continues its approach to DFW.

In 44C and D, the center set of seats in the last passenger section, Marilyn and Mike Steinberg are thinking ahead to their California vacation. The middle-aged couple from Miami—one widowed, the other divorced—had met five and a half years before through a dating service. The first time Mike came calling, they went to the bandstand in Hollywood, Florida. There was a concert—big bands. They danced beneath the moonlight to the strains of Miller and Basie. Later, like a couple of kids, they walked along the beach. "There is a Jewish expression," smiles Marilyn Steinberg—"*Beshert:* It was meant to be." After the date, "I asked him to come in and have some iced tea. He looked up and said, 'You know, I think it's *beshert*.' I laughed." They were married a year and a half later.

He is an insurance salesman, she an artist. Just before leaving Florida, Marilyn gave her framer two works for finishing. "What Marilyn does," says Mike, "is take fabric, either quilts it, embroiders it, or both. Then, she paints it or dyes it. After that, she puts it into a shape. You could call it a soft sculpture." One of the two pieces, "Upward Bound," was particularly striking. "Most of my works, even though I may plan them beforehand, never come out the way I want them to." From the beginning, this one seemed to take on a life of its own. And the form it took, in tones of mauve and pink, was of flight. After they returned from California she planned to pick up the piece and take it to a museum for showing. But that was in a few days. For now, there was vacation and freedom from routine. Enjoy.

17:59:37—FEEDER EAST: "Delta one ninety one turn right heading three four zero. Contact approach [the Arrival One controller] [on frequency] one one nine four."

17:59:42—CONNORS: —"Three four zero [degrees]. Nineteen four [frequency]. So long, thanks for the help."

At 17:59:44, two seconds after Connors switched frequencies, Feeder East broadcast: "The wind's zero six zero [degrees] at two [knots] and we are showing gusts to sixteen. There's a little bitty thunderstorm sitting right on the final [approach to Runway 17L]. It looks like a little rain shower."

A thunderstorm. The ATIS had said nothing about rain, much less a thunderstorm. Neither Connors nor Price heard the new report from Feeder East. From their perspective, it was just what they were told earlier: "A little rain shower just north of the airport."

17:59:47—PRICE TO CONNORS: "We're gonna get our airplane washed."

17:59:50—CONNORS: "What?"

17:59:51—PRICE: "We're gonna get our airplane washed."

It wasn't the shower that Johnny Meier saw, but something off to the right side. "I was sitting on the right side of the plane. I told the people behind me, 'It sure does look like a dust storm out there, or heavy rain.' Because it was solid black. It looked like it was always to the right of us." The woman behind him "was getting kind of scared. I turned around and told her, 'Don't worry about it. There won't be a problem. Planes always take off and land in the rain.'"

17:59:54—CONNORS TO ARRIVAL ONE CONTROLLER THOMAS WAYSON: "Approach. Delta one ninety one with ya at five [thousand feet]."

17:59:57—ARRIVAL ONE: "One ninety one heavy, expect [to land on] one seven Left."

17:59:59—PRICE: "Thank you, sir."

18:00:21—ARRIVAL ONE: "Delta one ninety one heavy, fly heading of three five zero [degrees]."

18:00:24—CONNORS: "Roger."

There are three airplanes preceding 191 on final to 17L: a Delta Boeing 737 (Flight 1061), an American Airlines 727 (Flight 351), and a corporate Learjet (N715JF). At 18:00:38, Arrival One asks the American flight: "American three fifty one, do you see the airport yet?"

18:00:38—AMERICAN 351: "As soon as we break out of this rain shower we will."

18:00:40—ARRIVAL ONE: "Okay, three fifty one. You're four [miles] from the [outer] marker [one of three that precedes the runway threshold]. Join the localizer [the ILS radio beam that will guide the plane to the ground] at or above two thousand three hundred [feet]. Cleared for ILS one seven Left approach."

18:00:46—AMERICAN 351: "Cleared for the ILS. American three fifty one."

Flight 191, on the same frequency as American 351, is monitoring the conversation. Connors and Price are now fourth in line to land. The pilot of the plane two aircraft in front of them is doing just fine. It's a rain shower, nothing more.

18:00:51—ARRIVAL ONE: "One ninety one heavy reduce speed [to] one seven zero [knots]. Turn left two seven zero [degrees]."

18:00:54—CONNORS: "Roger."

During the next couple of minutes, passengers in the rear rows of section B and the forward rows of C feel the landing gear being lowered: first, the rumble of the undercarriage fighting the 170-knot slipstream; then, a visceral "thunk" as ungainly assemblies of metal, wiring, and hydraulic lines lock into place, reaching for the earth below.

At 18:03:32 a Delta 737, Flight 963, has just cleared 17L after landing and is awaiting clearance to taxi to the terminal. Copilot David Davis, looking out his side window to the north, sees something extraordinary. He asks Captain J. A. Coughlin, "Is that a waterspout out there on the end [of the runway]?"

18:03:33—COUGHLIN: "I don't know. Sure looks like it, doesn't it. Looks like a tornado or something. I've never seen anything like it."

Neither man tells the tower about the observation, a sighting screened from others approaching 17L by the "dissipating rain shower."

18:03:34—AN UNIDENTIFIED VOICE IN THE COCKPIT OF 191: "Stuff is moving in—"

18:03:43—CONNORS TO PRICE: "One six zero's the speed."

18:03:46—ARRIVAL ONE: "Delta one ninety one heavy, reduce speed to one five zero. Contact tower [on frequency] one two six five five."

18:03:49—CONNORS: "One two six five five. You have a nice day. We appreciate the help."

At 18:03:58, 191 contacts Gene Skipworth in the DFW tower. The controller mans the Local East position, the last leg of the approach phase. Connors's voice is relaxed and light as he greets Skipworth:

"Tower. Delta one ninety one heavy. Out here in the rain. Feels good."

18:04:01—LOCAL EAST: "Delta one ninety one heavy. Regional tower. One seven left cleared to land. Wind zero nine zero at five [knots] gusts to one five."

18:04:06—CONNORS: "Thank you, sir."

Some seven seconds earlier, Ruben Encinas, the National Weather Service specialist manning Stephenville radar, called the agency's Fort Worth Forecast Office. Since returning from dinner a few minutes earlier, he had analyzed the mushrooming formation just north of DFW. Again, although it wasn't noted on the map overlaying his scope, Encinas knew the location of the giant airport. What he found was a very strong echo. When he picked up the phone to Fort Worth he called the return by name: a thunderstorm. The forecaster on the other end of the line said he would put out the word. Six minutes later, he issued a Special Weather Statement.

Jack Williams was another meteorologist at the Forth Worth Forecast Office. His responsibility was the aviation desk. He too knew of the echo; he'd been watching the rapidly building return on a remote display unit. Still, he decided it didn't require an Aviation Weather Warning and didn't warrant calling the DFW tower. After all, there had been no reports from law enforcement agencies or surrounding communities of actual thunderstorm impact, no "ground truth." Williams, a longtime veteran of the Weather Service nearing retirement, said he didn't want to "cry wolf."

And yet, at 18:04:18, an airplane was on the threshold of the lair:

PRICE: "Lightning coming out of that one [cloud]."

18:04:19—CONNORS: "What?"

18:04:21—PRICE: "Lightning coming out of that one."

18:04:22—CONNORS, LOOKING UP FROM THE INSTRU-MENT PANEL: "Where?"

18:04:23—PRICE: "Right ahead of us."

When Rufus Lewis turned his Learjet on final approach, he saw "nothing that alarmed him" ahead. As he continued down the ILS glideslope to 17L, turbulence began. The rain became extremely heavy, the world in front of the windscreen opaque. In an instant, his airspeed plummeted from 150 knots to 125 knots. The airplane fell. Lewis pushed the power levers forward and climbed, adding speed. He "had his hands full." And yet, he said nothing to the tower.

The lightning flash was the moment of decision for Ted Connors. We'll never know what went on in the man's mind, but, without too much conjecture, we can conclude that it is likely he did what any good pilot would. He weighed the evidence: others, right in front of him, had gone through the shower with no apparent problem; there were no reports of thunderstorms from the ground; Gene Skipworth had just cleared him to land. One ninety-one flew on.

The ride becomes bumpy. Johnny Meier sees rivulets of rain form a horizontal screen out the window of 41J. The lady behind him isn't getting any calmer. He turns around again to reassure her. They're over Grapevine Lake. "It won't be long now, 'cause when you take off and land [at DFW] you always see these two lakes. Don't worry about it. We're almost there now."

Just after 18:05, something strange begins to happen: a subtle, then sudden, mutation in airspeed. One ninety-one has been instructed to maintain 150 knots so as not to nudge too close to the Lear ahead. But the indicated airspeed is rising. No one is pushing the power levers forward; it's happening all by itself. By 18:05:07, it has ballooned to 157.46 knots. Eight seconds later, it's 162.42. Three seconds after that, at 18:05:18, IAS peaks at 173.20. Something's screwy. Connors's eyes are riveted on the indicator. Tick, tick—it flickers and drops to 171.65 knots. Flight 191 is 754 feet above the ground. The time is 18:05:19:

CONNORS TO PRICE: "Watch your speed."

A second later, the deluge begins. What has come before was mere precipitation. Now, the L1011's thin aluminum skin reverberates to the ca-cophonous beat of some unholy tattoo.

18:05:21—CONNORS TO PRICE: "You're gonna lose it [airspeed] all of a sudden. There it is."

Rudy Price's left hand grips the power levers, urging them forward. At the same time, the nose pitches up. One ninety-one is trying to climb. And yet, the altimeter unwinds. Inexorably. By 18:05:26, they are 699 feet above Texas and closing, the speed a dangerously low 137.88 knots. The airplane damn well better start responding soon because it's beginning to flirt with stall speed, the point at which this machine will simply cease to fly.

CONNORS: "Push it [the power] up. Push it way up."

The urging becomes more insistent, rising with the crescendo of the rain: "Way up. Way up. Way up!"

By 18:05:29, the L10's three Rolls Royce turbofans are screaming—gobbling up air, compressing it, heating it, expelling it back into the turbulent heavens from whence it came. The idea is to build thrust. Speed. Life-giving, God-blessed *speed*.

In the passenger cabin, the world is a darkened, roiling absurdity. Passengers grip armrests, hands, heads, each other, in white-knuckled, vise-like panic. "The plane really started rocking sideways," remembers Johnny Meier.

By 18:05:30, the airspeed starts to creep back up. Two seconds earlier, it had apparently bottomed at 129.36. Now it's 133.60.

CONNORS: "That's it."

The altitude: 635 feet above ground level, still falling, to be sure. But there's the merest hint of sanity. For a few rational seconds, the airspeed climbs. Then, at 18:05:35, all hell breaks loose. In one second, 20 knots are lopped off. Within another couple of heartbeats, the TriStar rolls wildly to port. The left wing dips by a full 20 degrees.

CONNORS: "Hang on to the ##!"

The roll is just a teaser, a prelude to unbridled terror. The bottom drops out. From the cabin, there is a collective gasp. Flight attendant Wendy Robinson says the Serenity Prayer. Other passengers pray aloud. By now, the screech of the engines is punctuated by the isolated cries of infants and adults alike.

At 18:05:44, there is another sound, that of the Ground Proximity Warning System. Its strident mechanical message is moot: "Whoop whoop. Pull up!"

A second later, Connors shouts a single word: "Toga!"—take off go around. Already trying to reach for the receding heavens, man and machine now strain with collective sinew against forces primeval.

18:05:46—GPWS: "Whoop whoop. Pull up!"

18:05:47—CONNORS: "Push it way up!"

The mechanical voice mocks the airmen. They are losing: "Whoop whoop—." The ground is barely 200 feet away.

"There was a lot of screaming and yelling going on. I mean everybody. I wasn't really scared. I was just sort of getting tensed up," recalls Johnny Meier. "Some of the babies that were sitting up front were crying."

At 18:05:52, 6,336 feet short of the threshold of 17L and at a ground-speed of more than 215 knots, the L1011's main mounts brush rich black soil. It is a sisterly kiss. Short. Light. Toward the end of the furrow, the impression fades as 191 bounces back into the sky, only to brush the earth again.

Meier looks out the window. Prairie flashes by, then pavement. He thinks it's the runway. It's not.

State Highway 114 is an arterial road skirting the northern end of DFW, a prime thoroughfare for commuters between the mid-cities area and Dallas. This Friday evening, at the tail end of rush hour, 114 is busy. A west-bound driver spots the TriStar some 500 to 1,000 yards to his right. It is very low, too low. How is it going to make the runway? In an instant, the question is answered. The car right in front of him, a Toyota, disintegrates. The driver dies in half a heartbeat.

"Ya know," Mike Steinberg says to his wife, "I've heard of rough land-ings, but this is ridiculous." The port engine ingests bits of Toyota. It's as if a giant had reached down to swat a fly and suffered a heart attack in the process. The impact slews the TriStar some 10 degrees left. The port wing dips. The conjunction of automobile and airplane dooms both.

The L10 hurdles the eastbound lane and snaps off two light standards. There's a downslope on the other side of the highway. For a second the craft skims above the ground, in mimicry of flight. Then the port wing lances the ground.

To their left, the Steinbergs see fire, then "flames roaring back." "Someone put a bomb on this plane!" screams Marilyn.

"Oh, #!" It is the last human sound from the cockpit.

"About the same time he hit on the left, he hit on the right—but a smaller version of it," says Meier. "That's when I ducked. I just put my hands on my knees and kind of locked them. I said, 'God, don't let me die. Let me live.'"

The disintegrating left wing grazes the northernmost of two water tanks near the airport's east freight area. The forward fuselage of the TriStar pivots, almost cartwheels, into the southernmost tank. In an instant, seats buckle, metal melds with flesh. Fire scourges what life is left in the forward three-fourths of the craft.

The tail cracks off as if propelled by a bullwhip and slides backwards. Fuel spills along the way, and a river of fire races to catch up.

Mike Steinberg never sees the tail shear off, never sees the carnage re-cede from their escape pod seemingly at the speed of light. When the tail snaps, he loses his glasses. But he feels it. Like some sort of crazy, netherworld carnival ride, he feels it. "When it finally stopped, I was on my side held by the seat belt. To my left, there was nothing. I was thinking 'We're still here!' I looked at Marilyn and she said, 'Please get me out of here.' I said, 'Wait a second and I'll get down and then help you out.'"

The tail was lying on its left side. Those like the Steinbergs had a relatively short drop to the ground, while people like Johnny Meier were literally dangling in space, held in place by seat belt and buckle.

When the plane dipped to the left and the fireball exploded, flight attendant Wendy Robinson ducked her head and closed her eyes. Her specially designed seat was equipped with a lap belt and shoulder harness. After the tail stopped, she opened her eyes: "The plane was tilted to the left. I was dangling to the left side of my jumpseat, at which point I struggled to release my seat belt. This was difficult because my body weight was on the buckle."

Quickly she changed from victim to functioning, capable crewmember, one of a handful left living. No longer was it "coffee, tea, or milk." It was, "Release your seat belts. Get up. *Get out!*" The first mandate of a flight attendant after an accident is to get passengers away from the plane. The possibility of fire is always present. Wendy's professional instincts took over. Then she saw where she was, specifically at what angle to the ground. She realized that if passengers released their belts, they would fall and injure themselves.

Johnny Meier felt the same way: "I said to myself, 'Man, I lived through this. I ain't gonna beak my neck falling.'" He saw "the stewardess leave her seat and kind of crawl down some seats. So I did the same thing." There was a row hanging virtually vertically. Meier used it like a ladder. When he got to the last seat, he jumped the remaining fifteen feet or so. "When I hit the ground, I just buckled. It was a pretty good fall."

When Meier hit the ground, he immediately noticed the wind. So did the Steinbergs. It was an ill wind, and it was blowing from the north. "As I was going down [to the ground] there were bodies going by me, debris going by me. It was *flying* by me." The source of the ghastly airborne effluent was the forward part of the plane, still burning despite being drenched by thousands of gallons of water from the breached southern tank.

Aside from the wind, it was quiet—"absolutely quiet," remembers Marilyn. Meier recalls the same sensation. There were no more screams, no more crying babies. Nothing.

Then the sirens began. And so did the rain. It came in horizontal torrents, lashing the few pitiful survivors who struggled, crawled, and were carried from the tail carcass. Meier and two other male passengers who had been sitting nearby collected themselves and started back toward the tail in search of other survivors. "I saw one woman," says Meier, "[who had been] sitting on the opposite side of the plane from where I was. She was crawling out from under the plane. She had a great big old gash on her head and was bleeding. I took my handkerchief out and held it up to her head. There was another lady to the left of the plane. She was trying to crawl away. Her legs were banged up real bad. She was crying, in a lot of pain.

"I got over to this girl, and about that time the hail started coming down—marble sized. I picked up some insulation from the plane, about two feet by four feet, and held it over her upper body. She was saying, 'It hurts, it hurts.'" Meier stayed with the young woman, shielding her from the storm. "She kept on saying, 'Hold my hand. I don't want to die. Don't let me die.'"

Rescuers had been on the scene for a few minutes. "It looked like the ambulance and fire trucks were already out there waiting for the crash," says Meier. "I couldn't believe it—they got there so fast. We kept yelling, 'Come over here. We need help, too.' But no one was hearing us." The reason was the rain. "The sky was blacker. The rain—it looked like it was harder than you *ever* saw rain. It felt like it was 30 degrees outside. Everything was going in slow motion."

While Johnny Meier's improvised rescue team was combing the wreckage, Wendy Robinson was leading others away from the scene. Mike Steinberg says, "There weren't that many people who were walking around." By now the rain had washed the fuel from his eyes. Although his glasses were gone, he could still make out what was happening to the north. "I'm standing in the field and I'm looking to the left and I see what looked like the front of the plane. It was in flames, and the fire fighters were pouring whatever it was to try to douse it. I was fascinated." A fire fighter came along, took a look at the couple, and said, "You're okay. You're still walking." Hand in hand, the Steinbergs were led to a police car for shelter.

And shelter was needed from the demonic wind. It began to whip the airport at peaks of 70 knots. So strong was it that those few who dared to look up from frightened crouches saw something remarkable: the tail of the TriStar, virtually the only recognizable piece of airplane left, righted itself with an agonized groan.

After the gusts subsided, Meier managed to carry half a dozen people to waiting ambulances. "You don't know how you're going to react to something like that until it hits you," he says quietly. After he had searched the tail, he headed toward the water towers. "I didn't want to leave anybody that was out there that was injured and just couldn't cry for help. I wanted to make sure I looked everywhere I could that people might be. You know, you always see it on TV that somebody got left behind. I didn't want it to haunt my memories that I left somebody that could've lived if I'd found them."

On the path toward the tanks were scattered pieces of passenger cabin, remnants shed by the tail as it rocketed backwards. "I walked up to this one guy I thought was all right. He had a cut on the left side of his face, and his jaw was kind of cut up, had blood on it. But his eyes were open," shivers Meier, "and he was still sitting there." The man was dead. Meier stumbled on, soot, fumes, debris, and stench assaulting him every step of the way. "I kept walking back, about 175 yards at the most, and a couple of guys—I don't know who they were, they had ID badges—asked me if I needed help." Meier said no. He just wanted to know what time it was. His flight for Temple left at eight. He didn't want to miss it.

Am I Alive?

BY SANDY PURL WITH GREGG LEWIS

(Excerpted from the book *Am I Alive?*)

The crash of Southern Airways Flight 242 at the village of New Hope, Georgia, on April 4, 1977, and the loss of seventy-two lives would have been bitter to contemplate under any circumstances. But in this case, the fact that an aviation disaster was taking place was no doubt obvious to the eighty-one passengers and four crew members on board. They endured agonizing, tortuous minutes before impact.

The DC-9, with twin rear-mounted engines, was forced to attempt an emergency landing on Highway Spur 92 after both engines failed during a hail-filled afternoon thunderstorm. Surviving flight attendant Sandy Purl went on to write the vivid book that brings the events of the crash right into your reading room and goes on to document her inspirational story of ultimately triumphing over the tragedy. This excerpt from *Am I Alive?* takes you to the last place on earth you would have wanted to be on that tragic afternoon—the cockpit and cabin of the doomed Flight 242.

Early on April 4, 1977, a cold front moved into the lower Mississippi River Valley and rumbled eastward like a huge bulldozer, shoving a mass of warm, moist air ahead of it. By afternoon, as the front rolled over northern Alabama and on into Georgia, the mixture of warm and cold air created towering clouds in a blackening sky.

Inside one of these giant cumulus clouds, rain started to fall. But before the moisture could reach the ground, it was caught in a violent updraft that channeled it more than five miles back up into the cold air. The rain froze and fell again. This process repeated itself again and again until the interior of this cloud became a vicious maelstrom of hailstones.

I rebuckled my seatbelt as Southern Flight 242 began a bumpy descent into Huntsville, Alabama. Since the beginning of our early morning shift, our crew had flown in and out of Atlanta all day on short hops. Cathy, the other flight attendant, and I had spent nearly all our air time strapped in our jumpseats because of rough skies.

Cathy, a 5-foot, 3-inch blonde, was my senior by eight months. We'd flown together a number of times. So we knew each other's preferences and could work together efficiently and well.

The cockpit crew was just about my favorite. The captain, Bill, always treated flight attendants as fellow professionals with an important role on the airplane. Apart from the professional respect I had for him, he was also a friend.

The first officer, Lyman, was a handsome, dark-haired, former Navy pilot, twenty years younger than the captain. His fun-loving appeal made him enjoyable company. He loved telling stories and jokes—in fact, at one of our stops just the day before he had had all the rest of the crew in stitches, laughing. Lyman said he'd applied as a pilot with another airline and the interviewer, wanting to test his response to an emergency situation, asked him what he'd do if he saw two trains running full speed on the same track and heading right at each other. Lyman said he thought for a moment and said, "Well, I guess I'd say a quick prayer to God and then run get my brother." The man said he certainly understood the prayer part, but he wondered why Lyman would run to get his brother. At this point in telling the story, Lyman had paused and looked around at each of us who were listening. Then he said, "I told the man, 'I knew God could stop it. But if He was busy, I knew my brother had never seen a train wreck.' "

Flying always seemed a lot easier when I worked with people I liked. And on a two-day stint of stormy up-and-down short trips, a compatible crew like ours made the tension and fatigue more bearable. We now only had two more stops to make and our exhausting thirteen-leg assignment would be over.

On the ground in Huntsville, Cathy stationed herself at the door to board and welcome new passengers while I prepared the cabin. My head count of sixty Huntsville boarders, with the twenty-one already on board, gave us a total of eighty-one passengers for the flight back to Atlanta. I was assigned to the rear of the aircraft, and quickly checked to see that all baggage was stowed, seatbelts fastened, and tray tables up. No sooner had I finished my inspection than we began a short taxi—so short that Cathy and I hardly had time to demonstrate the safety instructions before taking our own seats for takeoff.

The plane quickly climbed through the nasty weather—and then descended a couple of thousand feet. I assumed that the cockpit crew was checking radar and talking with ground control to find a smoother course through the storm.

Suddenly the whole sky broke loose. Hail that sounded like a million boulders battered the metal roof. Every passenger in the plane turned in unison and looked at me. The wave of raw emotions on their fearful faces seemed to push me back into my seat. I fought to keep the calm on my face.

Then, above the clattering din, I heard three explosions—Pow! Pow! Pow!—in the left engine. The cabin lights flickered, and the emergency lights kicked on for 15–20 seconds before power was restored in the cabin. But the familiar droning of the left jet was gone. I wondered if the engine had ripped away from the plane.

A few moments later, the hail stopped. I picked up the PA mike again. My own calmness surprised me as I reassured the passengers: "Keep your seatbelts on and securely fastened. There's nothing to be alarmed about. Relax, we should be out of the storm shortly—"

The lights flickered and went out again. When the power returned, I went on with the instructions I'd been drilled to give. "Please check to see that all carry-on baggage is stowed completely under the seat in front of you. In the unlikely event that there is a need for an emergency landing, we do ask that you please grab your ankles. If I scream instructions from the rear of the aircraft, there is nothing to be alarmed about. But in the event there is an emergency and you do hear us holler, please grab your ankles. Thank you for your cooperation and just relax. These are precautionary measures only."

As I finished my announcement, I strained to hear the sound of the right engine. I could hear a steady hum, but it didn't seem normal either. My ears had popped about the time the left engine went and the cabin temperature was quickly rising, so I knew we'd decompressed. Then I smelled smoke—like something electrical burning—I pictured the plane exploding in mid-air and scattering us over the north Georgia countryside.

In the cockpit, First Officer Lyman Keele piloted the crippled DC-9 while Captain William McKenzie raised Atlanta Center on the radio. "Okay, Flight 242 here. We just got our windshield busted, and we'll try to get back up to 15,000 feet. We're at 14,000," McKenzie reported. After checking the top set of gauges on the center console, he added, "Our left engine just cut out."

Less than 30 seconds passed. Suddenly Keele exclaimed, "My God, the other engine's going too"—

"Got the other engine going, too," the captain relayed to Atlanta.

"Southern 242, say again?" the ground controller asked.

"Stand by!" barked McKenzie. "We lost both engines!"

During the next five and a half minutes, the pilots tried desperately to restart the dead engines. Keele reset his course for Dobbins Air Force Base in Marietta. But the field was twenty miles away, and Flight 242 was losing altitude rapidly.

The instant I smelled fire, I knew we were in real trouble. But my first reaction was irritation: "Why today?" I thought. "I'm too busy today. Just one more leg after this, and I could be home! Mike's made special dinner plans."

But such thoughts quickly faded as the seriousness of the situation sank in. Any second I expected an emergency signal from the cockpit. It didn't come, so I finally acted on my own. I unbuckled my seatbelt, stood, and quickly made my way forward to Row 11. Trying to control my voice as much as possible, I reassured everyone that there was nothing to be alarmed about—storm situations such as the one we were experiencing simply called for an emergency briefing.

"This is standard procedure," I said, as calmly and convincingly as I could. "But I want you to listen carefully to all my instructions." I sat on the seat backs of Row 10. Stretching my legs to put my feet on the opposite aisle seat, I demonstrated the brace position for landing—hands on ankles, head between my knees. I individually briefed the people sitting at the window exits, showing them where to pull and how to lift the windows out, and making each one repeat my instructions so I knew he understood. I also told them they were to slide down the trailing edge of the wing and wait to help the passengers following. Everyone was to run 50 yards upwind of the plane. Then I told all the passengers to take off their shoes and stow their eyeglasses in the seat pocket in front of them.

Aware that Cathy had started briefing passengers in the front, I hurried to the rear, to show the passengers in the last seats, Row 20, their escape route. Asking them to get out their briefing cards, I showed how to open the rear bulkhead door, jettison the tail cone, and inflate the tail-cone slide. My jumpseat was the closest to the rear exit, but I didn't think I'd survive impact. I would be strapped into my seat between the engines, and if they didn't explode I was certain to be knocked out or have both legs broken when my jumpseat collapsed.

"If I'm incapacitated," I told the man in Seat 20B, "your job will be to drag my body out of the way and get to the door as quickly as possible. You won't have time to mess with me, so just pull my body over behind these sets and get out."

He nodded obediently. In fact, every passenger on board the plane responded unquestioningly to my every order without panic, as if I was some sort of goddess. I couldn't believe their calm.

As I returned to my jumpseat, three chimes sounded over the PA system. I quickly picked up my phone and heard Cathy's voice. "Sandy? They would not talk to me. When I looked in, the whole front windshield was cracked."

She paused for a response, but I didn't say anything. My mind raced over the possible implications of Cathy's report, none of them good. She jolted me out of those thoughts when she asked, "Okay, so what do we do?"

I still expected some instruction from the flight crew. "Have they said anything?"

"The captain screamed at me to sit down when I opened the door," Cathy said. "So I didn't ask him a thing. I don't know the results or anything. But I'm sure we're decompressed."

"Yes," I told her. "And we lost an engine." Then I asked her to make sure she had briefed the passengers up front and I reminded her to stow her shoes in a galley compartment.

"I took off my stockings so I wouldn't be sliding," I added.

"That's a good idea," she responded. "Thank you, bye-bye."

The intercom clicked off. I could only wait and wonder what was happening.

While Sandy and Cathy could only speculate on the danger, the cockpit crew knew all too well what was happening. Captain McKenzie eyed the plummeting altimeter and asked Ground Control if there was any airport closer than Dobbins. The first reply was negative.

"I doubt we're going to make it," McKenzie warned. "But we're trying everything to get something started."

"Roger," Ground Control answered. "Well, there is Cartersville. You're ten miles south of Cartersville, fifteen miles west of Dobbins."

"We'll have to go up there," Keele told the captain.

"Can you give us a vector to Cartersville?" McKenzie asked the ground.

"All right," came the instructions. "Turn left, heading of three-six-zero, direct vector to Cartersville."

Captain McKenzie requested the runway's heading and length. Ground Control asked him to stand by.

Ten seconds passed before Keele said, "Bill, you've got to find me a highway."

McKenzie spotted a blacktopped road.

"Is it straight?" Keele asked, as he checked his gauges and readied his controls for landing.

"No."

"We'll have to take it," Keele said. There was no longer any choice.

The ground controller finally came back to report on the runway: "At Cartersville, three-six-zero, running north and south, the elevation is 756 and uh . . . trying to get the length now . . . it's 3,200 feet long."

"We're putting it on the highway," the captain informed the ground. "We're down to nothing."

Keele called, "Flaps."

"They're at 50."

"I hope I can do it, Bill."

"I've got it. I got it!" Keele exclaimed. "I'm going to land right over that guy."

"There's a car ahead," the captain warned.

"I got it, Bill. I've got it now! I got it!"

"Okay. Don't stall it!"

"I gotta bug . . . We're going to do it right here!"

I knew we were low. But thinking we were descending into the Atlanta airport, I kept expecting to hear the five-bell emergency landing signal at any moment. When I noticed a male passenger get up and move to an aisle seat nearer the back, I ran forward and shouted, "Sit down. Now!" I returned to my seat and was standing, leaning forward to put my glasses in a pocket on the last row of seats when I glimpsed tree trunks out the window to my left. Still assuming we were landing on a runway, I started screaming, "Bend down and grab your ankles!"

Just before touchdown, I watched all the passengers go into the brace position. I wasn't yet buckled in my jumpseat on first impact. But when we bounced back up in the air, I yanked the belt across my hips and clamped it shut.

On second impact, a ball of flames flashed through the cabin. I saw one of my passengers catch fire at the same instant I was thrown forward into the brace position. I heard a woman scream. Then the whole world disintegrated around me. Flying bits of debris filled the cabin. I felt as if I were strapped into a big cardboard box rolling down stairs. My arms and legs flailed in front of my face as I tumbled over and over and over. Through it all I kept screaming, "Stay down! Grab your ankles. Grab your ankles . . ."

First Officer Keele had set the plane down right in the middle of Georgia Highway Spur 92—after bouncing once to avoid a car. The right wing had clipped a utility pole at an Amoco station. The left wing almost simultaneously wiped out the power and telephone lines to the New Hope, Georgia, fire station across the street. Careening on down the road, the DC-9's left wing smashed into gas pumps and a car parked in front of Newman's Grocery Store. The massive plane veered sideways and began to shatter and roll, tumbling past a number of homes in the tiny town of New Hope, before coming to rest in a cluster of pine trees off to the side of the road.

Suddenly everything stopped. Now the only sound invading the stillness was the crackling of flames spanning floor to ceiling in front of my jumpseat. For a second I wondered why I hadn't heard the emergency bells. But as my mind began to clear, I realized: "This is the real thing—a real emergency!"

With the wall of fire in front of me, I had only one way to go. I quickly unbuckled my seatbelt, stood, stepped back, and tried to force open the rear emergency exit between the bathrooms. But the handle wouldn't budge.

By now the smoke and fumes were searing my lungs. I began to cough and gasp for air. I knew that the toxic fumes were lethal within 30 seconds. I had to get out. By my vision began to go fuzzy. I watched my own hands slowly clawing their way down the emergency door as my body sagged to the floor.

There's no other way to explain what happened next except to say I felt myself slipping away. I slowly left my body through the back of my head and began to drift gently upward. About 10 feet above the floor of the plane, I looked down through the smoke and the flames and saw *my own body,* crumpled at the rear of the burning cabin.

At the same time, from my curled position on the floor, I was *watching* this happen. It was as if my eyes were turned around and looking out the back of my head. I could see myself hovering above the flames, waving goodbye to myself.

From above, looking down, I felt no real emotion until my father appeared beside me and took my hand. Instantly, I was engulfed by an overwhelming sense of peace and calm and a love that hadn't existed in my life since he had died six years before. But as I looked down at my flame-trapped body, fear returned. "I don't want to go back!" I pleaded, but my father gently answered, "It'll be all right, Sandy. You'll be okay."

The next thing I knew I was coughing on the fumes again and screaming out a prayer in my mind, "God! I survived impact—please don't let me die here by myself between these bathrooms."

The moment those words crossed my consciousness, I felt something grip me under the arms and jerk me to my feet. Certain I would die if I stayed where I was, I stepped forward, determined to go through the fire to the first window exit in the cabin. "If I don't make it," I thought, "at least I'll die with my passengers."

But as I shielded my face with an arm and moved forward, the wall of flames parted like a curtain, and I stepped out of the plane onto solid ground.

I had taken only a half-dozen dazed steps away from the aircraft when an explosion knocked me to the ground. Sputtering and gagging on the dirt in my mouth, I looked back over my shoulder to see the tail section—with my jumpseat between the two bathrooms—completely engulfed in flames. The rear of the cabin had broken away between my seat and the last row of passengers.

Large, twisted pieces of the plane lay scattered around me. The stench of jet fuel, pine smoke, and burning flesh filled the air. Another explosion rocked the tail wreckage. Finally I picked myself up and staggered to the road.

When I reached the pavement, an awkward weight pulled at my shoulders. I looked down to see a burn-blackened body under each arm. Both men were dead. As I laid them down at the edge of the roadway, I tried to remember how they'd gotten there. But I didn't know.

Then I spotted a blue pickup truck racing toward me on the highway. I ran out to the middle of the blacktop and waved my arms. As the truck screeched to a halt, I ran to the driver's door. "A Southern DC-9 just crashed!" I told him. "I'm a stewardess. You've got to help!"

"How many people?" he asked.

My mind went blank. "I don't remember. We were almost full—about a hundred."

The driver didn't wait for anything else. He shifted into gear and raced up the hill. I ran after him, screaming, "Come back! Come back! We need help!"

Finally giving up the chase, I stopped. And for the first time I turned and looked at the holocaust stretching for hundreds of yards along that little country highway.

At the crest of the hill ahead of Sandy, a downed utility pole at the Amoco station marked the spot the plane had first touched down. Halfway up the slope, on the right, long tongues of fire billowed into the sky from Newman's Grocery, fueled by the gas tanks in front of the small store.

Burning, broken pines and torn metal littered the entire scene. Incredibly, not one of the houses lining the road had been struck by the plane. The largest remains of the aircraft rested in the heavily wooded front yard of the one house built back away from the highway.

A man stumbled away from the biggest piece of wreckage and staggered toward the road. All his clothes had burned off him, except for his underwear and two strips of elastic around his ankles where his socks had been.

"If there's one survivor, maybe there are more." I ran back toward the two biggest sections of the cabin. The ground near the wreckage was covered with charred victims. One moved a leg, so I dragged him out of the wreckage to the road and hurriedly returned to search for anyone else who might still be alive. I checked every body I found for a twitch of the leg or a wiggling of a finger. The slightest movement offered a glimmer of hope.

But even as I dragged another passenger to the edge of the road, I felt so helpless. So alone. "I gotta get help." The thought became an instant obsession. I carefully eased the body I was carrying to the ground, looked around, and headed to the nearest home across the highway.

I heard a woman's voice screaming, "Help that boy. Somebody help that boy!" The lady stood in a nearby driveway and pointed back toward the crash. The passenger was still strapped in, and struggling to get out.

But I didn't have time to go back. I had to get to a phone. As I ran up the driveway across the road from the crash, an elderly couple came out to stand terrified on their porch. "Can I use your phone?" I asked. "I have to call for help."

"Sure," the man nodded. But he was looking past me. His wife gasped, and I whirled to see why.

A passenger came half running, half reeling toward us. The back of his suit still smoldered. His arms and face were black with burns. I knocked him to the ground and rolled him in the grass to put out the fire. Ordering the couple to cover him and keep him still, I ran into the house.

I found the phone with no trouble. But when I picked up the receiver I heard no dial tone. Instinctively I reached into my serving smock, pulled a quarter out of my liquor change, and desperately searched the phone for a slot to drop the money in. Seconds passed before I realized the phone was dead. I dropped the receiver and rushed back outside.

Rescue vehicles were beginning to arrive. "Help is coming," I thought. But now I had a new obsession: "I have to call Mike. I have to let him know I'm okay." A man with a service station shirt came running up. He took my arm and tried to lead me away.

"I'm a stewardess," I said, jerking free. "I have to stay. But you can help. You can call my husband and tell him I'm okay. You *have* to call him!"

I grabbed the man's pen and a piece of paper sticking out of his pocket and quickly wrote down my name, Mike's name, and our home number in New Orleans.

"But lady, the phones—"

"Just call him," I ordered as I started back down the road.

While I tried to decide what to do next, I paced along the road, clenching and unclenching my fists, saying out loud, "Stay calm. Stay calm."

My most overpowering thought was that number one fact, drilled into us in training and in all the safety courses I'd had. "You are responsible for your passengers." I had to keep searching for survivors.

I ran back to the largest cabin section of the plane and began moving sheets of hot metal and pulling more bodies out of the wreckage. I shouted instructions at the gathering bystanders—telling them to cover the people still alive and treat them for shock. And I did it all without thinking.

Something had snapped inside me; I was functioning purely on instinct and adrenaline. From the time I'd first reached the road and looked down to see the two corpses under my arms, I'd felt more like a dispassionate observer than a participant. I was an actress, acting out an unbelievable role in a gruesome disaster movie being projected in circle-vision.

Once, as I sifted through the jagged wreckage, I stopped and stared with surprise to see my own bare feet covered with blood from open, oozing cuts. But I felt no pain. When I looked into the blistered, agony-wrenched faces that were screaming in pain, I heard nothing.

Every time I touched or moved a body, big patches of burned skin stuck to my hands; I had to constantly wipe my hands on my slacks. Once I pulled on a hand protruding from under a seat, but nothing was attached to it. I just dropped the hand and continued my search.

I felt nothing. I guess it was a self-defense shield to keep me from being overwhelmed by the ordeal. There was no way my conscious mind could handle the horror; so it refused delivery on most of the sensory messages coming in.

The one thing my mind didn't block out was the screeching of dozens of sirens, which seemed to wail on and on for eternity. Perhaps my mind used the screaming sirens to overload and shortcircuit my senses and shut out everything else.

Within minutes I was aware of a host of emergency workers around me. They too sorted through the crumpled sheets of aluminum and dragged bodies away from the crash. "Help has arrived!" I realized, feeling my first tinge of relief.

Suddenly I felt the urgent need to go to the bathroom. So I headed toward another house. To get there I had to walk through a gathering cluster of spectators who stood on the road and gawked at the carnage.

"Don't just stand there," I screamed. "*Do* something! Pray! Do something!" Anger surged through me. But they only shifted their attention to me for a moment before turning to stare again at the rescue workers combing the crash.

A woman stood in her yard. I headed toward her. "I have to use your bathroom."

"Follow me, honey," she motioned, leading me into her home.

When I finished in the bathroom, the woman was waiting for me—with a cold wet washcloth and a hairbrush. She gently bathed my face, the washcloth turning black with soot by the time she finished. Then she brushed my hair back and fixed a pony-tail with a rubber band.

I thanked her and hurried back outside once again.

By this time even more ambulances had gathered at the top of the hill. Dozens of firefighters hosed down the larger pieces of the plane. Emergency workers carried stretchers at a trot toward the waiting ambulances. Others carried bodies toward a big yellow schoolbus that had been pulled onto a small gravel driveway among the trees near the top of the crash site. From overhead sounded the whomp-whomp-whomp of a rescue helicopter.

With no particular purpose in mind, I walked toward the line of emergency vehicles at the top of the hill. I couldn't escape the nagging feeling that there was something more I ought to be doing—some further responsibility I had to carry out. But I couldn't think.

"Am I alive?" I asked a nurse who was loading a passenger onto an ambulance.

Turning to look at me, she smiled. "Yes," she nodded. "You're all right."

I wasn't so sure. I made her touch me. The pressure of her hand on my arm felt warmly reassuring.

The nurse tried to talk me into one of the ambulances. A paramedic with her took my arm and tried to coax me to get on a stretcher. But I protested. "I'm okay. I'm okay. I'm one of the flight attendants, and I can identify pieces of the plane no one else would recognize. I can help you find bodies. I have to stay."

Insisting on my responsibility as a crew member, I finally thought of my undone duty. I realized what it was that I needed to do before I could leave the scene. "I have to find the rest of the crew." I didn't know I'd been thinking aloud until the nurse responded, "We'll go with you."

So she followed as two ambulance attendants picked me up and carried me through the woods, along a tiny stream and toward the far end of the crash site where I was sure I'd find the cockpit. My rescue workers wanted to protect my lacerated face, but I finally convinced them to let me down.

For several minutes I looked for something resembling the cabin door. But all I found were more bodies and gore. Rushing from one clump of smoldering wreckage to another, I happened on a large patch of charred earth. And there in the middle of the burned area was a body, charred stiff in a distorted U-shape. The arms, legs, head, and even the torso were off the ground, straining heavenward. And across the middle of the body lay the live power line that had created the scene.

Uncertain whether the corpse was male or female, I stared for a moment. Then, deciding it didn't matter, I backed away and resumed my own search for the crew.

Finally I spotted the nose of the plane. "Over there," I shouted and began to run. But the sight I found when I reached it stopped me cold.

The front of the plane had been ripped off in front of the cabin door. Lyman, the first officer, still sat strapped in his seat as it dangled from what was left of the cockpit. He was staring straight ahead with his eyes wide open, his face expressionless.

The captain lay on the ground just a little way away, his face mutilated almost beyond recognition. I fell to my knees beside him and cried, "Bill, Bill!" In what seemed like a natural reaction at that moment, I put my hands over his face and pressed the pieces back together in an attempt to lend a semblance of dignity to his death. He was my friend, and I couldn't leave him like that.

That's when I felt hands grabbing me. Four men pulled me away as I kicked and fought. "Cover him up!" I screamed. "At least have the decency to cover him up!"

Back on the road, the men set me down. But I took off again, running toward the yellow schoolbus, where a cluster of emergency personnel were working. There, among the trees, lay row after row of sheet-draped bodies. I walked between them to the rear of the bus, where men were loading stretchers through the emergency door. I looked in at the bodies already stacked inside. Then I turned and slowly surveyed the nightmare around me—the smoke, the broken metal, the shrouded dead, and the expressionless faces of the rescue workers who were also fighting to deny the horror of the scene.

Two more stretcher-bearers climbed the hill with a load. When they approached, I looked to see if I could recognize the body. But there wasn't any body—only *pieces* of bodies, collected from the wreckage.

That awful, gut-wrenching sight finally cracked my emotional shield. I felt strength drain out of me with a gush. And I put both hands to my head and began to shake.

"Get me out of here!" I wailed. "Get me out of here! I just can't look anymore."

Again I felt the firm hands. This time I surrendered to them. A man lifted me in his arms and trotted toward the road, clutching me to his chest. He handed me to another man who carried me a little farther and passed me to another, then another. Within moments I was transported back up the hill. The last man to take me wore the uniform of the Georgia State Patrol. He set me gently in the back seat of his patrol car before he climbed in, turned on his siren and flashers, and pulled quickly away from the crash site.

Racing down that little country road, we met what seemed like an endless procession of more police cars, ambulances and firetrucks, sirens screaming, heading for the crash. They and the rolling Georgia terrain whizzed past in a blur.

"What's your name?" I asked, in an attempt to be friendly.

"Phil," the trooper responded.

"Hi, I'm Sandy."

The conversation got no further. We reached the first houses in the town of Dallas, Georgia. Moments later the patrol car pulled to a halt at the emergency doors of Paulding Memorial Hospital.

Tragedy and Escape

BY ERNEST K. GANN

(Excerpted from *Fate Is the Hunter*)

As promised when we introduced Ernest Gann back in the "Dooley" chapter, this second Gann excerpt is taken from his now-classic 1963 nonfiction account of his career in commercial aviation.

First published in 1963, *Fate Is the Hunter* is the Ernest Gann book that took his readers from the cockpits of fiction—his novels—to the cockpits he had actually known during his career as an airline pilot, chiefly with American Airlines in the 1940s and 50s. If the title of the book gave readers a certain foreshadowing of what was to come in the pages ahead, Gann's dedication certainly clinched the deal: "To these old comrades with wings . . . forever folded." And then a listing of 351 pilots with the abbreviations of the airlines they flew with, followed by another Gann note: "Their fortunes were not so good as mine."

In introducing Ernest Gann with the "Dooley" excerpt from *Island In the Sky* earlier in this collection, I attempted to describe the many qualities that make his prose so engaging. Everything referenced there applies doubly, in my opinion, to *Fate Is the Hunter*. Arguably, it is Ernest K. Gann's finest book.

As he did with every other chapter in *Fate Is the Hunter,* Gann prefaced this piece with a short "tease line" of prose. The one for "Tragedy and Escape" reads: "There is a degree of mercy beyond which any man is rude to inquire."

Astute readers who have followed Ernest K. Gann's novels and films will recognize certain real-life events in this tale that eventually were used by Gann in his novel, *The High and the Mighty*.

Incidentally, the film version of *Fate Is the Hunter* serves Mr. Gann's aviation prose less well than did the much older film of *Blaze Of Noon* and the

Duke Wayne classics *Island In the Sky* and *The High and the Mighty.* The film version of *Fate Is the Hunter,* starring Glenn Ford, isn't half bad, in my view. But it isn't bonded Ernest Gann, either. Hollywood used the title and pretty much concocted its own story line.

To really enjoy Ernest Gann, you do not need Hollywood at all. His words are so good they will be the only film the camera in your mind needs.

Now there came a time of great aeronautical confusion. The long-established airlines were fighting for position and the postwar ventures were striving for recognition. Surplus airplanes were cheap and could be converted to airline use with relative ease.

Thus, innumerable small airlines came into being. Some of these possessed only one heavily mortgaged airplane and it was not uncommon for the captain and co-pilot to be the entire board of directors. Economics soon eliminated most of these pitifully brave little outfits.

Many of the new airlines were irresponsible and actually dangerous. Yet, particularly in America, it was discouraging to witness the willful choking of free enterprise by the clammy hands of government. It was also prophetic. Aviation had grown up. Airline officials were no longer to be found in shirt sleeves, and a universal seeking for respectability drained much of the color from the scene. Yet the natural forces behind such a revolutionary means of human communication was so tremendous neither the governments, unions, the pettiness of newly cautious officials nor the international jealousy for power and prestige could restrain this era any longer.

The transformation amazed the eyes and the minds of those who had pioneered the skies so short a time before. It was already affecting the affairs of all mankind, from savages to the most polished sophisticates. And all of this, as if to match the intrinsic nature of the development, occurred in such a short span of time that the same men who had flown helmeted and goggled in primitive open-cockpit mail ships were still flying the most modern airliners. Since most of them had barely reached middle age there was no reason to believe anything but ill-luck could ground them for years.

One of these men was Sloniger. Now, in the establishment of a new airline, he was as eager and enthusiastic as in the days when he flew mail between St. Louis and Chicago. He seemed to be forever young although he was now burdened with the complicated responsibilities of many airplanes.

The governmental vacillations and consequent uncertainties created serious morale problems in all the new lines wherever they flew. Sloniger had also to explain our maverick airplane. It was understandable that our steamship sponsors should wonder why airplanes which looked exactly alike should perform so differently. The mystifying reluctance of our maverick led some of them to wish they had never heard of airplanes. Expensive and frustrating attempts to operate this airplane now had all of us tugging at our lower lips, and the mechanics were alternately snappish and inconsolable. The experts, long since humbled, were drinking their whisky in straight, desperate gulps. At last, after countless changes and outbursts of profanity, the airplane was considered safe to fly the two thousand four hundred odd miles to Honolulu.

It seemed that I had personally inherited this monster and I soon came to regard it with the same mistrust once reserved for C-87's. Here are actual comments from my logbook, each having been set down on a different flight and day.

". . . All engines cutting out for six hours. Very uncomfortable ride. Quit until they fix it!"

My determination was not very strong. Three days later:

"Test hop with Sloniger . . . ship no better."

Ten days later on a flight to Honolulu:

"Engines cutting out again . . . certain it is plugs."

Four days later:

"Engines not getting fuel."

Two days later:

"Ditto."

And there follows a series of ditto marks for seven more flights.

The additionally curious events on one of these flights were not set down in the logbook. I had neither space nor heart to record how one flight from Honolulu to Portland, Oregon, was flown at two thousand feet because the engines simply would not perform at any other altitude. Nor did I wish to record that one of the stewardesses became drunk at the approximate point of no return, and the labor pains of a lady passenger were occurring every five minutes.

I could no longer approach the maverick with anything but hatred. But I continued to fly it because it had taken on the quality of a living thing, a well-dressed dandy of a villain, and I intended to conquer it.

Before we killed each other, cooler heads came to a reluctant and extremely expensive conclusion. It was not the fault of the engines. It was not the fault of the airplane . . . or me. It was the combination of the DC-4A type ship and the Dash–13 engines. The fundamental design of the fuel system simply would not continuously supply the new and hungrier engines with sufficient fuel—a circumstance aggravated by altitude. To correct the fault demanded the most elaborate alterations. So eighty thousand dollars' worth of beautiful new engines were removed and the older types set in place once again.

Thus equipped, the airplane flew quite happily ever after.

The visual aspects of the worlds aloft are as different as the corresponding areas below. And there are countless subtle variations for which there is no honest meteorological explanation. The mewling, drizzle-laden skies over the Bering Sea are not the same as the atmosphere to be encountered in the vicinity of Iceland, though the latitude of the observer may be nearly identical. The jungle skies of South America are lush and inviting and altogether lacking in the antagonism and elderly stink of skies over similar jungles in Africa. The skies of the Sudan are hostile and given to snarling phenomena, while the desert skies of Arizona and New Mexico seldom trouble an airplane. In many ways the North Atlantic skies suggest masculinity, while Pacific skies appear relatively feminine in character except over the Japanese archipelago. This is particularly true of the immense region from the West Coast of North America to the approximate longitude of the Philippines, where an occasional typhoon can destroy the illusion.

The steamship-airline proposed to operate all through these areas, which, in addition to the South Seas, had long been the domain of their surface ships. It was an enticing prospect, and with our earlier troubles defeated there was every reason to hope for growth and permanency. Month after month we flew the great triangle, gathering reputation and pride in the reliability as well as the luxury of the operation. We had no lack of passengers though the government had specifically warned against advertisement of any kind. Thus we were obliged to tiptoe around the increasingly concerned Pan American and rely on the praises of our passengers. Our flights were simply not supposed to be happening.

Another steamship-airline was similarly engaged over the Caribbean area and the operation was equally successful. It was inconceivable that any government agency could ever be persuaded to force such publicly convenient and well-founded organizations out of business.

During this time Abernathy and Holle were killed at Laramie, Wyoming, and later Anderson and Miner ended their brief careers at Richmond, Virginia. Both accidents occurred during attempts to land in bad weather. These were "hungry" airlines and unfortunately the airplanes had flown a great deal longer than the crews.

The same could not be said of Sprado and Weber, who soon afterward flew into Mt. Laguna, California. The official cause: "Let down below terrain without positive fix." The investigation carefully avoided any speculation on why two intelligent and experienced men should do such a thing.

It appeared that fate was on a rampage, taking the rich and highly skilled as well as those more hopeful than able. Now Tansey and Sparrow were killed at Shannon, Ireland. When the investigators discovered the cause, we found it harder than usual to shrug our shoulders. One of the finest and largest maintenance organizations in the world had made a tragic mistake and failed to discover it. They had somehow *reversed* the static pressure lines, which resulted in false altimeter readings. It was night. Circling the fog-shrouded airport, relying on his instruments as he should, Tansey flew into the ground.

Almost simultaneously Ham and Ring were lost at Michigan City, Indiana. The feeble technical explanation was "Probable carburetor ice." If Ham and Ring were amateurs, ill-schooled, and flying for an impoverished airline unable to maintain the equipment properly, this verdict might have been barely satisfying. But none of these shortcomings applied in their destruction. Whether they were many or few, occult or technical, there had to be other contributing factors to such an accident.

Fortune now also abandoned Haskew and Canepa, who flew into the Blue Ridge Mountains, Hearn and Day in a check flight, and Weeks and McKeirnan in another check flight. Cushing was lost in the most inexplicable of all appointments, a mid-air collision.

All of these men were gone in less than five months.

And there now occurred a combination of events which, if not so exactly proven by logbooks and records, would have been incredible. Afterward, I was at last thoroughly convinced that technical causes are only the active and recognizable instrument of a flying man's destiny. The fundamental cabal, the originating force of decision or postponement, remained exactly as it had been always—just an instant of distance beyond the reach of the human mind. But it was there, inscrutable, and only half hidden behind the comforting veil of our childish faiths. It did not linger for explanation, although it did allow the inquisitive to uncover the more obvious surface manipulations. It said to the boldest seekers, "Pry to the limits set for your intelligence but expect to retreat with bruises when you strike the veil. I am the final explanation. You could not long survive if you knew."

The date was the thirtieth of May.

Captain Coney and First Officer Willingham of Eastern Airlines were flying at four thousand feet in the vicinity of Bainbridge, Maryland. Forty-nine passengers attended by two stewardesses sat relaxed in the cabin. There was one infant aboard.

The weather was pleasant and routine radio reports were received from the flight by the various ground stations. Then suddenly the airplane went into a steep and uncontrollable dive. There was no call of distress. There was only silence. And the silence, except for the whimpering of metal fragments in the wind, continued forever because there were no survivors.

Again, the thirtieth of May.

Far to the west, over the Pacific, I was flying exactly the same type airplane bound from Honolulu to Burbank. The time was approximately the same, but since our fuel weight was greater, we carried only thirty-three passengers. This was the sole difference. Our weather was also most agreeable and the forecast for the California coast, plus unusually favorable winds, promised an easy arrival. All of this pleased me because I was due for my first vacation from the steamship-airline. It would commence when this flight was ended. I could not know how important the thought of this vacation would be to my continuing existence.

It had been my custom to leave the flight deck and visit the cabin after we had reached cruising altitude and were well established on course. As I had once schemed to preserve my captaincy, now I was determined to do all I could to help the steamship-airline stay in business. A lucky tour down the rows of seats could usually be managed in ten minutes of question-answering, and if it made future customers I was satisfied. Then I would stand in the galley, which was near the tail, and listen briefly to the woes or matrimonial progress of the stewardesses. Cardboard romance, culminating in official ritual, was an obsession with most of the girls who flew on any line, but Grimes was an unusually realistic young woman. She was petite, sparing of cosmetics, and forthright in manner. Now, as I reached the galley, she seemed to have been waiting for me.

"What's the matter with this airplane?"

"Nothing is the matter with it. Why?"

"Well, it sure shakes back here. It's a good thing I'm wearing a girdle."

I waited a moment. Here the sound of the engines was only a muted humming. It was not even loud enough to dominate the faint whisper of our slipstream along the fuselage. So we spoke softly, lest the passengers in the rear seats overhear us. I moved about the galley, touching at the bulkheads, listening, sensing the airplane as a man may question any unwillingness in a long-familiar love. But I could not detect the slightest quiver which might be thought unnatural. There was only the slight and regular yawing induced by the automatic pilot. The sensitivity control had only to be turned half an inch and the yawing would cease. I told Grimes that as soon as I returned to the flight deck I would ease off on the control and things would seem right again.

"No. That's not what I mean. It's a funny . . ."

"We passed through a little patch of choppy air about half an hour ago. Maybe that gave you ideas."

"I've flown long enough to know rough air when I feel it. This was something different. It sort of scared me."

"Relax. All airplanes shake a little back here."

I left Grimes with what I hoped was a reassuring smile and returned to the flight deck. There, everyone was busy about his duties. Drake and Hayes happily occupied both pilots' seats, and I was reluctant to pull rank. I asked them to reduce the sensitivity on the auto-pilot, then stood for a time between Smith, who was at his navigation table, and Vaclavick, who was, as usual, lost in a book. Snow occupied the jump seat, which cut down considerably on my forward line of vision. The view through the two small portholes alongside Smith's table and Vaclavick's radio gear was only of dappled clouds and splotches of sea between. What little I could see forward was exactly the same, interrupted at first by the less interesting side of human heads and the curving snout of the ship itself. Directly above me, the astrodome revealed a rather sour milk sky worthy of only a glance.

By now the normal characteristics of the route between the California coast and Honolulu were as familiar to me as the North Atlantic had once been and AM-21 had been in what already seemed the long ago. Here the seasons brought little variety. East of Honolulu and the Molokai passage, the small trade-wind clouds were almost a fixture and seldom rose above seven thousand feet. Then after a few hours' flying we would pass through a variable area which sometimes presented a weak front and, as often, clear skies. We were flying in this area now. Soon we would enter the almost dormant central area of the great Pacific "High" and it would hold for another two or three hundred miles. Then finally we would sight a very low layer of rumpled stratus which frequently extended as much as a thousand miles to the California coast. Except in the brief winter season this pattern was so regular that the slightest change called for comment. Even the forces and direction of the winds rarely surprised us. They appeared to blow or not to blow according to an eternal code.

I checked our progress and position with Smith and then went back to the crew compartment and sat down on the lower bunk because it was the only place available. I flipped through a magazine and thought more about my vacation, which was to be in a sailboat.

Anticipation of such pleasure lulled my senses and I had almost dozed off when Snow entered the crew compartment. He stood thoughtfully in the center of the small floor space with his head cocked toward one shoulder. I saw him do an about-face, but he remained in position. He bore the air of a man insulted, so I asked him of his trouble.

"I think we have a rough engine, but I can't figure out if it's number two or number three. Every time I make up my mind which one it is it seems to be the other."

Since we were directly in line with the four engines I thought it strange that I had failed to notice any roughness. Perhaps my thoughts were too much on sailing. I stood up beside Snow and listened. After a time he said, "It beats me. I can't notice anything back here, but up forward . . ."

The rumble of the engines was monotonously smooth, but I went forward and stood between Hayes and Drake.

"What's all this about rough engines?"

Drake spoke carefully. "Well, I *thought* I felt a little roughness a while ago, but nothing showed up on the instruments."

All three of us now instinctively looked back at the wings, although we knew very well the action was meaningless. Trouble would reveal itself first on the instrument panel. We checked the ignition on all four engines. Perfect.

Snow held the logbook open before me.

"Number three is about due for overhaul . . . might be a sticky valve. . . ."

We all now took turns at inspecting number three engine from the right window, and the repetitive dumb show struck me as sadly lacking in either originality or good sense. Since the engine was obviously as staunch and solid as anyone could wish, I was able to state my opinion with convincing nonchalance.

"Well, as long as it doesn't jump off the wing . . ."

Later in the afternoon I took over the flying, which at least gave me a comfortable place to sit down. The duties were almost embarrassing in their simplicity. With Hayes, I kept a casual lookout ahead. Otherwise my chief exertion was to reach forward every ten or fifteen minutes and turn a small knob on the auto-pilot to maintain exact course. Our altitude, varied only by the occasional movements of people in the cabin, could be corrected by a quarter turn of the stabilizer wheel. I performed this chore with mock delicacy, using the tip of my little finger.

The air was smooth and now the sun had descended enough to enliven the scenery ahead. There were great oblongs of brass upon the sea and the occasional bundles of clouds huddling along the southern horizon were also of brass. Far to the northeast I saw a lenticular cloud which was so unusual in this part of the Pacific we spent several minutes admiring its graceful, spooklike formation. Later, Smith reached in front of me and tore away the small slip of paper taped over the gyro compass. The figure 62° had been written on the paper—a reminder of our course for the zone in which we were flying. The figure 63° was written on the substitute paper which Smith now fixed in place. And there followed at once the usual badinage about who the hell did Smith think he was, the navigational genius of all time? And how could he fix our position so exactly that he dared change the course so little as one degree? We also made bitter comments on who he might think *we* were who could follow a course within one degree, and didn't he know that the compass was probably off more than one degree anyway? Then Vaclavick said that it was no wonder flight crews were so given to reading funny papers since their potential of concentration was constantly and thoughtlessly strained by existing in confined quarters with people who chattered like middle-aged ladies at a laundromat. And all of this exchange we found healthy and very pleasant and we had flown another hundred miles or so before we quieted again.

As the hours pass the subtleties of flying fatigue gradually reduce the mental alertness and physical smartness of any airline crew. Ties are pulled down and collars loosened; there is much stretching of limbs. It is at this point that some airmen become preoccupied with personal thoughts, their dreams transporting them over the horizons to more prosaic scenery. They create romantic visions of their homes in which their wives are continuously delightful, their children never cry, and all the bills are paid. Sometimes they become so sentimental they even trifle with thoughts of remaining permanently earthbound. Only the cynics know this is next to impossible.

We had achieved this condition when Smith announced a more immediate fact. We had passed the point of no return.

All long-range flights are planned according to zones, so that as fuel is consumed and the weight of the aircraft correspondingly reduced, the most economical speed can be maintained. Thus we intended to reduce power at approximate intervals of three hours. There was such a change now due, and after Snow had consulted his power charts for our present weight and obtained my permission, he eased back on the throttles and propeller controls. I noticed at once that the engines seemed rougher. This was not unusual since one power setting may create quite different harmonics of vibration than another; yet this new note strangely annoyed my basic flying nerves. I saw the others were equally uncomfortable.

Again we twisted around so we could see the number three engine. Because I was seated on the opposite side of the ship I had to stand and lean across Hayes for a clear view. None of us knew what we expected to see. Our preconception that something was not altogether right with number three simply bade us stare at it.

We were still twisting and posturing when Grimes came forward and plucked at my sleeve. "We've got the shakes again. I was just starting to serve some drinks when—"

"I know. I know. There's a little roughness in the number three engine."

This was far from a confirmed truth because there was still no indication on the instrument panel that number three was not performing as well as the other engines.

"Is it going to quit?"

"No."

I spoke positively because I wanted to be positive.

The slight vibration passed through the ship in cycles. I had never felt anything exactly like it. It was nothing and yet it was something because certainly all of our imaginations could not create the same mischievous genie.

There . . . there was the cycle again. . . .

My mind explored the vitals of the airplane which I knew so well. Perhaps the entire ship was slightly out of rig? It could happen. It had happened. Yet the cycles of vibration denied any such fault.

I told Snow to set the power where it had been. He complied and all was serene once more. It seemed, then, that this was purely a mechanical matter. I thought that one of the propellers must be a hundredth of an ounce out of balance and the result was a vibrational harmonic at certain critical power settings. I told Snow to leave things as they were. We would forget power reduction for the rest of this flight. Because of the favorable winds en route and weather along the coast, there was no demand to hoard our fuel. So we are fat, I thought. If number three runs smoothly at this power setting why trouble it? An early arrival won't hurt anyone.

I sat back to think of sailing but found it impossible. My theory about unbalanced propellers was so weak I simply could not nurse it to probability. I knew well enough that the weight relation of the giant blades was actually measured to a hair's degree.

Damn Grimes! Her skittish complaints had started all this.

I brooded on our mysterious ailment until the evening sky held my favorite moments. A violet hue had already seeped upward from the eastern horizon and after a while I knew the limits of my sight would become washed with purple and magenta until finally, all of this fading, it would be night. Usually I could find infinite peace in the upper skies at this time, but now I could not.

"Snow?"

He looked up from the logbook where he had been noting the engine instrument readings and I saw in his eyes that he was also dissatisfied.

"Go back and stand around in the tail for a while. See if you can find out what's troubling Grimes."

Snow buttoned his collar, put on his coat and white cap, and went back to the cabin. I waited easily enough. There was no vibration. No ... there. There it was again. Or was it? I looked at Hayes, who was preparing the weather report for the zone in which we were flying. He seemed unaware of anything unusual. I turned about to watch Smith. He was standing on his stool, his head projecting into the astrodome. I knew he would be waiting for the first stars of evening. Vaclavick? What about you? To see him, I had to half-rise in my seat, twisting my whole body around. Of course. His chin was cupped in his hands and his elbows sheltered a book as if it were a secret treasure. I rose still farther until I could see that the pages were mostly complex electrical diagrams. I could not imagine what Vaclavick could be studying since he already held the highest license obtainable. But from his absorption, it seemed certain he had not felt anything unusual.

Drake was resting back in the crew compartment. I could not see him.

I slumped unhappily in my seat and stretched my legs their full length. I placed my feet on the rudder pedals, a position of habit. I had long used the pedals as a convenient footrest when the airplane was on auto-pilot. Now very suddenly, as if the pedals had been waiting for me, the right one developed a tremor. There was no reason for this. When the auto-pilot was in operation the rudder pedals were practically locked in position. The trembling endured for several seconds and then was gone before I could be certain it bore any relation to our periods of vibration.

I disengaged the auto-pilot and flew manually for several minutes. It seemed the airplane had a tendency to yaw to the left. I adjusted the rudder trim tab, an incongruously large wheel below the magnetic compass. Why hadn't I checked this before? For a moment I thought the mystery was solved. The airplane was slightly out of trim, the auto-pilot tried to correct matters, and the consequent argument between two brainless entities set up a periodic vibration.

Snow returned. He pushed his cap far back on his head, blew out his cheeks, and took the pose of a farmer surveying a blighted crop.

"I'll be damned if I could feel anything back there."

Then he said that I knew how women were, which was assuming a great deal, and explained how he had even crawled into the tail cone. "It's colder than a whore's heart back in there."

"You didn't feel anything about five minutes ago?"

"Nary a thing. Smooth as silk all the time."

My theory on the auto-pilot was smashed. If it were fighting a directional correction the greatest force should be apparent in the tail.

I glanced out at the last of the evening. There would not be more than a few minutes even of this lilac light remaining.

"Take a look through the drift meter. See if there's anything hanging down."

I wondered if one of the various landing gear doors might have failed to close properly. It could be enough to create buffeting.

The drift meter was a periscopelike device fixed to the deck. It stood beside the navigation table. By adjusting and turning it properly we could examine the underside of the airplane.

Snow was bent over the device for several minutes. At last he straightened and shook his head. "Just fine. Everything sewed up tight."

There. Another spasm of vibration. It seemed to pass across the ship from one wing tip to the other and then it was gone. It was far from a violent reaction, yet it did seem to have more authority than any of the others. At least it was enough to bring Drake strolling from the crew compartment.

"Number three acting up again?"

We were back to thinking of number three. And I was momentarily ashamed because we were so like other men in blaming the understood, guilty or not. We had neither the courage nor the imagination to pry up the lid of anything unknown. We stared accusingly at the instruments, almost demanding they confirm our easy suspicions. But there was no signal.

"How about reducing power?"

Snow asked the question and with good reason. If there is something wrong with an airplane which cannot at once be detected, the specter of structural failure rises up, and it will not go away until the trouble is resolved. It just stands and stares . . . waiting. And it is axiomatic that the easiest way to postpone any such threat is to slow down.

But we all knew that cruising airplanes do not come apart in perfectly smooth skies, particularly after having been en route for several hours. A thunderstorm, a rough front, really excessive speed in a dive, there was always the remote possibility . . . but not under these conditions. It was something we not only had to believe but something supported by fact. So I spoke with conviction when I said that if number three was going to throw a master rod, or whatever it might conspire to do, the sooner we arrived over Burbank the better. Also, I thought without the slightest tinge of guilt, the sooner and better for sailing.

I asked Drake to take over the flying. Then I borrowed Snow's flashlight and went back to the cabin.

Grimes was in the buffet preparing to serve the passengers' dinner. She pointed to the glasses and dishes in the racks just above her head.

"Watch . . ."

As she continued with her work I studied the dishes and could find nothing wrong. I was about to turn away when the entire assembly came alive. The dishes clattered together and one glass which Grimes had placed upon the counter skittered across the metal surface until it came to rest against a low partition.

"How about that?" Grimes's manner was accusing, as if I had personally shaken the buffet. "See what I mean?"

I did. I sought through my whole aeronautical past for some logical explanation, needing it as much to reassure myself as Grimes. And I could find nothing. As she continued grimly about her work, I made several limping excuses. The airplane was perfectly all right, but it was far from new, of course. . . . I had noticed similar vibrations in other airplanes when the speed was just so. There was the number three engine to be considered . . . hem and haw. . . . Yes, it *had* to be number three engine and we would certainly give it a going over when we landed in Burbank. In the meantime Grimes should stop worrying about a very expensive airplane and try smiling. The passengers like it. "Think of your future. . . ."

"I am," she said.

I waited several minutes for another cycle of vibrations. The dishes remained still. So when Grimes went forward with a tray, I turned in the opposite direction and climbed through the small door leading to the tail cone. I closed the door after me and switched on the flashlight.

I directed the flashlight beam all around the circular compartment. Here, without benefit of heating or any upholstery, it was cold enough to see my breath. It was an eerie place, the relative silence somehow emphasized by the hushed murmur of the slipstream passing along the fuselage. There was no familiar sensation of flight because the area was windowless; it seemed rather that I stood in one end of a great projectile. We could be moving at fifteen miles an hour or fifteen thousand—the loneliness would have been the same. And it was difficult to remember that beyond such a thin skin there were only stars and ocean.

I straddled the harps of control wires which led through a multiple series of fiber sheaves. I plucked at several, not really expecting to find one loose, but more in curiosity because I had never been in this place before while the airplane was actually flying. In the flashlight beam I saw the various wires move ever so slightly, responding to corrections from the flight deck.

I waited patiently for any hint of vibration and at last decided that if I remained in the cold metallic cocoon any longer, the major shivering would be in myself. Before I returned to the cabin I examined every frame and all of the control wires. They appeared to be in perfect condition.

Grimes was waiting for me. I blew on my hands for warmth and told her that as far as I could determine the tail was still on the airplane.

"You're very funny."

I shrugged my shoulders. I did not like the look in Grimes's eyes because they were being far more honest than my own. She knew I was like a small boy whistling in a great and forbidding forest. If I ignored the goblins they might go away.

As I left her the dishes clattered again.

During the final four hours of the flight the vibration came unpredictably at intervals of five minutes or eight minutes or ten minutes and nothing we could do on the flight deck seemed to influence its arrival in any way. We called San Francisco on the radio and reported that we were experiencing unusual vibration.

"What is the nature of your vibration?"

"We don't know. It is not violent. One of our engines has been acting up."

"Are you going on three-engine operation?"

"Negative."

I wished we had never called in the first place. There was nothing anyone in San Francisco could do. We had, in the calling, instinctively sought assurance of our faith. We had succeeded only in raising eyebrows.

We were delayed a half hour in Burbank while Snow made a thorough inspection of the airplane. He joined with the regular ground mechanics and spent a long time atop a ladder examining the tail. Finally Snow came to operations office and I saw that he was much easier of mind. They had discovered only one very minor defect. An exhaust clamp on number two engine was broken. It had been replaced and was not worth further discussion since it could not possibly have caused any vibration.

Twenty-one of our passengers disembarked at Burbank. The twelve remaining would continue on to Oakland. I passed them at the boarding gate. They were clutching at their precious souvenir leis, and their midnight weariness seemed to have made a total conquest of their spirits. Even their palm-frond hats appeared to droop, so I told them the flight would be swift and they would soon be home in bed. It was a safe enough prediction. With so few aboard, the airplane would be very light and I intended to use the same power required for a full load.

The night was so benevolent and the air so tranquil I chose to fly the entire trip without surrendering the controls. For there was still a deep and sensuous joy in such personal dexterity, in selecting a star to guide upon and causing it to move up or down or slightly to either side at easy will. When we cared to look down there were the diadems of sleeping cities to invigorate our thoughts, and between them even the spaces of little habitation were enlivened by flashing airway beacons.

There was an unusual sensation of speed because the lights slid so quickly beneath, plus the magic awareness of flight, and none of us had ever seen the combination so rightly joined. Hence we were silent nearly all the way, each man wrapped in private counsel, and we found the visual environment subdued all sounds as other atmospheres had never done.

There was no vibration. Our notation in the logbook about a certain trembling now seemed baseless and even a little silly.

Grimes came forward for a cigarette and broke our silence momentarily. She said it was a heavenly night but, when she received no answer, said nothing more.

At eight o'clock in the morning the phone rang at my bedside. I answered with a yawn and at once a voice asked if I had been to church.

I said that I had been up most of the night and was trying to sleep . . . and furthermore I considered the joke ill-timed.

"Well, you'd better do some praying somewhere. Then come to the field."

I recognized the voice of our maintenance chief and protested a mistake. I was officially on vacation.

"It was almost permanent. The outboard hinge bolt of your left elevator was missing. Congratulations. See you shortly."

Because I resented losing even a few hours of a sailing vacation the journey to Oakland Airport seemed endless, and my grumpiness had overtones of self-pity by the time I arrived. I had some bitter things to say about time and tides as I joined the cluster of men now gathered near the tail of the airplane. They showed me the place where the hinge bolt should have been and was not, and it was easy enough to understand how anyone might have passed over its absence—particularly if the inspection occurred at night. I hoped that I had not been brought so far simply to look at the place where a metal pin of one-quarter-inch diameter was supposed to be, but since so many full-grown men affected concern about such a tiny object, I joined in the general head-shaking with all the enthusiasm of a social coward. Actually, the absence or the presence of one innocent little bolt failed to impress me. There were other bolts to keep the elevators where they belonged, and after having flown this same type airplane through all manner of violent turbulence and maneuvers, I was convinced its margin of strength was past my understanding.

When the head-shaking ceased and a written report was requested, I prepared one immediately as a sort of exit visa from the whole affair. Just before I left the airport I learned Coney and Willingham had crashed at Bainbridge. But the facts of the disaster were still few and on that morning it never occurred to any of us that there might be some relation between incident and accident.

Released at last, vastly and happily preoccupied with more gentle elements, I went directly to sea. There I remained blessedly beyond any communication for three weeks.

At the end of this time I put into Los Angeles and there, by chance, joined with Howard at a Chinese restaurant. And he revealed certain surface workings combined with incident which persuaded me there was either marvelous order or sheer anarchy in the house of fate. Otherwise how could I be alive when Coney and Willingham were dead?

Howard might have been Sloniger's twin since both men were wiry, leather-complexioned, and notably sharp of feature. They were so hawklike it was easy to envision them wheeling over any terrain with effortless grace and it was true that both men seemed capable of levitation when confined to a room. Howard was also a very early bird and he affected a thin and dapper mustache in the tradition of the clan. He had lost one leg when the propeller came off his plane in a Bendix race, but he had never lost either his courage or enthusiasm for flight. Indeed, his career had greatly prospered and he became one of the most respected test pilots in history. His flying skill was complemented by his engineering ability, and his combined talents were deeply involved in the design and proof of the fabulous DC-3 airplane. Later he went on to the DC-4 and was now wet-nursing the DC-6. Thus Howard was a unique figure in American aviation, and if he pronounced an airplane or any device of flight good or bad or indifferent, his views were regarded most solemnly. Fortunately, there was nothing solemn about Howard.

He approached me slowly and his eyes were so filled with mischief I wondered if he had preceded all of us to the Chinese wine. A hand flew upward in a gesture which might have been made by Sloniger. The hand executed the beginning of a chandelle and landed lightly on my shoulder.

"Let me touch you," Howard said. "When we eat I'd like to sit at your side. Maybe some of your luck will drip on me."

He caressed my shoulder and then my arm as if I were some pagan statue and I was exceedingly embarrassed.

"Yes, you're the living proof that it doesn't pay to be overly smart."

My embarrassment turned to bewilderment. I didn't know Howard well enough to exchange insults.

He led me to the table. "Please . . ." He pulled out a chair and bowed me into it.

Then he sat beside me and drew out a pencil. And while he talked, he made notations and diagrams on the tablecloth, each line and figure neatly set down after his hands had flown their interpretation. He began by saying that my written report on the suspected vibration had been a masterpiece of innocence. He stated flatly that if I had any training as an engineer I would never have had the opportunity to write it. It seemed that only a most remarkable series of causes and effects had kept us from duplicating the catastrophe of Bainbridge. The aura of fantasy was compounded when we considered both had occurred on the same day.

"Did you know we grounded every DC-4 in the world because of you?" he asked.

"I've been sailing. . . ."

"Never giving a thought to vibration, of course."

"No."

"Thank you for completing my picture of blessed ignorance." He frowned and his hands fluttered uncertainly. "But I will never understand your nonchalance. Listen to me very carefully. I've spent too much time on this investigation to miss the finale."

It soon became obvious that Howard's detective work had included my personal anticipations. Even what I had said to the crew and passengers had been remembered and considered.

"Although we can never be absolutely certain, we now believe the Eastern Airline crash at Bainbridge was caused by unporting. Do you know what that is?"

I confessed that I had never heard of it.

"Unporting is the balance destruction of the elevators by aerodynamic force. I won't confuse you with theory, but if enough separation between the fixed and the balance portion of your elevators occurs, your airplane will go into a vertical dive or even beyond the vertical, and no two men in the world are strong enough to bring it out. This can be caused by a missing hinge bolt."

He sighed heavily and drew wavelike lines on the table, then an airplane diving for the lines. He sketched another airplane more precisely and marked its approximate center of gravity. "Did you slow down when you first noticed the vibration? You did not because you had no fear of it. But if you *had* been the nervous type, if you *slowed* down, the center of gravity would have changed. That would have been quite enough to complete the process of unporting which had partially begun."

"The vibration really wasn't very bad."

"It doesn't take much. But let us assume another pilot would have reacted in the same way. It would only have postponed the inevitable. As soon as the time came for a normal power reduction and it was accomplished, unporting would begin. But not you. In the past you had lost all four engines so many times, the prospect of losing one gave you relatively little concern. So you sat there, fat, dumb, and happy, and you canceled all power reductions. This brilliant decision saved your life the *first* time that day."

I could think of nothing to say but a series of well . . . well's.

Howard held up one finger and then raised a second beside it.

"This was not enough," he said, and I saw that he was exasperated. "You landed at Burbank and disembarked twenty-one passengers. God alone knows why, but you took on just enough fuel to make up the difference in losing their weight. Even so your center of gravity would have been changed enough so that unporting was more likely than not. *But . . .*"

He moved a third finger up beside the others.

"You were in a hurry to reach Oakland so you could go about your silly sailing. As a result, and don't try to deny it because the figures are in the logbook, you used full gross weight cruising power all the way and your speed was correspondingly high. . . ." He paused, touched at his mustache, and stared at me incredulously. Then he spoke very slowly, clipping off each word as if he intended to impress them on my memory forever. "I would look at you quite differently if I thought you had planned what we eventually discovered. We had some long sessions with our slide rules and we found, my friend, that you had arranged the *only possible combination of power, speed, and weight* which would blockade the chances of unporting."

Later, when the wine had mellowed us both, I asked Howard if his slide rule could measure the fate of one man against another's.

About the Editor

LAMAR UNDERWOOD is the former editor in chief of *Sports Afield* and *Outdoor Life* and is presently the editorial director of the Outdoor Magazine Group of Harris Publications in New York. He is the author of the novel *On Dangerous Ground* and is the editor of several Lyons Press anthologies, including *The Greatest War Stories Ever Told*, *The Greatest Adventure Stories Ever Told*, *Tales of the Mountain Men*, *The Quotable Soldier*, *The Quotable Writer*, and others on hunting, fishing and the outdoors. His anthology *The Greatest Submarine Stories Ever Told* will be published by The Lyons Press in 2005. Lamar is an avid aviation buff who flies every chance he gets and reads about flying when stuck on the ground—between time spent running computer flight simulation programs and watching films, tapes, and DVDs. This book stems from his deep love of flying and his firm belief that one of the greatest things about flying is that not many people can do it—safely and with great skill. To Lamar, those that can are special people.

Permissions Acknowledgments